All-In-One Manual of Industrial Piping Practice and Maintenance

On-The-Job Solutions, Tips and Insights

K. K. Murty

Industrial Press, Inc.

Library of Congress Cataloging-in-Publication Data

Murty, K. (Kirshna)
Industrial piping practice and maintenance / K. Murty.
p. cm.
ISBN 978-0-8311-3414-3
1. Piping. 2. Pipe. 3. Pipelines--Maintenance and repair. I. Title.
TJ933.M87 2010
621.8′672--dc22

2009046979

Industrial Press, Inc.
32 Haviland Street, Suite 3
South Norwalk, CT 06854
Tel: 203-956-5593
Toll-Free: 888-528-7852
info@industrialpress.com

First Edition, February 2010

Sponsoring Editor: John Carleo
Interior Text and Cover Design: Janet Romano
Developmental Editor: Robert Weinstein

The author dedicates this book to the Supreme Being, Radhasoami

books.industrialpress.com
ebooks.industrialpress.com

10 9 8 7 6 5 4 3 2 1

Table of Contents

Index of Charts

PHILOSOPHY BEHIND THE BOOK

We live amid a myriad of industrial pipelines where even our domestic lives are intertwined by complex piping carrying a wide variety of liquids and gases. These vast networks process and transport their contents reliably, safely, efficiently, and economically - industrial piping being much more *galvanized* than the pipes that carry water to our kitchens.

Piping is a culmination of complex disciplines like metallurgy, standardization, manufacture, construction, and maintenance. Greatly-evolved technology and highly-engineered skills make the modern process plants viable and reliable. The environmental and safety record of pipelines is envious compared to that of other technology products. Pipeline systems are both the safest transportation mode and the most economical means, notwithstanding the millions and millions of miles of pipeline carrying everything from water and acids to crude oil.

In the late 1960s, George Christman and Neil Walton of Standard Oil Company, U.S., taught us young college graduates the rudiments of piping and made us feel at home in the complex piping systems, here in India at Coromandel Fertilizers Ltd (now Coromandel International Ltd.). That was the beginning of my lifelong journey around piping systems.

There are a host of books on design of processes, pipelines, fluid dynamics, and so on, as well as a good number of books on equipment like pumps, compressors, and turbines. But there are almost none on the piping that carries the fluid to and fro!

Back then, we had too many frustrating occasions looking for a book where all the descriptions, use, safety, techniques, and limits for a certain fitting, pipe, or valve could be found. We had to shuffle through sources like training notes, manufacturer manuals, or even the old scribbling pads. The experience of elders was (and is still) passed verbally down the line for maintenance engineers to fall back upon. Yet I have often met engineers and field personnel who lacked practical insight into many aspects of piping practice and do not know where to find it. They seemed to rely on the fitters or their contractors. I wished: if only I could lay my hands on a book with everything on piping, with all the information in one source!

Quite simply this book fulfills that wish! IT IS A BOOK ON INDUSTRIAL PIPING PRACTICE AND MAINTENANCE. I have prepared all those chapters I had wanted to learn from and use but could not obtain back in the day. There are no design theories, no calculations, but we have working principles, pictures, charts, engineering details, along with methods, problems, pitfalls, practical tips, and of course valuable notes in abundance. Theories and calculations can be readily found, all over the Web, all over the world, for modern engineers and technicians. Design details and complicated formulae which bog down the busy field engineer are avoided in this book. Maintenance engineers are expected to make miracles when a problem occurs. Who has time for calculations? Hence, charts are introduced wherever necessary. Complicated drawings and dimensions are kept away. Let us visualize first. So pictures are added. Precautions, practical points, problems, and solutions are given along with the text.

Reading this manual will be fluid and full; smooth rather than turbulent. Maintenance managers can walk along the “pipelines” of chapters, open the “valve” at the relevant branch or chapter. Out "flows" the information. I did not really conceptualize this book as a text for any specific course, but one that can serve as a guide and training manual for persons associated with process industries. And the book is meant not only for technical libraries, but especially for use on the shop floors and in field offices as a veritable maintenance manual on piping — not only as a handy ready-reference for finding a formula or size, but also for assimilating practical knowledge and experience.

The book is also effective for those who oversee plumbers and pipefitters. Throughout, I have focused primarily on ASME, ASTM codes, and the information most in demand.

The Big Bang of the Internet has led to a base of information that is expanding and exploding. Sites such as Engineeringtoolbox, Spirax Sarco, BDK valves, Expansion Joint Systems, Inc., Team Industrial Services, Corrview, Lamons Gasket Company, Mainland Valve and Fitting, RIDGID, Snap-on, ROYMECH, and Donadonsdd, among others, have added even more information, including some excellent pictures as listed in GIVING THANKS below.

A word of caution prevails, however. While I have made every effort to be 100 percent accurate, when in doubt, check to be sure — make yourself certain. That is the stuff engineers are made of.

GIVING THANKS

At the outset I would like express my indelible gratitude to Coromandel Fertilizers, Ltd., (now Coromandel International Limited) for making me what I am today. I wish to thank Mr. Anant Barbadikar, Managing Director, Pharmazel India, Ltd., (formerly Senior Director with Dr. Reddy's Labs) for his valuable guidance. My son, Mr. K. Prem Lahar B.E., (BITTS), contributed immensely by his strategic suggestions. Mr. V. Dayal Swarup developed some drawings for the book. I am also thankful to Mr. Hitesh Patni for his help. And friends and colleagues too numerous to mention lent me a hand by providing material and useful feedback.

I am thankful to the Industrial Press, Inc., for making my wish a reality. I am indebted to Mr. John Carleo for his continuous support of my project. My thanks are due to Ms. Janet Romano and Mr. Robert Weinstein for their superb help in transforming my manuscript into a finished book, and to Mr. Patrick Hansard for publicizing this book.

I am immensely thankful to the proprietors of Web sites who allowed me to capture images from their sites my book:

Engineering Toolbox (www.engineeringtoolbox.com) is an excellent base of engineering data mostly in the chart form and I am thankful to them for an immediate response and permission for use of their charts and figures in my book. All charts in Chapter 36, and Charts 3.1, 3.2, 3.3, 3.5, 3.7, 3.8, 3.9; 4.1, 4.2, 4.3; 6.1, 6.2, 6.3, 6.4, 6.5, 6.6; 7.1, 7.2, 7.3; 12.1;25.1, 25.2, 25.3, 25.4, 25.5, 25.6, 25.7, 25.8; 26.1; 27.1; 29.1, 29.2, and Figures 22.1B, 22.2, 22.3A, B, 22.5, 22.7, 22.8, 22.9, 22.10, 22.11, 25.3, 30.1.

Roymech is also great source of engineering information and I am indebted to Mr. Roy Beardmore for the permission for the Chart 17.4 and Figures 16.2 B, C, 16.4, 16.13, 16.16, 16.34, 16.60A&B and 17.7.

Spirax Sarco has almost all the information on steam and associated systems that you would normally ask for. I am gratified for their permission to capture some material from their superb website, 'Steam Engineering Tutorials' at http://www.spiraxsarco.com/resources/steam-engineering-tutorials.asp" .Their charts can be found at 16.1; 17.2, 17.3; 26.2 and Figures 16.15, 16.17, 16.20, 16.22, 16.23A &B, 16.24, 16.26, 16.28, 16.30, 16.43A&B, 16.45, 16.53A&B; 17.2, 17.3, 17.4 A, B, C, 17.5A, B, C, 17.6A, B, C; 18.1A, B, 18.2A, B, 18.3A, B, C, 18.4; 20.1A,B, 20.3, 20.4, 20.5, 20.6; 21.1, 21.2, 21.4, 21.5A, B, 21.6, 21.7, 21.8, 21.9, 21.10, 21.11, 21.12, 21.13; 22.1A, 22.2;24.1, 24.2, 24.3; 25.1, 25.2, 25.4, 25.5; 26.8, 26.9.

B. D. K. Engineering Industries Ltd., make high quality BDK Valves at their world class facility in India. I am thankful to Mr. Bharat Khimji, C.E.O. and Founder of the BDK Group of Companies, for his permission to use a number of valve pictures in my book, 16.1A, B, C, D, E, &F, 16.5, 16.7, 16.8, 16.9, 16.10, 16.11,16.18A &B, 16.19, 16.27, 16.29, 16.31B, 16.33, 16.37, 16.38, 16.39, 16.44A & B, 16.46, 16.47, 16.49A&B, 16.50A&B, 16.51A&B, 16.52A&B, 16.54, 16.55A&B, 16.56, 16.57, 16.61, 16.62, 16.63, 16.64, 16.65, 16.66.

Lamons Gasket Company make a wide range of industrial gaskets and a great resource at their website. They have readily agreed to share the information and even provided more. I am thankful for the Figures 12.1, 12.2, 12.3, 12.4A&B, 12.5, 12.6, 12.7, 12.8, 12.9, 12.10A &B, 12.11, 12.12, 12.13, 12.14, 12.15, 12.16, 12.17, 12.18.

Expansion Joint Systems, Inc. are a good starting place of knowledge in expansion joint systems at their website apart from very good place for their products. I am thankful for their permission for all the pictures in Chapter 24 except 24.17.

I am indebted for the kind courtesy of **Mainland Valve and Fitting Inc.,** for using their pictures, 8.1, 8.2C, 8.6, 8.8A, 8.8B, 8.11, 8.12, 8.13 in my book. They are manufacturers of a wide range of pipe fittings and valves.

Team Industrial Services offer services and information on emergency piping repairs and I am thankful for their permission for all the pictures in Chapter 32.

CorrView International is a rich source of knowledge on corrosion and its problems with a large stack of pictures at their website. I am thankful for their permission to capture images in Chapter 27 and 29 except 27.8 from their site into my book.

RIDGID specializes in high quality piping tools and I am thankful for The Courtesy of RIDGID in permitting me to use their pictures in my book and serve the piping industry better, all the pictures in Chapter 10 and Figures 9.1A&B, 9.9, 9.12, 9.15 A, B, C, &D, 9.16, 9.17, 9.18, 9.19, 9.20, 9.21, 9.22A &B, 9.26, 9.27A, 9.32A &B, 9.33, 9.34, 9.35, 9.36.

Snap-on Incorporated makes internationally well known Snapon tools and I am indebted for their permission for the use of Figures 9.10, 9.11.

I should say that **Donadon Safety Discs and Devices Srl** ("Donadon SDD") saves process plants by their rupture discs, and I am thankful for allowing me to share their all the figures in Chapter 29 except 29.1 from their website www.donadonsdd.com.

Figure 27.8 was taken from **NASA** website and Figures 9.2, 9.3, 9.4, 9.5, 9.6, 9.28; 22.4, 28.2, 28.3 are from Wikipedia. I thank them for the pictures and their service.

Introduction

Pipelines are the lifelines of almost every daily activity. Vast networks of pipelines provide, transfer, and process our needs reliably, safely, efficiently, and economically. In modern process plants, piping is not simple piping alone, but a culmination of complex disciplines like metallurgy, manufacturing process, codification, and strict standardization. Greatly engineered technology and highly-developed skills make the modern processes viable and reliable while using a whole range of pipes, fittings, and other components.

METALLURGY OF COMMON PIPING

The metallurgy of most used piping — like mild steel, stainless steels, non ferrous metal, and plastic piping — with their characteristics, uses, and limitations is discussed in Chapter 2, along with piping for high and low temperatures, high pressure, and corrosive chemicals.

COMMONLY USED PIPING MATERIALS

Modern process plants use a multitude of chemicals while operating with multi-faceted technologies. Very high pressures and temperatures are the order of the day. On the other hand, there are processes that require cryogenic temperatures or the lowest vacuum levels. Flow rates are simply incredible. The industries need highly compatible and cost effective materials and skills in making and maintaining the pipe lines and a host of other related components for the demanding process applications. Apart from A53 and A106 and Cast Iron grades, the discussion in Chapter 3 looks at special materials like P11, P22, etc.; stainless grades like 304, 310, 316, and 410; and super stainless steels like Alloy20, Hastelloy, and 904L.

MATERIALS FOR VALVES AND FITTINGS

Most of the present day valves have come a long way in the last hundred years, with mind-boggling codes, specifications, and standards. Unlike piping, valves and fittings are cast or forged. Thus in Chapter 4, we have the ubiquitous WCB to the more exciting varieties of cast and forged.

METHODS OF PIPE MANUFACTURE

Chapter 5 provides a peep into the manufacture of pipes.

SIZES, SCHEDULES, AND STANDARDS

In the jungle of standards like ASTM, ASA, DIN, and METRIC that govern the metallurgy, manufacture, and interchangeability of the large number of piping materials, fittings, and valves, Chapter 6 provides a stroll into more generally-used standards. Schedule gives out the wall thickness of the pipe, which in turn dictates the pressure and temperature rating of the line, like flange dimensions, materials, and ratings.

FLANGE TYPES AND MATERIALS

Flanges are classified by the construction design, mounting design, face, standard, and class, and by the material of construction of course. Chapter 7 is the main joining block of all piping systems.

PIPE FITTINGS

The weakest link in any piping system is the fitting like elbow, or tee. Chapter 8 describes how they are made stronger by screwed, welded, and flanged joints.

HAND TOOLS IN PIPING

We got used to a veritable lot of tools while working on piping, though many do not even bother to know the basics. There is a technique for every tool, a number of special tools meant only for piping work, and a bundle of bad practices. For example, one of the most misused tools, a pipe wrench is used as a hammer and adjustable spanner and what not. Chapter 9 deals with tools used in piping.

CUTTING, THREADING, AND WELDING PIPES

The process of pipe joining has evolved highly specialized but standardized procedures. Chapter 10 discusses cutting and threading pipe, flange types and materials, and flange joint assembly. This chapter addresses the very basics of piping, cutting, and threading. How to and how best? What are the pitfalls?

With intricate materials now being used for process applications, welding pipelines has become both an art and a technology, more so than maintenance welding. Welding on pipe, fittings, flanges, and pressure vessels must only be carried out by someone with proper skill and credentials; welding on pressure piping must be carried out with a qualified welding procedure. Chapter 10 also deals with the basics of these welding and cutting techniques.

BOLTS AND STUDS

The right choice of bolts and studs is vital for reliable and safe operation of modern processes. Chapter 11 deals with sizes, standards, and effective use of the very bolts and nuts of a piping system.

GASKETS

Gaskets are necessary to make leak-tight joints. Packing is the gland packing that seals leak across the spindles of valves and other moving parts. Modern gasket and packing materials are engineered to suit temperatures, pressures, and corrosivity of the fluids handled and should be used appropriately for reliable and safe operation. Chapter 12 deals with just that.

OPENING LINES

Opening lines is one of the frequently undertaken jobs in chemical plants. If proper care and technique are not practiced, this job could prove to be extremely hazardous in view of the exceedingly corrosive and dangerous chemicals used in the modern process plants. Chapter 13 makes that operation safe and fast.

BOLTING-UP

Chapter 14 discusses the reverse operation of opening of flanges, which needs to be more stringent. With the right materials and methods, we can ensure that we have a perfectly leak proof and quality joint.

BLINDING AND NORMALIZING LINES

Again, one of the more demanding maintenance activities! Pipelines are sometimes deliberately blinded for a positive shutoff of any ingress of fluids into the upstream or downstream systems. The pipelines need to be opened and closed under trying conditions. Chapter 15 discusses safety and skill in this crucial maintenance activity. The chapter advises to test, check, or maintain the line, stop the liquid to the other side by all means. Don't believe in valves, they "pass!"

PROCESS VALVES

Valves play a major role in stopping, enabling, mixing, and controlling fluid flows in processes. With a large variety of types, classes, shapes, and sizes, they deserve a judicious treatment. Gate valves, globe valves, check valves, butterfly valves, plug valves, and ball valves are all here in Chapter 16. Pressure Temperature Ratings of valves are some of the most misunderstood aspects, even by experienced engineers.

CONTROL VALVES

Processes commonly used in industries a little over half a century ago were controlled manually. But now in the highly complex industries, processes are remote controlled precisely and fast, albeit automatically. Chapter 17 covers them — valves in control.

RELIEF VALVES

Here you have them, the real saviors of processes. Relief and safety valves are used to control, relieve, or safely vent the excess pressure, when process systems go haywire. Chapter 18 deals with various aspects of working with and operating these special valves.

RUPTURE DISCS (SAFETY HEADS)

Chapter 19 discusses rupture discs, which are expendable safety valves that are used extensively in the industry because of capital and maintenance savings. They do not demand any maintenance, though very fast acting.

STRAINERS

Chapter 20 deals with strainers which are used vitally in the piping systems. They do some good; allow quality fluid in, and stop or strain foreign material.

STEAM TRAPS

Chapter 21 addresses steam traps, the real energy savers in a world of ever expanding expenses. Proper selection, installation, and maintenance of steam traps are important for effective utilization of the power of the steam traps.

FLOW MEASUREMENT

What is control without measurement? In the wide variety of processes and their precision management, we have an equally wide range of flow measurement systems. Chapter 22 browses various methods like Orifice Plates, Venturi, Target Flowmeters, good old Rotameters, turbine meter, and Coriolis meter.

STRESS ANALYSIS

Present demands in modern chemical plants are for higher pressures, temperatures, and flows need systems designed, built, and operated with intrinsic reliability. Nowadays, a number of operating conditions, pipe line stress conditions, and worst case scenarios are evaluated, much before the pipe line is designed and built. Basing on these conditions, stress analysis is made; in turn, pipe line is designed and built. Chapter 23 deals specifically with stress analysis.

EXPANSION JOINTS

The pipeline tends to expand and contract depending on the actual temperatures. It should be sufficiently flexible to accommodate the movements of expansion and contraction. In these days of high technology and consequent high pressures and temperatures, expansion joints carry the stresses of the line. Chapter 24 addresses expansion joints.

EXPANSION LOOPS

100 feet of carbon steel pipe at an initial temperature of 100°F expands by about 1.6 inches for about 300°F rise. This catastrophic effect can be averted by expansion elbows, loops, or Z bends. Expansion loops are a cheaper and easier alternative to expansion bellows. They can be fixed in line with a length of piping made out of the same piping materials. Chapter 25 discusses expansion loops, supports, etc.

PIPE SUPPORTS

The life of a pipe line is as strong as its supports. Let us support our life lines. Chapter 26 briefly deals with pipeline supports, restraints, anchors, and guides.

CORROSION

Corrosion is the greatest pain for maintenance engineers and a major challenge for the designers. Metallurgical engineering has developed sustainable products for most difficult applications. But still the problem persists. Chapter 27 reflects on a maintenance worker's solutions to corrosion.

MISCELLANEOUS PRACTICES

Chapter 28 has some miscellaneous but very valuable pipe line practices such as fillers and spacers, jacketed and gutted piping, traced piping, double block and bleed, and pipeline pigging.

INSULATION

Insulation conserves energy and money. Chapter 29 provides a small discussion on these useful practices and problems with bad insulation.

PIPE LAYOUT

At this point, we have fully covered piping materials, standards and schedules, and metallurgy. We have reviewed the methods of joining piping with the piping tools. We also have suitable bolts and studs, packing, and gaskets. Let us draw the lines now. Chapter 30 deals with pipe sketches, layout, and some finer points.

TESTING OF EQUIPMENT

Pressure tests are carried out to determine that the equipment is safe for operation. Normally hydrostatic tests are carried out on the equipment, before it is initially put in service, at scheduled intervals as per statutory controls, or after repairs or alterations. Chapter 31 deals with hydrotesting of equipment or pipe lines, safety, and other techniques.

EMERGENCY PIPE REPAIRS

A small leak in a piping system could force a shutdown in a factory. However, if maintenance can manage the leak without shutting down the system, plant operations can continue, saving huge amounts of money. Chapter 32 suggests little techniques that affect in a big way critical plant operations. Any process plant may need them some time or the other, but safety is important. Emergency repair techniques included online leak sealing, and hot tapping to keep the plants running!

INSPECTION

The environmental and safety record of pipelines is excellent compared to other technology products. However, pipelines are susceptible to attack by internal and external corrosion, cracking, third party damage, and manufacturing defects. Unfortunately, they do not give any audible early warnings like rotating equipment does. Hence, it is imperative that periodic inspections are carried out to detect defects and prevent damage.

A large number of inspection methods like visual, radiography, magnetic particle, and ultrasonic are discussed in Chapter 33 to evaluate the integrity of the piping and joints.

PIPELINE RIGGING

Lifting operations are inherently dangerous, even when proper training is conducted, equipment is properly maintained, and employees work in a safe manner. Accidents can still occur. You are the only controlling influence and can minimize hazards.

Pipelines fall into the more serious category of this. Pipelines are slender beings often lifted into position where there is no access. Being round they may roll off and they are prone to slip, as a number of them are often lifted. More frequently, pipeline supports are built after the pipeline is positioned. Chapter 34 discusses rigging in the piping practice with its tools, tackles, and resources for lifting and lowering pipelines and components, with the help of cranes, pulley blocks, come-alongs, etc.

FLEXIBLE HOSES

Flexible hoses are extensively used all over the industry, primarily for vibration isolation or where the upstream / downstream lines have considerable movement or distance. A small discussion on metallic and non-metallic hoses is found in Chapter 35.

MORE INFORMATION

Chapter 36 has more useful charts such as pressure drop in steel pipes in imperial and SI units, pressure drop and velocity for water in imperial and SI units, specific gravity of some common gases, specific weight of other materials, and corrosion charts.

METALLURGY OF COMMON PIPING

MATERIAL SELECTION

Pressure, temperature, flow, type and requirements of process, and corrosion characteristics of the piped chemical are some of the most important factors to be considered when selecting the material of construction, size, and wall thickness of the pipe. External environmental conditions and all the process parameters should be considered in conjunction with each other when selecting or reselecting the size and material and lay out of the piping.

Pressure is the paramount consideration in deciding the size and material of the piping as higher pressures impose higher stresses in the piping. In general, the strength of metals decreases as temperature increases. In Cast Steel, strength reduces at 653°F/345°C and becomes critical in the range of 932°F/500°C to 1022°F/550°C. For example, the strength of low carbon steel is reduced by 22% when temperature is raised from 932°F/500°C to 977°F/525°C. Temperature also has a profound effect on corrosion rates. For example, in streams containing sulfuric acid, an increase of merely 6°C temperature can double corrosion rate.

Needless to repeat, the amount of flow decides the size of the piping and its layout. Flow characteristics like pulsating or uniform, and intermittent or continuous also should be considered for proper selection of the piping layout.

Another important parameter to be taken into account is the corrosion characteristics of the fluid and ambient conditions. Even though the whole range of materials is currently developed to combat corrosion, adequate precautions must be taken in selecting the material size and corrosion allowance.

Let us browse through the whole range materials available to us for piping.

STEEL PIPING

Steel is the most widely used material for piping in the process industry. An assortment of steels and its alloys are available for extremely diversified services and are continuously tailored to suit varied requirements of industrial processes. The list is exhaustive; any attempt to review all of them will be borne with the risk of leaving one more. Steel is used for high pressures and temperatures and, similarly, in sub-zero temperatures and vacuum services. It is extremely strong and tough, and shows good resistance to piping strains, vibration shock, and low and high temperatures.

LOW CARBON STEEL

Fortunately, low carbon steel is a very versatile material for most plant applications. It is relatively inexpensive, yet provides the strength, ductility, workability, and welding properties required. The steel used for equipment is low in carbon (0.3% or less), sulfur, and phosphorus; it contains small quantities of silicon or aluminum and also sufficient manganese to offset the effect of sulfur. Carbon steel is used from –18°F/–28°C to around 797°F/425°C. Though it is the most used material in the industrial piping, its corrosion resistance is found wanting against most acids, alkalis, and salts.

A number of other alloys have been developed to cope up with the severe conditions of the chemical processes. However, not one single material in this large list is suitable for all services. One may have to select the right material considering various requirements.

LOW ALLOY STEELS

Low carbon steel is a tough and ductile material. But beyond 653°F/345°C, the strength of low carbon steels decreases; it develops a tendency toward scaling and suffers loss of material due to corrosion by sulfur compounds. It becomes brittle in hydrogen service beyond 500°F/260°C. The addition of Molybdenum (moly) of about 0.5% greatly pushes its strength up to 896°F/480°C. Chromium up to 9% combats the tendency to oxidize at high temperatures and resists corrosion from sulfur compounds. But these steels need heat treatment when welded. Steels with chrome and chrome molybdenum are used in pressure vessels, piping, furnace tubes, and exchangers operating at high temperatures and pressures. Low alloys of carbon, moly, and chrome are used in high temperature service over 797°F/425°C. Below –18°F/–28°C, steel loses its resistance to sudden shock. An addition of 3% to 5% nickel will produce steels that remain tough to –148°F/–100°C.

HIGH ALLOY STEELS

Larger quantities of alloying elements are necessary to produce the desired characteristics. Steels that contain 5% or more of alloying metals are generally called high-alloy steels. Chromium steel and stainless steel fall into this category. At the medium temperature end, stainless steels hold the fort, but at higher temperatures a complex variety of materials are developed.

CHROMIUM STEELS

Process chemicals with considerable amounts of sulfur compounds become quite corrosive to steel at temperatures ranging from about 550°F/288°C to 842°F/450°C. Chromium steels withstand this type of attack very well, but in some cases the low chromium alloys previously described are not resistant enough. In these cases, alloys containing from 12 to 17% chromium are used, but they have a tendency to become brittle after extended heating cycles in the

698°F–1022°F (370°C–550°C) range. Their primary use is now largely confined to pump and compressor parts.

Chrome-Moly alloy steel piping is the backbone of high-pressure, high-temperature piping in process plants like power, fertilizer and petro chemical plants. These grades contain molybdenum and chrome as their alloying elements. These pipes are referred to with a P-number, such as P-11 or P-22, with corresponding forged fittings with an F–number. They are often referred to as 1 1/4 CR-1/2 moly or 2 1/4 CR-1/2 moly, depending on the percentage of chrome and molybdenum content.

STAINLESS STEEL PIPING

Stainless steels demand a little more attention than others as they are extensively used in the most challenging applications. There are several types of stainless steels used, depending on the corrosion and temperature burden of the process. The most common types are 304 and 316.

Stainless steel is basically a low carbon steel with chromium at 10% and other alloying elements like nickel, vanadium, columbium, etc. Chromium gives corrosion-resistance properties. Stainless steels develop a thin corrosion-resisting chromium oxide film on their surfaces. The secret is that this film is self-healing if oxygen is present in its surrounding atmosphere, even in very small amounts. To improve this "passive" oxide layer, stainless steel is treated generally with 10% nitric acid plus 2% hydrofluoric acid chemical bath or circulation. This process is known as *passivation*. Steels containing more than 3.99% are classified as Stainless Steels.

The most important advantage is its corrosion resistance, with its various grades withstanding most acids, alkaline solutions, and chloride-bearing fluids to a certain extent. Stainless steel has excellent fire and heat resistance properties; certain grades with high chromium and nickel maintain strength at high temperatures and resist scaling. They are highly suitable for applications in temperatures ranging up to about 1800°F/982°C and down to cryogenic temperatures of about –320°F/–200°C. Stainless steel renders itself to easy fabrication and forming as it can be cut, welded, and machined as other steels. Austenitic grades exhibit work-hardening properties.

Stainless steel has a modern, attractive, and aesthetic appearance with its bright, highly polished and easily maintainable surface. Stainless steel is hygienically safe for use in hospitals, kitchens, and other food and pharmaceutical plants. Though initial costs may be higher than conventional steels, life cycle costs are low and are cheaper in the long term. Stainless steels have high strength-to-weight compared to other steels. They exhibit stress-corrosion cracking and a high coefficient of expansion which make them impractical for certain applications. Stress-corrosion cracking is a mechanical–chemical type of deterioration. The most familiar occurrence is the cracking of stainless steels in chloride environments.

More of this discussion is in Chapter 27 on Corrosion.

When stainless steel is heated, it expands at a rate approximately 150% of that of steel. This expansion becomes a problem whenever stainless steel is used in close contact with other metals. At high temperatures, great internal strains can be produced when two adjoining materials expand at different rates.

It is believed that in 1821 a French metallurgist Pierre Berthier, alloyed iron with chromium and found it to be resistant to certain acids. In 1872, Messrs. Woods and Clark applied for a British patent for an alloy containing 30 to 35% chromium and 1.5 to 2% tungsten which was supposed to be acid and weather resistant. In 1904 and 1906 Leon Guillet published research papers on the present day 300 series and some 400 series stainless steels. Harry Brearley, chief of the research lab run jointly by John Brown & Co. and Thomas Firth & Sons, pioneered the industrialization of stainless steel.

GRADES OF STAINLESS STEEL

Chromium, nickel, molybdenum, titanium, and other elements are alloyed in varying quantities to produce a wide range of stainless steel grades, each with its unique properties. The suffix L after the grade number, i.e., 304L, means that stainless steel has low carbon content to a maximum of 0.03% (normal level is 0.08% max.) This gives better corrosion resistance, particularly where welding is involved, by preventing depletion of chromium at the weld zone. Similarly the suffix H denotes higher carbon content which yields a little higher strength.

There are more than 60 grades of stainless steel, divided broadly into five classes basing on their microstructure.

- *Austenitic*
- *Ferritic*
- *Duplex*
- *Martinsitic*
- *Precipitation Hardening*

AUSTENITIC Austenitic stainless steels are the most widely used with SS304 in the lead. When nickel is added in sufficient quantities to stainless steel, the crystal structure changes to *austenite* — hence, the name austenitic. SS304 contains 18% chromium and 8% nickel, whereas SS316 contains 16% chromium, 10% nickel and 2% molybdenum. *Moly* is alloyed to resist corrosion to chlorides (like sea water). They have excellent corrosion resistance and weldability, and lend themselves well to a variety of forming and fabrication techniques. They offer good high temperature and excellent low temperature characteristics. They are non-magnetic and can be work hardened.

FERRITIC These are plain chromium stainless steels with chromium content between 12 and 18%, but with low carbon content. Typical grade: 430. They offer a moderate corrosion resistance, not hardenable by heat treatment. They are magnetic. Weldability is poor and formability not as good.

DUPLEX Duplex SS have high chromium content (between 18 and 28%) and a reasonable amount of nickel (between 4.5 and 8%). These stainless steels exhibit a combination of ferritic and austenitic structure, hence called duplex. Some duplex steels contain molybdenum from 2.5–4%. Typical grade: 2205. They offer excellent resistance to stress corrosion cracking, better resistance to chlorides. They are better than austenitic or ferritic steels in tensile and yield strength, while offering good weldability and formability.

MARTENSITIC Martensitic stainless steels exhibit relatively high carbon content (0.1–1.2%) with 12 to 18% chromium. They were the original commercial stainless steels. They offer moderate corrosion resistance and can be heat treated. They have high strength, but weldability is bad. They are magnetic. Typical grade: 410.

PRECIPITATION HARDENING Precipitation hardening stainless steels are hardened, after fabrication. They have moderate to good corrosion resistance and are magnetic. They offer very high strength and good weldability.

CAST IRON PIPE

Sewer lines and water mains are some of the principal uses of cast iron. Another most important use for cast iron is in the concentrated sulfuric acid at ambient temperatures. The major drawback of cast iron is its brittleness. It is now possible to weld some grades of cast iron, but not to a point of leak-proof. Cast iron fittings could crack if the bolting is bad, the line is misaligned, or supports are not proper, and in case of a fire or increase in line pressure. Cast iron pipes are seldom used beyond 150 class.

There are four basic types of cast iron

- White iron
- Malleable iron
- Gray iron
- Ductile iron
    - White cast iron is characterized by high compressive strength, hardness, and resistance to wear because of the carbides.
    - Malleable cast iron is white cast iron heat-treated to improve higher ductility.
    - Gray cast iron is characterized by good machinability, and wear resistance because of the micro-structural graphite.
    - Ductile iron develops high strength and ductility with the addition of small amounts of cesium or magnesium to gray iron.

GALVANIZED PIPE

A protective coating of zinc is made on steel pipe to prevent it from corrosion and increase its life. Reaction of zinc with steel makes successive layers of these reactants, but the outermost layer is all zinc. It has an amazing quality of healing itself after minor mechanical damages; it continues to protect from atmospheric corrosion and even certain salt water to some extent. GI pipes start with a nice shiny look, which changes to a dull gray finish over a year. The finish should be smooth and continuous.

Most of the pipes are threaded because welding of GI pipes is not recommended. Galvanized steel pipes have a tendency to build up rust in the inside bore, which may eventually choke it. Sometimes flakes of zinc may come out and so this pipe is not recommended for gas and process lines. However, it is extensively used for water applications even though the plastic pipes are taking over. Long-term exposure beyond 392°F/200°C may result in peeling of the coating, loss of mechanical properties, and reduction in corrosion resistance.

NONFERROUS ALLOYS

A metal or alloy that contains little or no iron is called a nonferrous material.

NICKEL ALLOYS

In specific conditions where extreme resistance to chemicals is required and where stainless steels do not fit the bill or the service, alloys containing large amounts of nickel are used. These alloys usually hold additions of iron, copper, aluminum, chromium, cobalt, and molybdenum. Some typical examples of these alloys are Monel, Hastalloy, and Inconel. These alloys are used in a variety of services that involve acids and caustics. For example, Monel is used in hot sodium hydroxide or hydrochloric acid service and severe sea water applications.

TITANIUM

Titanium offers extremely high resistance to corrosion of salt water and marine environment with exceptional resistance to a wide range of acids, alkalis, and chemicals. It is about twenty times more erosion resistant than the copper-nickel alloys. Its excellent erosion resistance permits use of higher pipeline velocities. It has a high strength-to-weight ratio. Due to its high heat transfer efficiency and higher strength, thinner wall tubes can be used. Commercially pure or unalloyed titanium is available in ASTM Grades 1 through 4, and 7, with Grade 1 being the most pure. It maintains corrosion resistance even at high temperatures with maximum limit at 2000°F/1095°C. Brinell hardness is about 215.

COPPER ALLOYS

Brass is a family of alloys of copper and zinc with copper content from 90% to about 60%, with the balance zinc. Some brasses have small amounts of other elements such as lead, tin, antimony, arsenic, or phosphorus. Because of their resistance to corrosion from water containing various impurities, brasses fit where steel fails. They are weaker than steel and not normally used at temperatures above 450°F/232°C.

Yellow brass has 66% copper and 34% zinc. Though it shows good corrosion resistance in most environments, is not recommended for acetic acid, acetylene, ammonia, and salt. Maximum recommended temperature limit of 500°F/260°C. Brinell hardness is 58.

Aluminum and silicon bronzes are more resistant to salt water than simple brass and are widely used as condenser tubing when salt water is the cooling medium. Brinell hardness of copper is about 80.

Arsenical Admiralty contains 71% copper, 28% zinc, and 1% tin and traces of arsenic. It offers extremely good corrosion resistance against salt and brackish waters, and water containing sulfides. Maximum working temperature of 500°F/260°C and Brinell hardness is about 64.

There are a number of copper-nickel alloys, the most common being cupro-nickel and monel.

Cupro-nickel contains 69% copper, 30% nickel, and small quantities of manganese and iron. It is used in condenser tubing when the cooling water has extreme concentrations of salt, like saltwater. It is specifically meant for high temperature, pressure velocity applications. Maximum temperature limit is in the order of 500°F/260°C. Brinell hardness is about 70.

Monel contains 67% nickel and 30% copper and has a maximum temperature range of 1500°F/815°C. It offers excellent resistance to most acids and alkalis, but is not recommended for fluorosilic acid, mercuric chloride, and mercury. Brinell hardness is about 120.

ALUMINUM ALLOYS

Aluminum has good resistance to corrosion from sulfur compounds and to atmospheric oxidation. Its action on corrosion is similar to stainless steel, but it is a very low strength material with a Brinell hardness of approximately 35. It is rarely used in industrial piping except in the certain typical service of urea. However, aluminum coated steel is used in certain equipment to save it from sulfur compounds and oxidation at high-temperature. It is extremely light in weight and melts at 1220°F/660°C. Its specific corrosion resistance characteristics make it non toxic and finds applications in drug, food, and beverage industries. It has good thermal and electrical conductivity and machining properties.

LEAD ALLOYS

Lead is a heavy, extremely ductile, extremely weak material with a low melting point. It is used as lining over steel against sulfuric acid corrosion, but Teflon is taking over its place.

Commonly Used Piping Materials

Low carbon steel is the most widely used piping in the industry. It is made in a variety of grades to meet various process requirements of the industry. Commonly used grades are ASTM A106 and A53. Material composition and other small mechanical details are given below.

ASTM A106/A53 — PRESSURE/TEMPERATURE CHARTS

ASTM A106 is the black carbon steel pipe for high-temperature, high-pressure service of seamless pipe in three grades A, B, and C of varying strength. Grades A and B are available in most sizes and schedule numbers. Grade B permits higher carbon and manganese contents than Grade A. Although the physical and chemical properties for ASTM A106 Grades A and B are comparable to those for A53 pipe, A106 is preferred for more stringent services.

The following charts show

- Pressure and temperature ratings of A106 grade B (**Chart 3.1**)
- Maximum working pressure of carbon steel pipes ASTM A53 B. (**Chart 3.2**)
- Operating temperatures and allowable stresses on the pipewalls of seamless pipes (**Chart 3.3**)

SPECIAL MATERIALS FOR HIGH TEMPERATURE PIPING

To satisfy the demands of processes operating at higher temperatures with extremely difficult chemistries of fluids, suitable materials need to be continuously developed.

Carbon steel loses all its stress resistance at 653°F/345°C. It is extremely difficult to select material at 1200°F/650°C satisfying the needs of pressure and corrosive properties of the fluid in the line. Tensile strength of 60,000 psi reduces at 797°F/425°C to 10,800 psi, at 896°F/480°C to 6500 psi, and at 1004°F/540°C to 2500 psi.

When specified amounts of chromium, molybdenum, and vanadium are alloyed into steel, they increase its creep resistance for high-temperature strength, which in turn allows lighter and thinner pipes. Some of the more popular materials are discussed here.

However, a careful welding procedure should be followed with proper pre-heat and post-weld heat-treatment to avoid cracking. Chrome-moly steels should be preheated from becoming brittle,

Chart 3.1
Pressure (psig) and Temperature (deg F) Ratings of A106 Grade B Carbon Steel Pipes, Based on ANSI/ASME B 31.1

Maximum Allowable Pressure (psig)										
Pipe Size (inches)	Pipe Schedule	Temperature (°F)								
		100	200	300	400	500	600	650	700	750
1″	40[1)]	2857	2857	2857	2857	2857	2857	2857	2743	2476
	80[2)]	3950	3950	3950	3950	3950	3950	3950	3792	3423
	160	5757	5757	5757	5757	5757	5757	5757	5526	4989
1 1/2″	40	2116	2116	2116	2116	2116	2116	2116	2032	1834
	80	2983	2983	2983	2983	2983	2983	2983	2864	2585
	160	4331	4331	4331	4331	4331	4331	4331	4157	3753
2″	40	1783	1783	1783	1783	1783	1783	1783	1712	1545
	80	2575	2575	2575	2575	2575	2575	2575	2472	2232
	160	4217	4217	4217	4217	4217	4217	4217	4049	3655
3″	40	1693	1693	1693	1693	1693	1693	1693	1625	1467
	80	2394	2394	2394	2394	2394	2394	2394	2298	2074
	160	3600	3600	3600	3600	3600	3600	3600	3456	3120
4″	40	1435	1435	1435	1435	1435	1435	1435	1378	1244
	80	2075	2075	2075	2075	2075	2075	2075	1992	1798
	160	3376	3376	3376	3376	3376	3376	3376	3241	2926
5″	40	1258	1258	1258	1258	1258	1258	1258	1208	1090
	80	1857	1857	1857	1857	1857	1857	1857	1783	1610
	160	3201	3201	3201	3201	3201	3201	3201	3073	2774
6″	40	1143	1143	1143	1143	1143	1143	1143	1098	991
	80	1794	1794	1794	1794	1794	1794	1794	1722	1554
	160	3083	3083	3083	3083	3083	3083	3083	2960	2672
8″	40	1006	1006	1006	1006	1006	1006	1006	966	872
	80	1586	1586	1586	1586	1586	1586	1586	1523	1375
	160	2976	2976	2976	2976	2976	2976	2976	2857	2579
10″	40	913	913	913	913	913	913	913	876	791
	80	1509	1509	1509	1509	1509	1509	1509	1448	1308
	160	2950	2950	2950	2950	2950	2950	2950	2832	2557

STD (standard) = schedule 40, [2)] XS (extra strong) = schedule

Chart 3.2
Maximum Working Pressure of ASTM A53 B Carbon Steel Pipes Manufactured According to ASME/ANSI B 36.10

1) CW — continuous weld — a method of producing small diameter pipe (1/2-4")

2) ERW — electric resistance weld — most common form of manufacture

Nominal Size (inches)	Pipe Outside Diameter OD (inches)	Schedule Number or weight	Wall Thickness - t - (inches)	Inside Diameter - d - (inches)	Working Pressure ASTM A53 B to 400°F		
					Manufacturing Process	Joint Type	Psig
¼	0.540	40ST	0.088	0.364	CW[1)]	T	188
		80XS	0.119	0.302	CW	T	871
3/8	0.675	40ST	0.091	0.493	CW	T	203
		80XS	0.126	0.423	CW	T	820
1/2	0.840	40ST	0.109	0.622	CW	T	214
		80XS	0.147	0.546	CW	T	753
3/4	1.050	40ST	0.113	0.824	CW	T	217
		80XS	0.154	0.742	CW	T	681
1	1.315	40ST	0.133	1.049	CW	T	226
		80XS	0.179	0.957	CW	T	642
1 1/4	1.660	40ST	0.140	1.380	CW	T	229
		80XS	0.191	1.278	CW	T	594
1 1/2	1.900	40ST	0.145	1.610	CW	T	231
		80XS	0.200	1.500	CW	T	576
2	2.375	40ST	0.154	2.067	CW	T	230
		80XS	0.218	1.939	CW	T	551
2 1/2	2.875	40ST	0.203	2.469	CW	W	533
		80XS	0.276	2.323	CW	W	835
3	3.500	40ST	0.216	3.068	CW	W	482
		80XS	0.300	2.900	CW	W	767
4	4.500	40ST	0.237	4.026	CW	W	430
		80XS	0.337	3.826	CW	W	695

Chart 3.2 Continued

6	6.625	40ST	0.280	6.065	ERW[2)]	W	696
		80XS	0.432	5.761	ERW	W	1209
8	8.625	30	0.277	8.071	ERW	W	526
		40ST	0.322	7.981	ERW	W	643
		80XS	0.500	7.625	ERW	W	1106
10	10.75	30	0.307	10.136	ERW	W	485
		40ST	0.365	10.020	ERW	W	606
		XS	0.500	9.750	ERW	W	887
		80	0.593	9.564	ERW	W	1081
12	12.75	30	0.330	12.090	ERW	W	449
		ST	0.375	12.000	ERW	W	528
		40	0.406	11.938	ERW	W	583
		XS	0.500	11.750	ERW	W	748
		80	0.687	11.376	ERW	W	1076
14	14.00	30ST	0.375	13.250	ERW	W	481
		40	0.437	13.126	ERW	W	580
		XS	0.500	13.000	ERW	W	681
		80	0.750	12.500	ERW	W	1081
16	16.00	30ST	0.375	15.250	ERW	W	421
		40XS	0.500	15.000	ERW	W	596
18	18.00	ST	0.375	17.250	ERW	W	374
		30	0.437	17.126	ERW	W	451
		XS	0.500	17.000	ERW	W	530
		40	0.562	16.876	ERW	W	607
20	20.00	20ST	0.375	19.250	ERW	W	337
		30XS	0.500	19.000	ERW	W	477
		40	0.593	18.814	ERW	W	581

Chart 3.3
Operating Temperature and Allowable Stresses in Pipe Walls for Seamless Carbon Steel Pipes

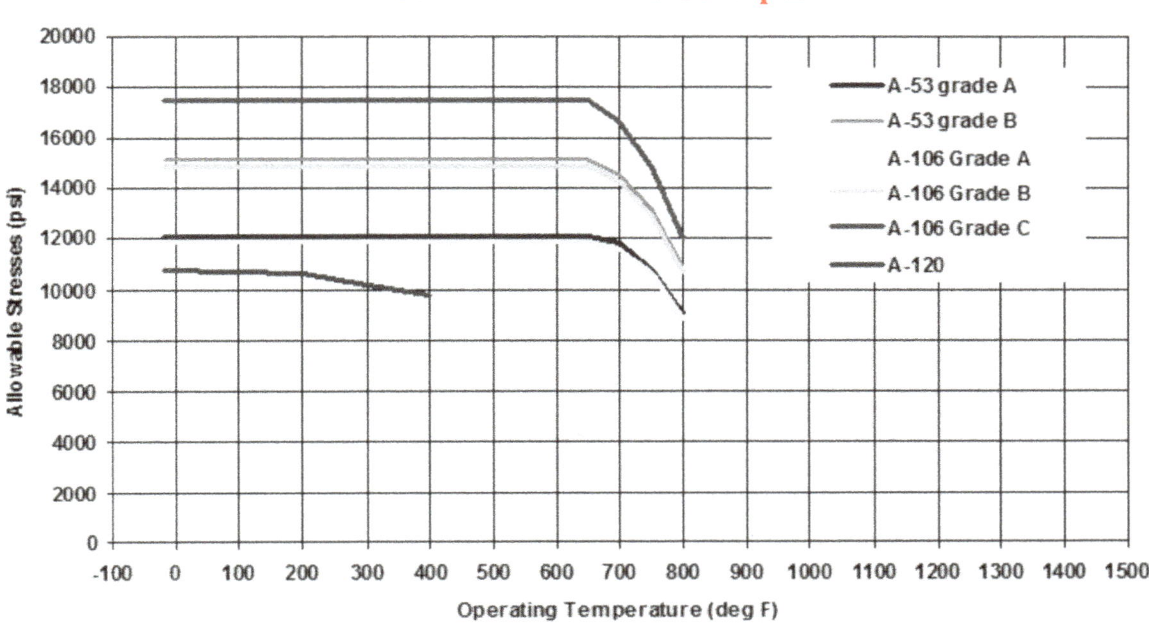

and cracking requires preheating, typically between 248°F/120°C and 500°F/260°C, to drive off moisture, thereby reducing hydrogen and slowing the cooling rate. Hydrogen embrittlement leads to cold cracking of the finished weld. Slowing the cooling rate reduces thermal stresses and further allows hydrogen to diffuse from the weld. Maintaining a minimum interpass temperature is important to maintain low hydrogen level and reduce thermal gradients. But overheating disturbs the mechanical properties and renders the weld puddle too fluid. It will then be difficult to control the weld and will end up in hot cracking. Chrome-moly steels are more sensitive to this. Post-weld heat treatment from 1094°F/590°C to 1247°F/675°C reduces residual stress and hydrogen within the material.

Most common alloys of this category are P5, P9, P11, P22, P91, etc. Normally piping is standardized as, say, P5, weld fittings as WP5, forged fittings by F5, etc., in this class of materials. Chemical composition and tensile strength details for these alloys are given here.

P5 material has higher mechanical properties than 1 1/4 chrome steel, which makes it more suitable to higher temperature and pressure applications particularly in certain corrosive atmospheres. It is resistant to hot sulfide corrosion cracking and useful from –18°F/–28°C to 1145°F/645°C. This class of alloy falls into standards of ASME/ASTM A234, A182, and A335 for piping and forgings, etc. It is widely used in refineries and petrochemical plants in catalytic reformers, hydrocrackers, etc.

P9 material has higher mechanical properties than 400 series stainless steels like 410, but exhibits equivalent corrosion resistance. 9% chromium makes it more suitable to higher temperature and pressure applications, particularly in certain corrosive atmospheres. It is used in ranges from –18°F/–28°C to 1292°F/700°C in refineries and petrochemical plants.

P11 is the choice material in steam lines operated at slightly higher temperatures, where carbon cripples steel at elevated temperatures. It is used in ranges from –18°F/–28°C to 1300°F/704°C in refineries and petrochemical plants for crude distillation systems and extraction steam lines and associated piping.

P22 is used at elevated temperatures for its creep strength and stress rupture properties within a typical range of –18°F/–28°C to 1094°F/590°C. It is widely used in refineries and petrochemical and power plants.

ALLOY 400

Alloy 400, also known as Monel 400, has excellent corrosion resistance against sea water and mono ethanolamine, fluorine, hydrofluoric acid, and hydrogen fluoride. Corrosion resistance — coupled with strength, ductility, and weldability — makes it suitable material for some of the most trying services. It maintains mechanical properties from subzero to 1022°F/550°C. It resists stress corrosion cracking induced by chlorides. Heat exchangers and pumps, valves, and such other mechanical components in the salt water, MEA service, and in nuclear power plants find extensive use of this material.

Its varieties like piping, forgings, etc., are governed by the codes ASTM B165, B366, B127, B164, B564; ASME SB165, SB366, SB127, SB164, and SB564.

ALLOY 600

A rare combination of mechanical strength and corrosion resistance, Alloy 600 is virtually immune to chloride stress corrosion cracking. It maintains its mechanical properties right from subzero to 2012°F/1100°C. It resists oxidation at high temperatures and offers very good resistance to many organic and inorganic compounds. It is not precipitate hardenable. It finds applications in super heaters and steam generators, and food and chemical process plants. Brinell hardness is about 150.

Its varieties like piping, forgings, etc., are governed by the codes ASTM B166, B366, B167, B564; ASME SB166, SB366, SB167, and SB564.

ALLOY 800

This alloy was specifically designed to meet high temperature applications with good strength, ductility creep, and rupture qualities and to resist corrosion, oxidation, and carburization. It can stand well in atmospheres that alternate between oxidizing and reducing atmospheres. It is used in steam hydrocarbon reforming, ethylene pyrolysis, and high temperature reactors.

Its varieties like piping, forgings, etc., are governed by the codes ASTM B407, B366, B514, B408, B564; ASME SB407, SB366, SB514, SB408, and SB564.

Chart 3.4 shows the chemical composition of high temperature steels **Chart 3.5** looks at seamless ferritic alloy steel pipes (high temperature steels) and summarizes the operating temperature and allowable stresses in their pipe walls.

Chart 3.4 Chemical Composition of High-Temperature Steels

MATERIAL	CHEMICAL COMPOSITION PERCENTAGES, BALANCE Fe												Tensile Strength KSI	
	C	Cr	Mn	Mo	Ni	P	S	Si	Ti	Al	Mg	Cu	YS	TS
P5	0.15	4.0-6.0	0.3-0.6	0.44-0.6		0.025	0.025	0.5					60	30
P9	0.15	8.0-10.0	0.3-0.6	0.9-1.1		0.025	0.025	0.25-1.0					60	30
P11	0.05-0.15	1.0-1.5	0.3-0.6	0.44-0.6		0.025	0.025	0.5-1.0					60	30
P22	0.05-0.15	1.9-2.6	0.3-0.6	0.87-1.13		0.025	0.025	0.5					60	30
ALLOY 400	0.15		1.25		63		0.02	0.5				28.0-34.0	81	36
ALLOY 600	0.15	14.0-17.0			72		0.015	0.5		0.15-0.60	1	0.5	80	35
ALLOY 800	0.06-0.1	19.0-23.0			30-35				0.15-0.60				72.5	25

Chart 3.5
Operating Temperature and Allowable Stresses in Pipe Walls for Seamless Ferritic Alloys Steel Pipes

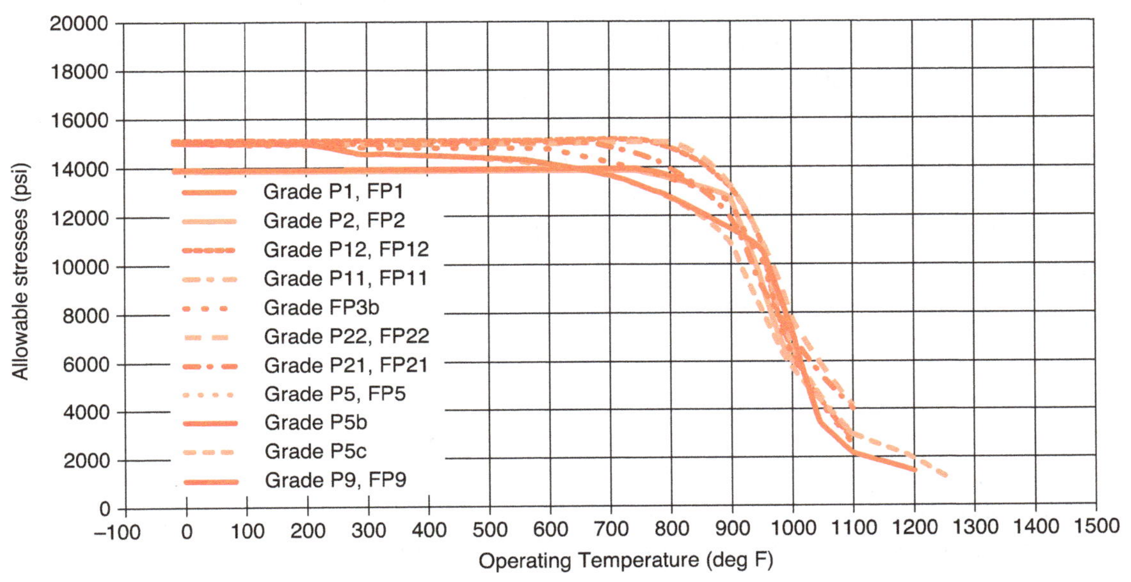

STAINLESS STEELS

Chart 3.6 summarizes the chemical composition and tensile strength for the commonly used stainless steels discussed in this section.

304/304L

304 stainless is the basic, most widely used stainless steel. Excellent fabrication and welding performance with general corrosion resistance attracts it to many industrial and non industrial purposes. Brinell hardness is approximately 160. Upper temperature limit is 1400°F/760°C.

304L has a carbon content of 0.03% or less. 304L is less prone to intergranular corrosion than 304 because low carbon content tends to reduce carbide precipitation along grain boundaries. However it may exhibit chloride stress corrosion cracking. Brinell hardness is about 140. Its varieties like piping, forgings, etc., are governed by the codes ASTM A312, A376, A358, A269, A249, A403, A182, A351; ASME SA312, SA376, SA358, SA269, SA249, SA403, SA182, and SA351.

316/316L

316 is more resistant to atmospheric and other corrosion conditions than SS304 because of the addition of molybdenum. It has better tensile, creep and stress–rupture strengths at higher temperatures. It resists dilute sulfuric acid (1–5%) up to 122°F/50°C. It is not hardenable and non magnetic. It has superior corrosion resistance to chlorides. It is an industry standard for pharmaceutical, fertilizer, nuclear plastic, and paper and rubber industries. Though it has temperature range up to 1400°F/760°C, it is not recommended within the carbide precipitation range of 800° to 1650°F (425°C to 900°C). Brinell hardness is approximately 160.

316L has much less carbon content in the limit of .03% and hence is less prone to stress corrosion cracking and intergranular corrosion than SS304 It has a continuous maximum temperature range of 1400°–1500°F (760°C to 815°C). Brinell hardness is about 140.

Its varieties like piping, forgings, etc., are governed by the codes ASTM A312, A376, A358, A269, A249, A403, A182, A351, A479, A276; ASME SA312, SA376, SA358, SA269, SA249, SA182, SA276, SA403, SA479, and SA351.

Chart 3.7 shows selected pressure ratings for standard seamless A312–TP316/316L stainless steel pipes.

317

SS317 has excellent corrosion resistance far superior to SS316, particularly in phosphoric acid service, and a host of other critical applications. 317L effectively resists intergranular corrosion with its low carbon content. It has higher tensile strength, creep, and stress rupture strength than other stainless steels. "M" and "N" suffixes indicate higher levels of molybdenum and nitrogen respectively, which are effective in improving resistance to pitting and crevice

Chart 3.6
Chemical Composition and Tensile Strength Details for Commonly Used Stainless Steels

MATERIAL	CHEMICAL COMPOSITION PERCENTAGES, BALANCE Fe											TENSILE STRENGTH KSI	
	C	Cr	Mn	Mo	Ni	P	S	Si	Ti	Co	Ta	Y S	T S
304	0.035	18.0-20.0	2		8.0-13.0	0.04	0.03	0.75				70	25
316	0.035	16.0-20.0	2	2.0-3.0	10.0-15.0	0.04	0.03	0.75				70	25
317	0.035	18.0-20.0	2	3.0-4.0	11.0-15.0	0.04	0.03	0.75				75	30
310	0.08	24.0-26.0	2	0.75	19.0-22.0	0.045	0.03	0.75				75	30
321	0.08	17.0-20.0	2		9.0-13.0	0.04	0.3	0.75	Trace			75	30
347	0.08	17.0-20.0	2		9.0-13.0	0.04	0.3	0.75		Trace	Trace	75	30
410	0.15	11.5-13.5	1		0.5	0.04	0.03	0.75				60	30

Chart 3.7
Pressure Ratings for Standard Seamless A312–TP316/316L Stainless Steel Pipes — Temperatures 100°F to 750°F, based on ANSI/ASME B 31.1

Pressure Rating (psig)										
Pipe Size (inches)	Pipe Schedule	Temperature (°F)								
		100	200	300	400	500	600	650	700	750
1″	40	3048	2629	2362	2171	2019	1924	1867	1824	1810
1″	80	4213	3634	3265	3002	2791	2659	2580	2528	2501
1″	160	6140	5296	4759	4375	4068	3876	3761	3684	3646
1.5″	40	2257	1947	1750	1608	1496	1425	1383	1354	1340
1.5″	80	3182	2744	2466	2267	2108	2009	1949	1909	1889
1.5″	160	4619	3984	3580	3291	3060	2916	2829	2772	2743
2″	40	1902	1640	1474	1355	1260	1201	1165	1141	1129
2″	80	2747	2369	2129	1957	1820	1734	1682	1648	1631
2″	160	4499	3880	3486	3205	2980	2840	2755	2699	2671
3″	40	1806	1558	1400	1287	1196	1140	1106	1084	1072
3″	80	2553	2202	1979	1819	1691	1612	1564	1532	1516
3″	160	3840	3312	2976	2736	2544	2424	2352	2304	2280
4″	40	1531	1321	1187	1091	1014	967	938	919	909
4″	80	2213	1909	1715	1577	1466	1397	1355	1328	1314
4″	160	3601	3106	2791	2566	2386	2273	2206	2161	2138
5″	40	1342	1158	1040	956	889	847	822	805	797
5″	80	1981	1709	1535	1411	1312	1250	1213	1189	1176
5″	160	3414	2945	2646	2433	2262	2155	2091	2049	2027
6″	40	1219	1052	945	869	808	770	747	732	724
6″	80	1913	1650	1483	1363	1267	1208	1172	1148	1136
6″	160	3289	2836	2549	2343	2179	2076	2014	1973	1953
8″	40	1073	926	832	765	711	678	657	644	637
8″	80	1692	1459	1311	1205	1121	1068	1036	1015	1005
8″	160	3175	2738	2460	2262	2103	2004	1944	1905	1885
10″	40	974	840	755	694	945	615	596	584	578
10″	80	1609	1388	1247	1147	1066	1016	986	966	956
10″	160	3147	2714	2439	2242	2085	1986	1927	1880	1868

corrosion at higher elevated temperatures. Nitrogen also increases the strength of these alloys. These alloys are widely used in the production, concentration of phosphoric acid, and in flue gas desulfurization (FGD) systems and in pharma, nuclear, plastic, paper, and rubber industries.

Its varieties like piping, forgings, etc., are governed by the codes ASTM A312, A403, A182; ASME SA312, SA182, SA403, and SA479.

310

SS310 exhibits a lower coefficient of expansion than most of the 300 series stainless steels. Its creep strength is better than the others at elevated temperatures. It resists oxidation at temperatures 1100°C/2012°F (1895°F/1035°C) under cyclic conditions. High chrome and nickel allows it to be used in sulfur atmospheres. It lends itself to easy fabrication. Ducting, piping, and furnace linings in sulfur bearing gases, and heat exchanger tubes are some of the common applications.

Its varieties like piping, forgings, etc., are governed by the codes ASTM A312, A403, A182; ASME SA312, SA403, and SA182.

321

SS321 is stabilized against carbide precipitation and can operate in the severe temperatures where carbide precipitation occurs. It is alloyed with titanium, which combines with carbon and nitrogen to form titanium carbide and nitrides. This leaves chromium in solution free to fight corrosion. It is good in continuous services up to 1472°F–1652°F (800°C–900°C). It may exhibit chloride stress corrosion cracking. 321H has better high temperature creep resistance with higher amounts of carbon (.04–10). SS321 is used in heat exchangers operated at higher temperatures, and high temperature piping in refineries, power plants, and fertilizers. Brinell hardness is about 150.

Its varieties like piping, forgings, etc., are governed by the codes ASTM A312, A403, A182, A479, A276; ASME SA312, SA403, SA182, SA479, and SA276.

347

SS347 resists intergranular corrosion by eliminating carbide precipitation because of the addition of columbium. Hence it has better corrosion resistance than Type 321 and much better resistance than 304. SS347 can be used in high temperature services in continuous service at 1652°F/900°C and intermittent services at 1472°F/800°C. SS347H has better high temperature creep resistance due to higher amounts of carbon (.04 – 10). SS347 is used in heat exchangers operated at higher temperatures, and high temperature piping in refineries, power plants, and fertilizers. Brinell hardness is approximately 160.

Its varieties like piping, forgings, etc., are governed by the codes ASTM A312, A403, A182, A479, A276; ASME SA312, SA403, SA182, SA479, and SA276.

Alloys 310, 321, and 347 stainless steels are used in high temperature service because of their higher creep and stress rupture properties. 321 and 347 alloys have maximum temperatures of 1500°F/816°C for specific applications whereas Alloy 304L is limited to 800°F/425°C.

410

SS410 is used in services requiring abrasion and wear resistance combined with general corrosion resistance. It is martensitic stainless steel with high mechanical properties. It is magnetic and resists oxidation and scaling up to 1200°F/650°C. Because of its hardness, it is best used in valve seatings, hardened steel, balls, pump parts, springs, and in slurry piping and pipelines handling solids like coal. It can be heat treated to increase strength, hardness, and wear resistance. It is not recommended in severe corrosive services. Brinell hardness is around 155.

Its varieties like piping, forgings, etc., are governed by the codes ASTM A268, A815, and A182.

It is difficult to identify which stainless steel is what grade.

Chart 3.8 indicates the operating temperature and allowable stresses in pipe walls for seamless austenitic alloy steel pipes. **Chart 3.9** shows different methods for distinguishing stainless steel.

Chart 3.8
Operating Temperature and Allowable Stresses in Pipe Walls for Seamless Austenitic Alloy Steel Pipes

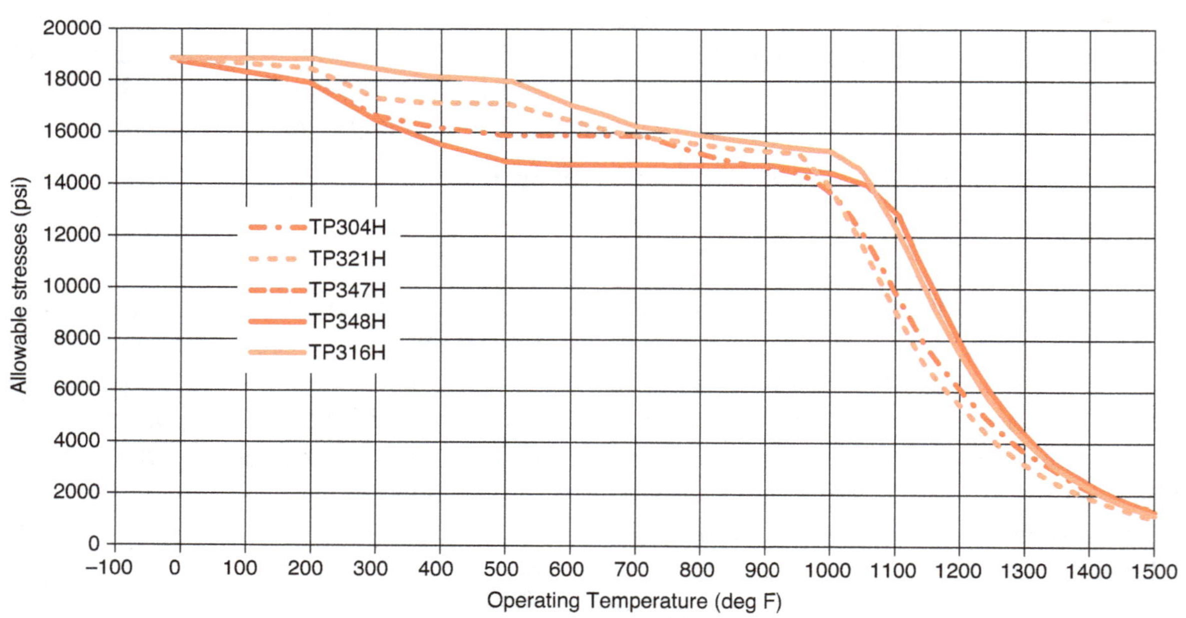

Chart 3.9
Different Methods for Distinguishing Stainless Steel

<table>
<tr><th>AISI Type</th><th>Grade</th><th>Group</th><th>Magnet Test</th><th>Spark Test</th><th>Hardness Test</th><th>Sulfuric Acid Test</th><th>Hydrochloric Acid Test</th></tr>
<tr><td>302</td><td rowspan="11">Chromium-Nickel</td><td rowspan="11">Austenitic</td><td rowspan="11">Non-Magnetic</td><td rowspan="4">Few forks - short, reddish</td><td rowspan="11">>165 Brinell after heated to 1800°F and water quench</td><td>Strong attack - green crystals and dark surface</td><td>Rapid reaction - pale blue - green solution</td></tr>
<tr><td>303</td><td></td><td>Spoiled egg odor - heavy black smudge</td></tr>
<tr><td>303Se</td><td></td><td>Garlic odor</td></tr>
<tr><td>304</td><td>Strong attack - green crystals and dark surface</td><td>Fast attack - gas formation</td></tr>
<tr><td>308</td><td rowspan="3">Full red without many forks</td><td></td><td></td></tr>
<tr><td>309</td><td></td><td></td></tr>
<tr><td>310</td><td></td><td></td></tr>
<tr><td>316</td><td>Few forks - short, reddish</td><td rowspan="2">Slow attack - tan surface turns brown</td><td>Very slow attack</td></tr>
<tr><td>317</td><td></td><td></td></tr>
<tr><td>321</td><td></td><td></td><td rowspan="2">Fast attack - gas formation</td></tr>
<tr><td>347</td><td></td><td></td></tr>
</table>

Chart 3.9 Continued

<table>
<tr><td>410</td><td rowspan="11">Chromium</td><td rowspan="7">Martensitic</td><td rowspan="11">Magnetic</td><td rowspan="3">Long white with few forks</td><td rowspan="7">> 280 Brinell after heated to 1800°F and water quench</td><td></td><td>Rapid reaction - dark green solution</td></tr>
<tr><td>414</td><td></td><td></td></tr>
<tr><td>416</td><td></td><td>Spoiled egg odor - heavy black smudge</td></tr>
<tr><td>416Se</td><td></td><td></td><td>Garlic odor</td></tr>
<tr><td>420</td><td>Long white-red with burst</td><td></td><td></td></tr>
<tr><td>431</td><td>Long white with few forks</td><td></td><td></td></tr>
<tr><td>440 A, B, C</td><td>Long white-red with burst</td><td></td><td></td></tr>
<tr><td>430</td><td rowspan="4">Ferritic</td><td>Long white with few forks</td><td rowspan="4">180 - 250 Brinell after heated to 1800°F and water quench</td><td></td><td></td></tr>
<tr><td>430F</td><td></td><td></td><td>Spoiled egg odor - heavy black smudge</td></tr>
<tr><td>430FSe</td><td></td><td></td><td>Garlic odor</td></tr>
<tr><td>446</td><td>Full red without many forks</td><td></td><td></td></tr>
</table>

SUPER STAINLESS STEELS

Chemical composition and tensile strength details are given for some of the more specific and specialized stainless steels, sometimes called super stainless steels. **Chart 3.10** tabulates them.

Chart 3.10
Chemical Composition and Tensile Strength Details for Super Stainless Steels

MATERIAL	CHEMICAL COMPOSITION PERCENTAGES, BALANCE Fe													TENSILE STRENGTH KSI	
	C	Cr	Mn	Mo	Ni	P	S	Si	Co	N	V	W	Cu	YS	TS
254SMO	0.02	19.5-20.5	1	6.0-6.5	17.5-18.5	0.03	1.01	0.8		0.18-0.22			0.5-1.0	94	44
2205	0.03	21.0-23.0	2	2.5-3.5	4.5-6.5	0.03	0.02	1		0.08-0.2				65	90
C276	0.01	14.5-16.5	1	15.0-17.0	54-55	0.04	0.03	0.08	2.5		0.35	3.0-4.5		100	41
C22	0.01	22	0.5	13	56			0.08	2.5		0.35	3		100	41
ALLOY 20	0.07	19.0-21.0	2	2.0-3.0	32.0-38.0	0.045	0.035	1					3.0-4.0		
904L	0.02	19.0-23.0	2	4.0-5.0	23.0-28.0	0.045	0.035	1					1.0-2.0	70	25

254SMO

This alloy is used where other steels fail, in severe corrosive conditions normally seen in high chloride environments such as bleaching plants in paper and pulp mills, and sea water and brackish water systems. It is much stronger and more ductile, and it has higher impact strength than other austenitic steel varieties. It shows far better characteristics than 317 alloys in pitting, stress corrosion cracking, and crevice corrosion. It adapts to easy fabrication and welding. It is proprietary of Avesta-Sheffield.

Its varieties like piping, forgings, etc., are governed by the codes ASTM A312, A403, A182; ASME SA312, SA403, and SA182.

2205

Avesta-Sheffield 2205 Duplex steel combines the advantages of ferritic and austenitic steels. It combines high mechanical strength with excellent corrosion resistance against pitting, stress, corrosion, and cracking. Balanced composition and microstructure yield to good weldability without risking the weld-affected zone to corrosion. It exhibits higher strength and low thermal expansion, but higher thermal conductivity. It is exclusively used in difficult services for heat exchangers and pipes in fertilizer, desalination plants, pressure vessels, tanks and piping, and other components. Its varieties like piping, forgings, etc., are governed by the codes ASTM A790, A815, A182; ASME SA790, SA815, and SA182.

C276

This steel resists stress corrosion cracking, and pitting, far better than other stainless steels. It has exceptional corrosion resistance in oxidizing atmospheres and resists oxidation up to 1900°F/1038°C. It has better fabricability and does not risk the weld-affected zone to corrosion, thus maintaining corrosion resistance in welded joints. It is used in severe conditions in fertilizer, pharma, and paper industries. Brinell hardness is about 210.

Its varieties like piping, forgings, etc., are governed by the codes ASTM B619, B336, B564, B574; ASME SB619, SB336, SB564, and SB574.

HASTELLOY® C-22®

This steel resists corrosion much better than C-276 and C-4. It has excellent resistance to oxidizing aqueous media including wet chlorine and mixtures containing nitric acid or oxidizing acids with chloride ions. C-22 alloy has outstanding resistance to pitting, crevice corrosion, and stress corrosion cracking.

Its varieties like piping, forgings, etc., are governed by the codes ASTM B-574, B-575, B-619, B-622, and B-626; ASME SB-574, SB-575, SB-619, SB-622 and SB-626.

904L

904L is a super stainless steel used under severe corrosive conditions, such as in dilute sulfuric acid and severe phosphoric acid applications. It exhibits good resistance to pitting and crevice corrosion and intergranular corrosion; it has very good resistance to stress corrosion cracking. It allows itself for easy fabrication, but service temperature is limited to 840°F/450°C. It is used in sulfuric acid and phosphoric acid plants, paper, and pharmaceutical industries and in sea water service for its excellent corrosion resistance.

Its varieties like piping, forgings, etc., are governed by the codes ASTM B677, B366; ASME SB677, and SB366.

ALLOY 20

The combination of nickel, chromium, molybdenum and copper in Alloy 20 has a profound effect on its resistance to chloride stress corrosion cracking and pitting attack. Columbium minimizes carbide precipitation during welding. It shows excellent corrosion resistance to phosphoric acid, boiling sulfuric acid at 20–40% concentration. It is one of the earliest super stainless steels which also exhibit good mechanical properties and fabricability. It was originally designed with sulfuric and phosphoric acid industries in mind; it now finds extensive use in pharmaceuticals, food, paper, and plastics industries. Maximum temperature range of 1400°–1500°F (760–815°C). Brinell hardness is about 160.

Its varieties like piping, forgings, etc., are governed by the codes ASTM B729, B464, B366, B473, B462; ASME SB729, SB464, SB366, SB473, and SB462.

NON METALLIC PIPING

PLASTIC PIPE

Plastic pipes are also brittle but they are better than cast iron. These pipes are slowly invading steel and cast iron domains. They must be properly supported and the temperature and pressure should be kept in limits. There are several varieties of plastic pipes used in the industry, like PP, PTFE, PVC, CPVC, HDPE, etc. Plastics are easy to handle because of their light weight. They are indispensable for certain services because of their specific corrosion resistance. Light weight, availability of fittings, easy joining methods, ability to withstand sizable shock are some of the features. Most of them can be welded by heat or by solvents. These pipes are available now which can handle about 15kg/cm^2.

Plastic pipes are joined by one or more of these methods: solvent cement, heat welding, mechanical joining like screwed and flanged fittings, and IR welding. One kind of a plastic cannot be joined to the other except by mechanical means.

These pipes are manufactured by the extrusion method. However they exhibit a high expansion characteristic which must be kept in mind while fabricating or repairing these lines. A general comparison of expansion values of various materials is shown in **Chart 3.11**.

Chart 3.11
Comparison of Expansion Values of Various Materials

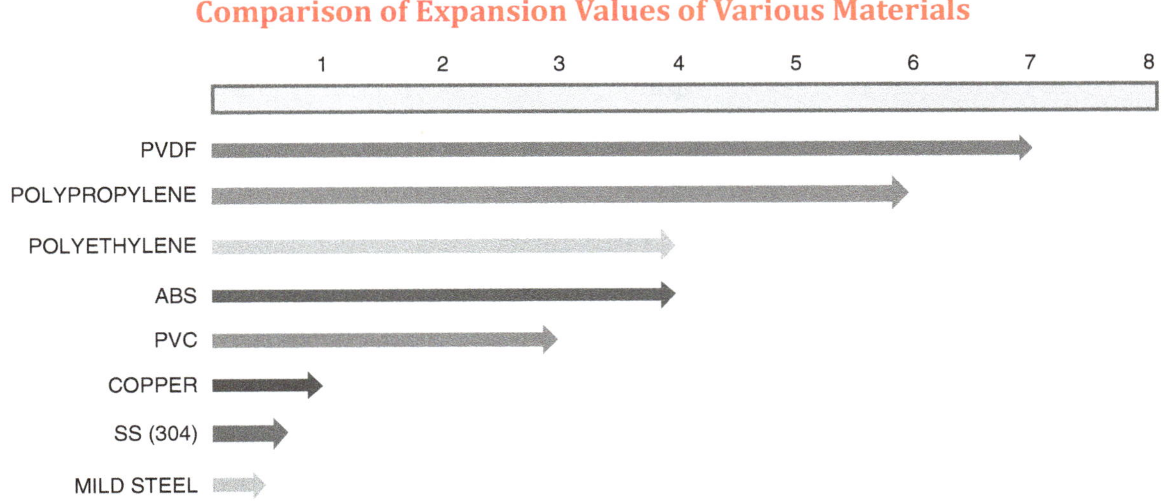

PVC-U PIPE

PVC-U (unplasticized polyvinyl chloride) is the most commonly used thermoplastic piping material. It has excellent chemical resistance and resists atmospheric corrosion. It has virtually replaced other materials in domestic potable water piping. It is light in weight and generally grey in color. Flow through PVC is smooth and PVC resists abrasion. Methods of joining PVC are much simpler and may not always require a skilled plumber to make domestic piping. The joints are made by threading or by solvent cement. Only schedule-80 pipes can be threaded.

These pipes should not be used for compressed air or gas. They are extremely corrosion resistant and lightweight. They offer low resistance to flow with temperature range up to 200°F/93°C. PVC-U is not recommended for aromatic and chlorinated hydrocarbon services and cannot operate beyond 210 psi or 15 kg/cm^2. Like most of the plastics, PVC exhibit high coefficient of expansion of five times that of steel. It is imperative that this be kept in mind while installing PVC lines.

CPVC

CPVC (chlorinated polyvinyl chloride) pipes are identical to PVC in corrosion resistance. CPVC is better suited for handling corrosive chemicals at temperatures 39–59°F (4 to 15°C) higher than PVC.

ABS

ABS (Acrylonitrile Butadiene Styrene) is widely used in food and beverage industries and in water and sewerage treatment. Its impact strength is high, but it exhibits a high coefficient of expansion like other thermo plastics. ABS piping and fittings can be joined by solvent cement. Piping is light in weight and comparatively rigid over PVC. It is not resistant to alcohol, petrol acetic acid, and other organic solvents although it is good against most organic and diluted inorganic acids, salts, animal fats, and oils. ABS can operate from –40 to 76°F (–40°C to +80°C) up to 210 psi or 15kg/cm^2. It is important to reduce the allowable pressure suitably when used at higher temperatures.

HDPE

HDPE (high density polyethylene) pipe has very good corrosion resistance against many chemicals and the choice material for municipal water transport, as it will not support biological growth. Pipe ID is smooth and maintains its flow capacity over a long period of time. HDPE pipe can be bent to a radius 25 times the nominal pipe diameter, a feature which can do away with many fittings required for directional changes. With its flexibility, HDPE pipe adopts itself on rough terrain. But like other plastics, these pipes are light in weight and also deflect. Hence the pipe needs to be restrained adequately and properly.

Black HDPE pipe stabilized with carbon black, the most effective ultraviolet stabilizer, can be used effectively outdoors where normal plastic pipes develop cracks. HDPE pipe can be used in a wide temperature range of –76 to 140°F (–60°C to +60°C). It is a cost-effective pipeline with an estimated service life of conservatively 50–100 years.

The coefficient of thermal expansion for unrestrained polyethylene is very high, about 10 times that of steel. It is a common experience for technicians working in HDPE pipes of this "unusual behavior." Pipeline measurements taken in the morning would be found totally wrong when the temperatures rise due to the sun. Suppose you try to change a length of pipe or fitting and take the corresponding measurements in the morning. Then you come back with the fitting or pipe to fix it after a few hours. You will find to your dismay that the pipe ends have come closer than you can fix the pipe or fitting. Exactly the opposite happens in the afternoon. HDPE may elongate by 1" in a 100' length for a 10°F rise in temperature.

HDPE pipe is easy to install. It is normally joined by heat fusion, which creates a leak-free joint stronger than the pipe itself. Joining time for the HDPE lines is small compared to steel, though dependent upon the pipe wall thickness and diameter, and the weather at the time.

METHOD OF JOINING HDPE PIPE

- *Pipe lengths and fittings to be joined should be secured rigidly and should move only as and when required.*
- *The mating ends of the pipe must be clean and parallel. They should be aligned so that the outside diameters of the pipes match, maintaining minimum required distance.*
- *Heat the ends of the pipe with a heating mirror between the pipes to the required temperature and pressure. By this time, a bead of molten material should be visible at the pipe ends.*
- *Then the heating mirror is removed and immediately the molten pipe ends are brought together with sufficient pressure. Pressure must be applied sufficiently and for the required time so that a homogenous joint is made.*
- *The pipes in the joints should not be moved at all until the joint is strong enough and achieved its integrity.*

POLYPROPYLENE (PP)

Polypropylene is also economical and offers good resistance to many acids and bases, but not to solvents. It can be used up to temperatures of 225°F/107°C, but it becomes brittle at low temperatures. It is not recommended beyond 21psi or 1.5kg/cm^2.

It is the lightest of all the thermoplastics, but strong and durable. It is resistant to most acids and alkalis but should not be used with active oxidizing agents like nitric acid and aromatics. They are joined by heating. They are socket welded by inserting a PP wire into it with the help of a hot air gun. Outer diameter is often covered with a FRP layer for additional protection.

PVDF (KYNAR®)

PVDF (polyvinylidene fluoride) is in the race with Teflon in corrosion protection. It offers excellent resistance to halogens and strong acids. It is tough and abrasion resistant and has a good temperature range of –0°F to 275°F (–18°C to 135°C). Chemicals with PH greater than 11 are best avoided for this material. It is used as a liner rather than as a straight pipe. It is not cheap like other plastics but boasts of exceptional qualities like resistance to ozone, UV and nuclear radiation and weather, and fungi. It offers high dielectric strength and high abrasion resistance. It is resistant to weathering and fungi. It has low permeability to most gases and liquids, and low flame and smoke characteristics. However, at higher temperatures it can dissolve in polar solvents such as organic esters and amines.

PTFE (TEFLON®)

PTFE (polytetrafloroethylene) (Teflon) is one of the boons given to the chemical industry which is virtually inert to most chemicals with a wide temperature range of 500°F/260°C to cryogenic. It cannot be processed by melting or by solvent cements, whereas PFA (TEFLON®) with the same characteristics is melt processable. It is a wonderful lining material.

PFA (TEFLON®)

PFA (perfluoroalkoxy) has all the excellent chemical resistance qualities of PTFE plus higher mechanical strength at higher temperatures. It resists stress cracking. PFA tubing is widely used in the services of ultra pure chemicals like in the semi-conductor, laboratory, environmental, and pharmaceutical industries. It is less abrasive resistant than PP and PVDF. It has a temperature range of –325°F to 500°F (–198°C to 260°C).

LINED PIPING

Metallic pipe is strong, and easy on cost and work. A lot of them are developed tailored to individual needs. Carbon steel is the most common material, but its corrosion resistance is very poor. Stainless steel scores better in corrosion resistance but is expensive.

Lined pipes combine the mechanical strength of steel and the corrosion resistance of plastics. They have wide operating range from –18°F to 500°F (–28°C to 260°C), offer low pressure drop in the industry, and are available in classes 150 and 300 and in other standards, in sizes ranging from 1/2" through 12" (15 to 300mm) and beyond. Liners can be bonded and locked or loose lined. Bonded plastic lined pipe can operate at full vacuum. Pipes are generally flanged in lengths of 6 meters (about 20ft) with the lined material protruding onto the flange faces.

A wide range of linings are available to choose from.

Polypropylene (PP) is the least expensive and is very common in the industry. It can be used from –0°F to 225°F (–18°C to 107°C) with a tensile strength of about 4000 to 4500 PSI (27000–31000 kpa). Pipe and fittings are not vented for safety. Polypropylene plastic lined pipe and fittings are the most common in the industry.

Polyvinylidene Fluoride (PVDF) (Kynar®) has a temperature range of –0°F to 275°F (–18°C to 135°C) with tensile strength of 4500 to 6500 PSI (31000–45000 kpa). It is resistant to most chemicals. Pipe and fittings are not vented for safety.

EthyleneChlorotrifluoroEthylene (ECTFE) (Halar®) has excellent mechanical properties (tensile strength 6000 to 7000 PSI (41000–48000 kpa)) coupled with exceptional chemical resistance (a wide variety of acids, bases, and solvents) and has a wide temperature range of –325°F to 300°F (–198°C to 149°C). ECTFE is resistant to abrasion with its smooth finish and good anti-stick properties. Permeation is not a problem with ECTFE and piping is not vented for safety.

EthylenetetraFluoroethylene (ETFE) (Tefzel®) has tensile strength of about 6000 to 7000 PSI (41000–48000 kpa). Its chemical resistance is close to Teflon with a wide temperature range of –18°F to 300°F (–28°C to 150°C). ECTFE is resistant to abrasion with its smooth finish and good anti-stick properties. Permeation is not a problem with ECTFE and piping is not vented for safety.

Polytetrafluoroethylene (PTFE) (Teflon®) Much has been written about Teflon, which is virtually inert to all chemicals except elemental fluorine and molten alkali metals. Temperature range is –325°F to 500°F (–198°C to 260°C). But it is softer and has bad cold flow characteristics. Microporosity is another problem, resulting in higher permeation, which requires higher thickness liners. This of course increases the overall strength (Tensile strength about 3000 PSI/21000 kpa).

Perfluoroalkoxy (PFA) (Teflon®) has higher tensile strength of about 4000 to 4500 PSI (27000–31000 kpa) with a temperature range of –325°F to 500°F (–198°C to 260°C). It does not creep or cold flow like Teflon but has the same corrosion resistance. Pipe and fittings are vented.

PVC and CPVC are rarely used as lining material because of low temperature range. However, they are widely used as a straight pipe in most domestic water applications.

Rubber is a rather loose term about a variety of natural and synthetic rubbers like natural rubber, Neoprene, Hypalon, Butyl, Chlorobutyl, Nitrile, EPDM, etc. It is the oldest lining system to be employed. Each kind of rubber has it is own characteristic advantages and disadvantages. Rubbers are available as soft for abrasive applications, semi-hard for general applications, and hard for severe applications. Temperature (200°F/93°C for neoprene) and pressures (rarely beyond 150 class) are limited. Multi-layer lining with soft and hard combination increases the temperature limitation.

Glass lined pipes are also used though they are shock prone, because of their extreme corrosion protection. They do not tolerate any thermal stresses, let alone mechanical shocks. They are joined by split flanges preferably with PTFE gaskets. They stand a temperature range of –18°F to 450°F (–29° to 232°C), though sudden fluctuation will surely crack the lining.

Cement mortar lined pipes were used traditionally for water and sewerage transport. HDPE has taken over in water transport, though older systems with cement mortar lined pipes are in service.

HANDLING, STORAGE AND MORE OF PLASTIC-LINED PIPING

- *Pipe line can be hydrostatically tested 1.5 times the design pressure, typically with water or with any other suitable liquid if water is contra-indicated for the process. A pneumatic test is not recommended for the inherent dangers involved with it.*
- *Main pipe is generally carbon steel schedule 10 through 80 as required, in lengths of 6 meters. Pipe grade may be ERW (electric resistance welded) ASTM A-587, or A-53 or A-106 seamless with corresponding flanges and fittings. Ductile iron, or galvanized or stainless steel are also made on special order.*
- *No welding should be carried out on the piping after the lining is carried out and needless to repeat, the same after the piping is installed.*
- *After lining, all pipe and fittings must be individually tested with a minimum of 10,000 volt non-destructive electrostatic spark test.*
- *Inspect all pipe and fittings visually for any damages and/or imperfections before installation.*
- *Gaskets are not used on the lined piping and fittings. However, if the pipe line is reused, it is a good practice to use compatible gaskets. If they are connected to unlined pipes, it is better to use gaskets.*
- *Over-tightening the lined flanges is not a good practice as plastics creep and will eventually crack.*
- *Long straight runs of pipes, even when they are flanged, should be avoided. Loops allow for expansions.*
- *Do not block weep holes if there are any on the lined pipes with paint or insulation.*
- *To obtain maximum performance from plastic-lined piping products, it is important that the end faces and surfaces of the plastic are protected from damage during storage, handling, and installation.*
- *These pipe and fittings are supplied with protective end caps which should be left in place until ready for installation. Do not damage the plastic lining while removing the end caps.*
- *Plastic and plastic-lined pipes are best avoided for use and storage in sub-zero temperatures unless well known of their feasibility. They can become brittle in low temperatures.*
- *Take adequate care when using mechanical means to transport them. Rough handling can definitely crack or break these pipes.*
- *Ultra violet light can spoil them when stored in the sun for a long time.*

PROBLEMS

The major reason for piping and flange leaks in the plastic piping is improper supports. Pipe deflection is the contributory factor, which is limited to 6mm accepted as an industry standard.

Improper supporting can straightaway lead to the failure of the pipeline at its weakest point, but it can also induce environmental stress cracking (ESC). Plastic-lined pipe under mechanical or thermal stress can fail below the tensile strength of the material.

The suggested support distance for most plastic lined pipes are summarized in **Chart 3.12**.

Chart 3.12
Support Span of Plastic-Lined Pipes
Based on 6mm Pipe Deflection

Lined Pipe Size mm/inch	Only Lined Pipe Mtrs/ft	Lined Pipe with Water Mtrs/ft
25 / 1	4.3 / 14	3.9 / 12
40 / 1 ½	5.2 / 17	4.9 / 16
50 / 2	5.8 / 19	5.5 / 18
80 / 3	7.0 / 22	6.7 / 22
100 / 4	7.9 / 26	7.3 / 24

Permeation is another problem with plastic-lined pipes with increased temperature, pressure, and concentration contributing to the problem. Weep holes are provided to release the accumulated liquid, but they can be blocked by paint, rust, etc. But again these holes can also structurally weaken the pipe, more so when the leaked liquid further corrodes the piping.

Another concern is the grounding or making an electrical continuity of plastic piping. This is essential to counter the buildup of static electricity. It is normally achieved by connecting a strip of electrical cable across each flange joint.

Materials for Valves and Fittings

Valves are indisputable leaders that control the fluid power and organize to enable, stop, direct, prevent backflow, or control flow or pressure. There are different types of valves like gate, globe, ball, check, butterfly, and needle valves that serve a variety of purposes. Gate, globe, and check valves are most widely used.

Valves have come a long way since the days when every manufacturer made its own kinds of valves which were often drilled at the user end to suit the flanges made by someone else. A huge number of conflicting pressure ratings and test pressures were listed by the early valve and fitting manufacturers. But slowly, increased concern for environment, general public, plant and worker safety, technological growth, and production demands forced creation of standards.

That places a heavier demand on the valve engineer in selecting, operating, and maintaining the valve and pipe fittings. Although there are a whole range of metallurgies available in front of us to choose from, a dizzying number of materials available to suit each and every application of the process industry make the selection a tough task. All aspects of valve manufacture, design, intricate details, design, functionality, inspection, and testing are standardized by various international organizations, foremost being ASME, ASTM, and ISO.

Each type of valve is classified by the pressure, MOC, and the construction type — flanged, screwed, socket welded, or butt welded. Pressure and temperature ratings are interdependent. Valves, which generally have their pressure rating marked on the body, should be at least equal or better than the design. The ASME codes specify which types of valves and fittings are permitted to be used and their proper service applications in pressure piping. Temperature and pressure of the service are interrelated. As the temperature goes up, the operating pressure must invariably be derated. Valves are manufactured with a large number of different materials of construction suiting to the demanding applications.

A more detailed description of valves is available in Chapter 16, Process Valves.

CAST IRON AND MALLEABLE IRON

CAST IRON

Cast iron has been the mainstay of the most valve bodies; ASTM A126 Classes A, B, and C are three classes of gray iron for castings intended for use as valve pressure retaining parts, pipe fittings, and flanges. Gray iron is economical to produce; it has been in production for centuries. However cast iron is brittle and will not tolerate temperatures beyond 446°F/230°C. Large valves for general service like water are cast iron valves and are rarely used beyond class 150.

Ductile iron (A395, A536) is sometimes referred to as spheroidal or nodular iron. Ductile iron (65,000 psi / 450,000 kpa tensile strength, 45,000 psi /31000 kpa yield, 12% elongation) is sometimes preferred to WCB as it has 50% higher yield strength properties and is more cost effective. However ductile iron castings have a maximum temperature rating of 652°F/345°C.

WCB is the most commonly used valve body material. (ASTM A216, WCB is used for carbon steel castings found in valves, flanges, fittings, and other pressure-containing parts for high-temperature service; these castings are of quality suitable for assembly with other castings or wrought-steel parts by fusion welding.) It contains a maximum of 0.3% carbon. Three grades — WCA, WCB, and WCC — are covered in this specification. Selection will depend upon design and service conditions, mechanical properties, and the high temperature characteristics. Grade B is generally used for non-corrosive fluids at temperatures higher than cast iron. It has yield strength of 70–95,000 psi, (482.000–655,000 kpa) tensile strength of 36,000 psi (248,000 kpa) and elongation 22.00%. It can be used for non-corrosive applications including water, oil, and gases at temperatures between –18°F and 800°F (–28°C and 425°C).

MALLEABLE IRON`

Screwed fittings are also made with malleable iron, particularly in service subject to expansion, contraction, stresses, and shock. Fittings made with MI are characterized by pressure tightness, stiffness, and toughness. Some valves are also made with malleable iron.

HIGH AND LOW TEMPERATURE STEELS

Chrome moly-steels used for higher temperatures and pressures are designated as the ASTM A217 series.

WC6 (1 1/4 Cr – 1/2 Mo) is used for non-corrosive applications including water, oil, and gases at temperatures between –18°F and 1100°F (–28°C and 593°C).

WC9 (2 1/4 Cr – 1 Mo) is used for non-corrosive applications including water, oil, and gases at temperatures between –18°F and 1100°F (–28°C and 593°C).

C5 (5% Chrome 1/2% Moly) is used for mild corrosive or erosive applications as well as non-corrosive applications at temperatures between –18°F and 1200°F (–28°C and 649°C).

C12 (9% Chrome 1% Moly) is used for mild corrosive or erosive applications as well as non-corrosive applications at temperatures between –18°F to 1200°F (–28°C and 649°C).

LCB Carbon steel for low temperature is designated with A352 LCB for low temperature applications to –51°F/–46°C and not recommended above 645°F/340°C. There are LC1, LC2, and LC3 grades for still lower temperatures.

STAINLESS STEELS

Stainless steel (ASTM A-351) grades cater to more corrosive services. They have carbon levels even lower than WCB (0.08% maximum.)

Some of the predominant grades are:

CF8, equivalent to 304 SS, is used for corrosive or extremely high temperatures non-corrosive services between 514°F and 1200°F (268°C and 649°C). For higher temperatures above 800°F/425°C, carbon content should be greater than 0.04%. CF3 is of low carbon content equivalent to SS304 L.

CF8M, equivalent to 316 SS, is used for corrosive or extremely low or high temperatures non-corrosive services between 514°F and 1200°F (268°C and 649°C). For higher temperatures above 800°F/425°C, carbon content should be greater than 0.04%. CF3M is of low carbon content equivalent to SS316 L.

CN7M is equivalent to Alloy 20 used for more severe services like phosphoric acid and hot sulfuric acid.

N12M (ASTM A494 Grade N12M) is the corresponding grade to **Hastelloy B** and is excellent for services in hydrofluoric acid at all concentrations and temperatures. It has good resistance to sulfuric and phosphoric acid atmospheres up to 1200°F/649°C.

CW12M (ASTM A494 Grade CW12 M) is the corresponding grade to **Hastelloy C** and has excellent resistance to sulfuric and phosphoric acid atmospheres 1200°F/649°C. It maintains properties at high temperatures.

CD4MCu (ASTM A351) is a duplex grade material excellent for sulfuric and phosphoric acids.

NON-FERROUS METALS

M 35 (ASTM 743 Grade M3-35-1) is the corresponding grade to Monel, which is extremely resistant to organic acids, sea water, and many alkaline solutions up to 750°F/400°C. It is weldable.

Bronze (ASTM B62) is generally used for salt water and other mild applications. B61 grade is used for valve trim. It should not be used for temperatures exceeding 550°F/288°C.

FORGED STEEL

Most of the smaller size valves (1/2" through 2" NPS) are of forged construction with ASTM A105 material. It is a low carbon, manganese, and silicon steel. Machinability is good and it is weldable. This material is used for the bonnet of the valves with bodies made up of WCA, B, C grades for higher size valves. A105 is a forged material whereas A216-WCB is cast. The chemical composition for both is essentially the same but with minor variations. WCB and A105 materials are not recommended above 800°F/427°C. Alloy steel valves must be considered beyond the limits of cast steels. Larger size valves are generally made with cast steel or its alloys. If an all-forged valve is required in those sizes, it may have to be specially made and supplied.

Chart 4.1 provides a list of cast materials and their corresponding forged class. Generally in valves, the bodies are cast while the bonnets are forged and other parts are made with similarly

Chart 4.1
Cross Reference of Materials and Corresponding ASTM for Cast, Forged, and Wrought Fittings, Flanges, Unions, and Valves

Material	Forgings	Castings	Wrought Fittings
Carbon Steel	A105	A216-WCB	A234-WPB
Cold Temperature Service	A350-LF2		A420-WPL6
Carbon-1/2 Molybdenum Alloy Steel			
High Temperature Service	A182-F1	A217-WC1	A234-WP1
		A352-LC1	
3-1/2 Nickel Alloy Steel			
Low Temperature Service	A350-LF3	A352-LC3	A420-WPL3
1/2 Cr-1/2 Mo Alloy Steel	A182-F2		
1/2 Cr-1/2 Mo-1 Ni Alloy		A217-WC4	
3/4 Cr-1 Mo-3/4 Ni Alloy Steel		A217-WC5	
1 Cr-1/2 Mo Alloy Steel	A182-F12 CL2		A234-WP12 CL2
1-1/4 Cr-1/2 Mo Alloy Steel	A182-F11 CL2	A217-WC6	A234-WP11 CL2
2-1/4 Cr-1 Mo Alloy Steel	A182-F22 CL3	A217-WC9	A234-WP22 CL3
5 Cr-1/2 Mo Alloy Steel	A182-F5		A234-WP5
5 Cr-1/2 Mo Alloy Steel	A182-F5a	A217-C5	
9 Cr-1 Mo Alloy Steel	A182-F9	A217-C12	A234-WP9
13 Cr Alloy Steel	A182-F6	A743-CA15	
304 Stainless Steel (18 Cr-8 Ni)			
Standard	A182-F304	A351-CF3	A403-WP304
Low Carbon	A182-F304L		A403-WP304L
High Temperature Service	A182-F304H	A351-CF8	A403-WP304H
310 Stainless Steel (25 Cr-20 Ni)	A182-F310H	A351-CK20	A403-WP310
316 Stainless Steel (16 Cr-12 Ni-2 Mo)			
Standard	A182-F316		A403-WP316
Low Carbon	A182-F316L		A403-WP316L
High Temperature	A182-F316H	A351-CF3M	A403-WP316H
		A351-CF8M	
317 Stainless Steel (18 Cr-13 Ni-3 Mo)			A403-WP317
321 Stainless Steel (18 Cr-10 Ni-Ti)			

Chart 4.1 Continued

Standard	A182-F321		A403-WP321
High Temperature Service	A182-F321H		A403-WP321H
347 Stainless Steel (18 Cr-10 Ni-Cb)			
Standard	A182-F347	A351-CF8C	A403-WP347
High Temperature Service	A182-F347H		A403-WP347H
348 Stainless Steel (18 Cr-10 Ni-Cb)			
Standard	A182-F348		A403-WP348
High Temperature Service	A182-F348H		A403-WP438H

Chart 4.2
Pressure Rating PSIG vs. Temperature F in ANSI Pressure Classes

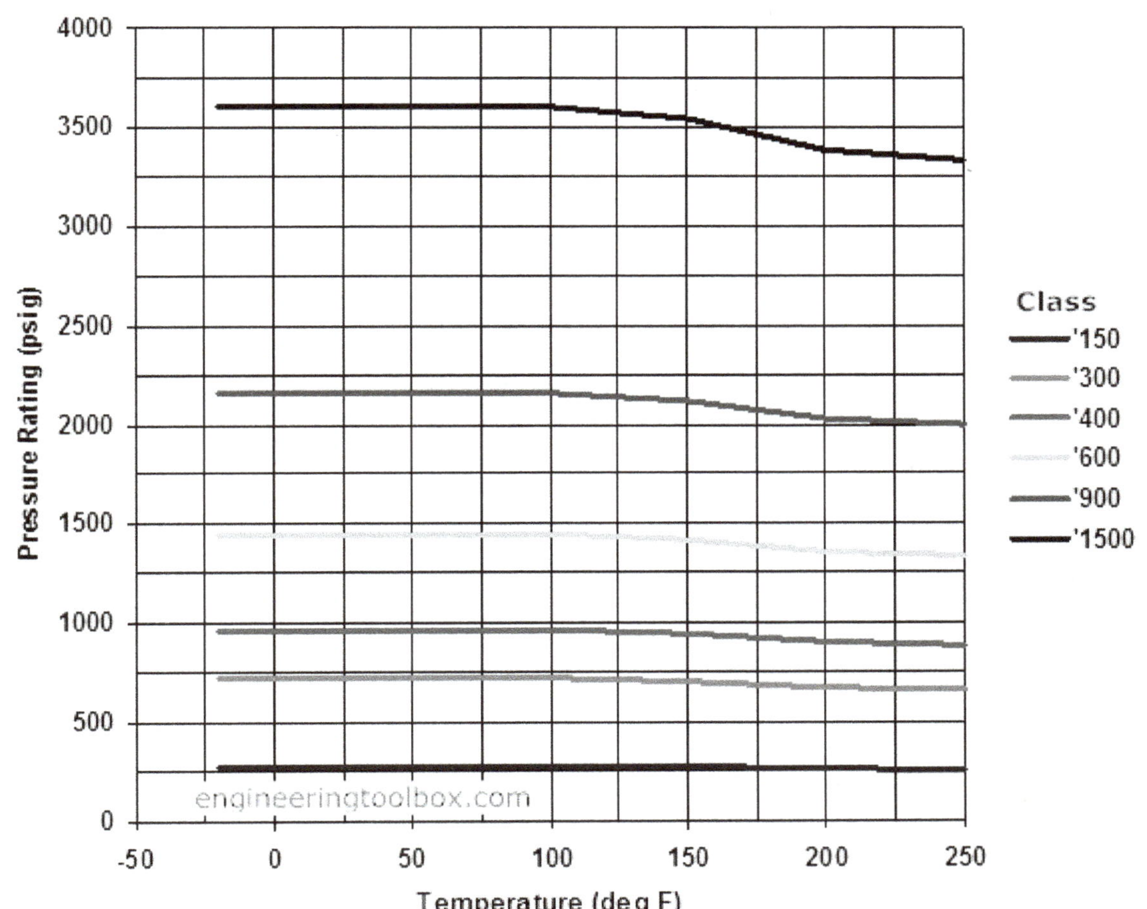

Chart 4.3
Pressure Rating Kg/cm² vs. Temperature C in DIN Classes

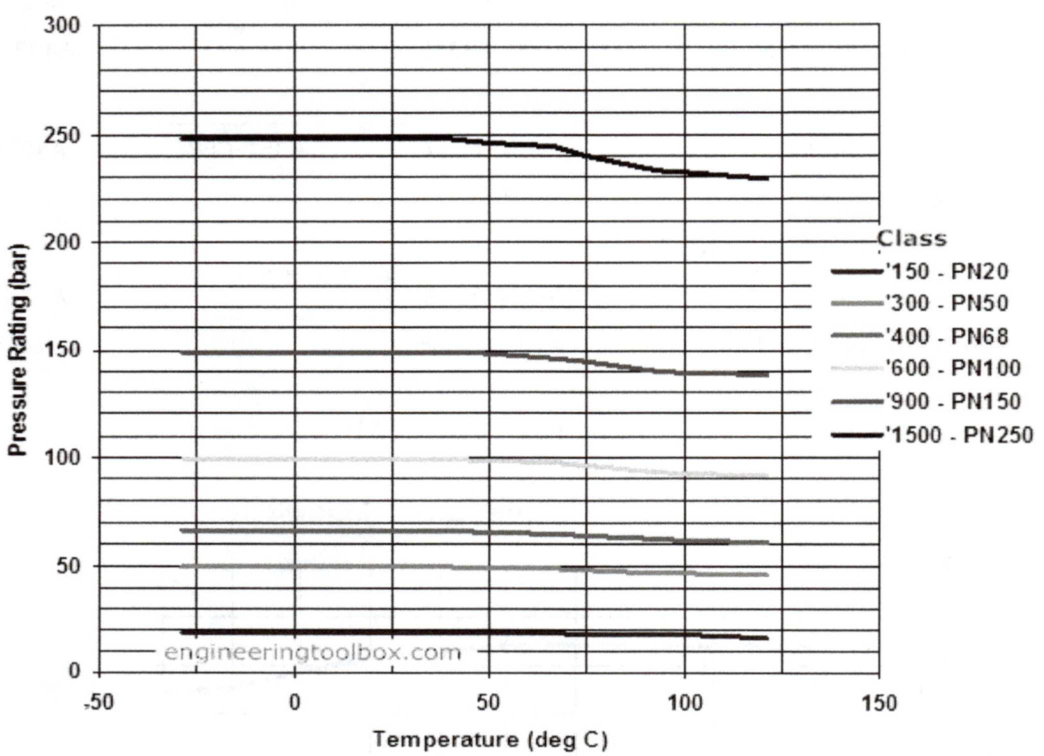

Chart 4.4
Temperature Limits for Valve Materials

TEMPERATURE LIMITS FOR VALVE MATERIALS			
VALVE BODY MATERIAL		DEGREES F/C	
		UPPER	LOWER
Carbon Steel	Grade WCB	1000/538	-20/-28
Carbon Steel	Grade LCB	650/343	-50/-46
Carbon Moly	Grade WC1	850/454	-20/-28
1 1/4 Cr – 1/2 Mo	Grade WC6	1000/538	-20/-28
2 1/4 Cr – 1 Mo	Grade WC9	1050/566	-20/-28
5 Cr – 1/2 Mo	Grade C5	1100/593	-20/-28
9 Cr – 1 Mo	Grade C12	1100/593	-20/-28
3 1/2 Ni	Grade LC3	650/343	-150/-100

Chart 4.4 Continued

Type 304 SS	Grade CF8	1500/816	-425/-260
Type 316 SS	Grade CF8M	1500/816	-425/-260
Type 347 SS	Grade CF8C	1500/816	-425/-260
Alloy 20		300/150	-50/-46
Hastelloy B		700/370	-325/-198
Hastelloy C		1000/538	-325/-198
Aluminium		400/204	-325/-198
Bronze		550/288	-325/-198
Inconel		1200/649	-325/-198
Monel		900/482	-325/-198
Cast Iron		410/210	-20/-28
Ductile Iron		650/343	-20/-28

suitable or better grades. For corrosive services, stainless steels must be considered and, if the temperatures and pressures are not high, thermoplastics should be used.

Chart 4.2 and **Chart 4.3** graph the pressure rating vs. temperature, ANSI classes in Chart 4.2 and DIN classes in Chart 4.3. **Chart 4.4** then summarizes the temperature limits for various valve materials.

PLASTICS

Plastic valves have virtually taken over the metals in domestic water piping because they are cheaper, offer tight shut off, and are easily fixed and maintained. Even though all-plastic valves are occasionally available, metals are still used somewhere in the plastic valves. Plastics may be reinforced with metal inside, and the fasteners are almost always metal, often stainless steel. Springs may be metal, sometimes encapsulated, but sealing elements are thermoplastics in these non-metallic valves. Within the wide choice of thermoplastics available, it is sometimes difficult to select the right material, even after weighing advantages and disadvantages.

PVC (Polyvinyl Chloride) is a proven plastic with over 30 years standing and is the most widely used thermoplastic. PVC offers very good chemical resistance to a wide range of acids, alkalis, salt solutions, and chemicals, but not for some solvents, aromatics, and chlorinated organic compounds. The maximum service temperature of PVC is 140°F/60°C. PVC is joined by solvent cementing, threading, or flanging. PVC may develop embrittlement after several years of exposure to sunlight. Ultraviolet light used in the sterilizers for deionized water can affect PVC.

CPVC (Chlorinated Polyvinyl Chloride) is very similar to straight PVC, but can handle corrosive liquids within its range up to 180°F/82°C. CPVC is also joined by solvent cementing, threading, or flanging.

Polypropylene has low specific gravity and offers good chemical resistance to a wide range of acids, alkalis, and organic solvents, but not for strong oxidizing acids, chlorinated hydrocarbons, and aromatics. The maximum service temperature is 180°F/82°C. Polypropylene is joined by heat fusion, threading, or flanging. Glass-filled polypropylene provides higher mechanical strength and consequently higher pressure and temperature limits.

FLUOROPOLYMERS

PVDF (Polyvinylidene Fluoride) has far better mechanical strength, corrosion, and abrasion resistance — even better than PTFE and PFA — as well as a higher temperature limit than all other thermoplastics. PVDF is an excellent material for transport of high purity deionized water and is resistant to halogens like chlorines, bromine, and many strong acids, mild alkalies, and organic solvents. Service temperature limit is 280°F/138°C with its melting point at 338°F/170°C. PVDF is joined by heat fusion, threading, or flanging.

PTFE (Polytetrafluoroethylene) is a wonder material that is resistant to most chemicals and solvents; it can be easily formed into various shapes and sizes. It is an excellent material for valves, diaphragms, and bellows because of its corrosion resistance, low friction, and non-sticky characteristics. It can safely be used up to 300°F/149°C. PTFE is joined by threading or flanging. It is often used to line or encapsulate other components. PTFE rings are extensively used as seat rings in the ball valves and as gland packing in many valve applications. PTFE composites can handle higher pressure temperature, typically up to 500°F/260°C. Carbon filled or glass filled Teflon materials are extensively used to impart lower wear rate, reduce abrasion, and creep.

PFA (Perfluoroalkoxy resin) is equal to PTFE in chemical resistance, is melt processable, and has formability with the same dielectric constant.

Borosilicate glass is used in glass piping systems in the pharmaceutical, food, and chemical industries for specific applications. It is extremely fragile but virtually resistant to acids and alkalis, except hydrofluoric acid and sodium hydroxide. It is routinely used on sight glasses and level indicators.

Acrylics are extensively used as sight glasses and level indicators, though some valves are also available. Low abrasion and opacity, and resistance to UV light are its qualities.

METHODS OF PIPE MANUFACTURE

Piping in modern process plants is not simple piping alone, but a culmination of complex disciplines like design, metallurgy, manufacturing process, codification and strict standardization, and maintenance. Most of the modern processes depend largely on piping using a whole range of pipes, fittings, and other components. Greatly evolved technology and highly developed skills make them viable and reliable.

The quality and reliability of pipes have improved a lot because of the continuous developments in materials, metallurgy, welding, and inspection techniques. Pipe is now subjected to continuous structural and dimensional stability in the manufacturing process and non-destructive tests such as ultrasonic and x-ray soon after. Each section of pipe is also pressure-tested before leaving the manufacturer. Furthermore, as a good engineering practice, new pipelines are also pressure-tested at the user end.

The manufacture of steel pipe dates from the early 1800s; pipe was originally manufactured by hand. Since then, manufacturing processes have gone through a tremendous change after about 1812 when automatic processes were introduced.

Three principal manufacturing methods are in use: electric resistance welding (ERW), Butt (continuous) welding, and seamless. Welded pipes use flat-rolled steel as raw material whereas the seamless pipes use steel billets. Presently most of the process industries like oil, gas, power, and fertilizer use seamless pipe for process applications. Longitudinally-welded or ERW pipes are more common in other chemical industries like pharmaceutical, food, and diary. Though initially discounted, the use of spirally-welded pipe is again slowly increasing in the industries, particularly for larger diameter range.

LAP WELDING

One of the earlier methods of pipe manufacture — lap-welded pipe — is no longer used, as the joints proved to be defective and the joint strength was inferior. (As per the American Society of Mechanical Engineers, ASME, seamless pipes have a joint factor of 1.0 and lap-welded pipes 0.6.) Steel was rolled into the shape of a cylinder as a pipe after heating in a furnace. The edges were overlaid or lapped together in a process called *scarfing* and then the joint was welded.

ELECTRIC RESISTANCE WELDED PIPING

Most process plant engineers consider ERW as a second-grade pipe and use it only for less risky applications. But the cost and availability of seamless pipe puts ERW pipe back into process lines. Improved manufacturing techniques — with better strips of steel, heat treatment, and automatic inspection — are yielding ERW pipes that are better than they were once.

Originally, electrical resistance was used to make a longitudinal pipe joint with straight electrical heat and weld (ERW), but now high frequency induction heating (HFI) is used by most pipe manufacturing plants for better quality. However this piping is still called ERW pipe. Butt joint is welded using double submerged arc welding (DSAW).

Pipes start their life at a steel mill in sheet form. They are manufactured from basic flat-rolled materials known as skelp, plates, and strip. A sheet of steel is cold formed into a cylindrical shape. Low frequency A.C. current is used to heat the edges in this manufacturing process. Current is passed between the two edges of the steel to heat the steel to a point at which the edges are forced together to form a bond without the use of welding filler material. Weld flash is removed and the weld zone is then heat treated to get homogeneity. The pipe is processed through an on-line nondestructive testing. The pipe is then sized for dimensional accuracy through a series of sizing rolls. The welds with low frequency were found to be susceptible to seam corrosion, hook cracks, and insufficient bonding of the seams.

From the 1970s, a process of high frequency welding was developed which produced a higher quality weld. Presently most of the pipe manufacture is by high frequency induction heating (HFI) though the pipes are still called ERW pipes. The pipe seam is then welded on both sides with electrodes while the pipe is double submerged under flux. This results in a strong joint as weld penetration is full. Even pipes that are 18 meters long are made by this process.

Alternatively, sheets are rolled into helical form of suitable size and welded spirally. This process helps large diameter pipes to be produced from narrower plates or skelp. But the pipe may suffer from poor dimensional accuracy, as the pipe ends are out of round and not true. This may lead to difficulty in proper field fit up.

SEAMLESS PIPE

Seamless pipe is used extensively as standard specification for petrochemical, fertilizer, and other similar process industries. Seamless pipe is highly tolerant of internal pressures. Even though the cost is higher, seamless pipe does away with weld and other seam defects. In the manufacture of seamless pipe, a mandrel is used to pierce a hot round steel billet. The tube obtained is then further rolled and stretched to get the finished length and diameter.

However larger size pipes of 6 to 16 inches (150 to 400 mm) diameter are made in a plug mill. Smaller sizes of 1 to 6 inches (25 to 150 mm) diameter are manufactured in a mandrel mill.

PLUG MILL

An ingot of steel of about two tons is heated to 2372°F/1300°C and pierced with a mandrel. This hole is further enlarged on a rotary elongator. The result is known as a bloom and the tube wall is still thick. A plug with approximately the same diameter as the finished inside diameter of the pipe is then pushed through the bloom. Then the entire bloom and plug are passed between the rotating rolls of the plug mill and the tube is also rotated by 90° for each pass. This step effectively reduces wall thickness and roundness is ensured. A reeling mill smoothes the internal and external surfaces and gets the proper finish and final dimensions are made by a sizer.

The rotating action of the rolls makes the hot metal flow over and the mandrel makes a hollow pipe shell. The pipe grows considerably in length during the piercing and sizing processes and the diameter is reduced as required. The tube is then cut to length and subjected to heat treatment, finish straightening, NDT, and hydro test.

MANDREL MILL

The steel ingot is heated to 2372°F/1300°C and pierced. A mandrel is inserted into the tube and both are passed through a rolling mandrel mill. Here pipe wall thickness is continuously reduced by a number of curved rollers. These rollers are fixed to each other in pairs at 90° angles. The tube is reheated and the pipe is passed through a multi-stand stretch-reducing mill to finish the pipe to required diameter. The tube is then cut to length and subjected to heat treatment, finish straightening, NDT, and hydro test.

EXTRUSION

Extrusion is used for small diameter tubes. The bar stock of suitable length is heated to 2282°F/1250°C and then extruded through a steel die. The tube is then passed through a multi-stand reducing mill to obtain finished dimensions and surface quality.

The development of seamless pipe led to the growth of process piping as now it is possible to load the piping to higher capacities, pressures, and temperatures. It reduced pipeline construction and maintenance costs, thereby reducing the overall transportation and processing costs.

CAST IRON PIPE

In the past, cast iron pipes were cast vertically in sand molds. But presently most cast iron pipe is manufactured by centrifugal casting, which yields longer pipes at faster production rates. Metal has better quality and smother surfaces, and the process can achieve thinner wall thickness.

Cast iron pipe has the disadvantage of low tensile strength. Hence it can break easily if not properly supported or bolted. Because of this, it is manufactured in short lengths, which results in higher fabrication costs. It is also comparably heavy. But it has excellent corrosion resistance which makes it indispensable for a few applications like sulfuric acid service.

GALVANIZED PIPE

Galvanized pipe is one of the most widely used piping materials, particularly in the domestic sector until plastic pipes stole the lead. Often steel is immersed in molten zinc at about 450°C.

A number of iron zinc alloy layers with a top layer of free zinc are formed due to the reactions. The coating has a characteristic gray appearance whereas the top layer may shine for some time soon after the coating.

Most of the larger diameter pipes are hot dip galvanized in batches, whereas smaller sections are galvanized in semi-continuous or continuous processes. Before hot dip galvanizing, pipes are acid pickled in batches.

In a semi-continuous process, lengths of cleaned tube are passed continuously through a bath of molten zinc at 842°F/450°C, resulting in a coating thickness of around 65 microns. In the continuous process, strip is formed into tube and then passed through a bath of molten zinc at 842°F/450°C. Coating thickness is around 12–25 microns on the exterior of the tube only. Coating ranges from 65 microns to over 300 microns.

Cast iron pipe was first used in Sugarland, Germany, in 1455. In 1664 King Louis XIV ordered cast iron pipe be laid from Marly-on-Seine to Versailles for 15 miles to pump water to a fountain and the town. It is still in service.

Sizes, Schedules, and Standards

Until the mid nineteenth century, pipe manufacturers made their own sizes. There was no standardization or interchangeability. Many products were designated with WOG, working pressure for water, oil, or gas service or other manufacturer's standards. Pressure classes as we are accustomed to today were not seen at all. Plant safety, worker safety, demands of higher technology, and concern for the environment and the public paved the way for stringent standards. Developments in technology continue to improve the safety and efficiency of pipeline operations.

Numerous codes were slowly developed and standardized, covering the whole range of manufacturing procedures, material content, strength, dimensional tolerances, joining methods, inspection and testing. But now we are amidst organized chaos in the standards with each nation or powerful geographical entity developing its own standard. The American Society for Testing Materials (ASTM), American National Standards Institute (ANSI) (formerly American Standards Association ASA), American Petroleum Institute (API), German Institute for Standardization (DIN), International Organization For Standards (ISO), and Japanese Industrial Standards (JIS) by Japanese Standards Association are some of the organizations covering piping standards. Modern piping standards are the outcome of the best of the engineering practices, experience, industry demands, and developments in technology and manufacturing processes.

SIZES

Various standards have been developed and are used throughout the world. For instance, API 5L, ANSI/ASME B36.10M is used in the United States, BS 1600 in the United Kingdom, and DN (Diameter Nominal) and ISO 65 in Europe. Furthermore, in Europe and some places as well, the DN system is used but, they are still using the same pipe IDs and wall thicknesses as Nominal Pipe Size. DN is generally equivalent to NPS multiplied by 25, but 1/2 is 15 and 3" is 80. DN conforms to International Standards Organization (ISO). Japanese standard pipe is known as JIS pipe. Presently piping is available from about 1/8" (3mm) to very large diameters of 48" or more. Process requirements dictate the size of the pipe, thickness, metallurgy quality, and allowances.

Pipe sizes are designated by two numbers: pipe diameter and thickness. Nominal pipe diameter is generally associated with the inside diameter. Outside diameter is the same for a given size in order to maintain certain interchangeability of pipe fittings. The inside diameter of a pipe will vary with the thickness or schedule of the pipe. But from sizes of 14" and beyond, the NPS is equal to the outside diameter (OD) in inches. NPS (Nominal Pipe Size) designates the pipe in inches. This meaning of NPS should not be mistaken with the U.S. use where NPS denotes National Pipe Thread Straight just as NPT means National Pipe Thread. NB (Nominal Bore) along with schedule (wall thickness) is used in British standards classifications.

We are not attempting to compare or correlate various standards. Be careful about the standard of the pipe size, whether it is British, U.S., German, or Japanese. They may be identical but may not be the same. **Chart 6.1** compares DN and NPS pipe sizes.

Chart 6.1
Comparison of DN and NPS Pipe Sizes

Diameter Nominal - DN - (mm)	Nominal Pipe Size - NPS - (inches)
6	1/8
8	1/4
10	3/8
15	1/2
20	3/4
25	1
32	1 1/4
40	1 1/2
50	2

Diameter Nominal - DN - (mm)	Nominal Pipe Size - NPS - (inches)
65	2 1/2
80	3
100	4
150	6
200	8
250	10
300	12
350	14
400	16

Diameter Nominal - DN - (mm)	Nominal Pipe Size - NPS - (inches)
450	18
500	20
550	22
600	24
650	26
700	28
750	30
800	32
900	36

SCHEDULE

The schedule of pipe refers to the wall thickness of pipe in the American system. Schedules are standardized from 5 to 160. Normally, the pipe schedule is determined by the service requirements like pressure, temperature, flow, and corrosion of the process. Using thicker piping gives a greater factor of safety. B31.1 and ASME code, Section I, gives the calculations for the design of safe working pressure of the piping based on its type, thickness, and minimum diameter.

Thickness of the pipe increases as the schedule number goes up. For example, schedule 10 pipe has less wall thickness than schedule 20 pipe of the same diameter. The wall thickness for typical 6 inches schedule 40 pipe is 0.280 inches but it is 0.432 inches for schedule 80. These differences also mean that schedule 80 steel pipe is stronger than schedule 40 pipe. Schedule also varies with the diameter of pipe, which means that the thickness at Schedule 40 in a 2" pipe is not the same as the thickness at Schedule 40 in a 3" pipe. Pipes of a particular size all have

Chart 6.2
Size, Schedule, Thickness — ID & OD — of Pipe Sizes from 1/8" to 42"

Pipe Size (inches)	Outside Diameter (inches)	Identification			Wall Thickness *t-* (inches)	Inside Diameter *d-* (inches)
		Steel		Stainless Steel Schedule No.		
		Iron Pipe Size	Schedule No.			
1/8	0.405	-	-	10S	.049	.307
		STD	40	40S	.068	.269
		XS	80	80S	.095	.215
1/4	0.540	-	-	10S	.065	.410
		STD	40	40S	.088	.364
		XS	80	80S	.119	.302
3/8	0.675	-	-	10S	.065	.545
		STD	40	40S	.091	.493
		XS	80	80S	.126	.423
1/2	0.840	-	-	5S	.065	.710
		-	-	10S	.083	.674
		STD	40	40S	.109	.622
		XS	80	80S	.147	.546
		-	160	-	.187	.466
		XXS	-	-	.294	.252
3/4	1.050	-	-	5S	.065	.920
		-	-	10S	.083	.884
		STD	40	40S	.113	.824
		XS	80	80S	.154	.742
		-	160	-	.219	.612
		XXS	-	-	.308	.434
1	1.315	-	-	5S	.065	1.185
		-	-	10S	.109	1.097
		STD	40	40S	.133	1.049
		XS	80	80S	.179	.957
		-	160	-	.250	.815
		XXS	-	-	.358	.599
1 1/4	1.660	-	-	5S	.065	1.530
		-	-	10S	.109	1.442
		STD	40	40S	.140	1.380
		XS	80	80S	.191	1.278
		-	160	-	.250	1.160
		XXS	-	-	.382	.896

Chart 6.2 Continued

1 1/2	1.900	-	-	5S	.065	1.770
		-	-	10S	.109	1.682
		STD	40	40S	.145	1.610
		XS	80	80S	.200	1.500
		-	160	-	.281	1.338
		XXS	-	-	.400	1.100
2	2.375	-	-	5S	.065	2.245
		-	-	10S	.109	2.157
		STD	40	40S	.154	2.067
		XS	80	80S	.218	1.939
		-	160	-	.344	1.687
		XXS	-	-	.436	1.503
2 1/2	2.875	-	-	5S	.083	2.709
		-	-	10S	.120	2.635
		STD	40	40S	.203	2.469
		XS	80	80S	.276	2.323
		-	160	-	.375	2.125
		XXS	-	-	.552	1.771
3	3.500	-	-	5S	.083	3.334
		-	-	10S	.120	3.260
		STD	40	40S	.216	3.068
		XS	80	80S	.300	2.900
		-	160	-	.438	2.624
		XXS	-	-	.600	2.300
3 1/2	4.000	-	-	5S	.083	3.834
		-	-	10S	.120	3.760
		STD	40	40S	.226	3.548
		XS	80	80S	.318	3.364
4	4.500	-	-	5S	.083	4.334
		-	-	10S	.120	4.260
		STD	40	40S	.237	4.026
		XS	80	80S	.337	3.826
		-	120	-	.438	3.624
		-	160	-	.531	3.438
		XXS	-	-	.674	3.152
5	5.563	-	-	5S	.109	5.345
		-	-	10S	.134	5.295
		STD	40	40S	.258	5.047
		XS	80	80S	.375	4.813
		-	120	-	.500	4.563
		-	160	-	.625	4.313
		XXS	-	-	.750	4.063

Chart 6.2 Continued

		-	-	5S	.109	6.407
		-	-	10S	.134	6.357
		STD	40	40S	.280	6.065
6	6.625	XS	80	80S	.432	5.761
		-	120	-	.562	5.501
		-	160	-	.718	5.187
		XXS	-	-	.864	4.897
		-	-	5S	.109	8.407
		-	-	10S	.148	8.329
		-	20	-	.250	8.125
		-	30	-	.277	8.071
		STD	40	40S	.322	7.981
8	8.625	-	60	-	.406	7.813
		XS	80	80S	.500	7.625
		-	100	-	.594	7.437
		-	120	-	.719	7.187
		-	140	-	.812	7.001
		XXS	-	-	.875	6.875
		-	160	-	.906	6.813
		-	-	5S	.134	10.482
		-	-	10S	.165	10.420
		-	20	-	.250	10.250
		-	30	-	.307	10.136
		STD	40	40S	.365	10.020
10	10.750	XS	60	80S	.500	9.750
		-	80	-	.594	9.562
		-	100	-	.719	9.312
		-	120	-	.844	9.062
		-	140	-	1.000	8.750
		-	160	-	1.125	8.500
		-	-	5S	.156	12.438
		-	-	10S	.180	12.390
		-	20	-	.250	12.250
		-	30	-	.330	12.090
		STD	-	40S	.375	12.000
		-	40	-	.406	11.938
12	12.75	XS	-	80S	.500	11.750
		-	60	-	.562	11.626
		-	80	-	.688	11.374
		-	100	-	.844	11.062
		-	120	-	1.000	10.750
		-	140	-	1.125	10.500
		-	160	-	1.312	10.126

Chart 6.2 Continued

14	14.00	-	-	5S	.156	13.688
		-	-	10S	.188	13.624
		-	10	-	.250	13.500
		-	20	-	.312	13.376
		STD	30	-	.375	13.250
		-	40	-	.438	13.124
		XS	-	-	.500	13.000
		-	60	-	.594	12.812
		-	80	-	.750	12.500
		-	100	-	.938	12.124
		-	120	-	1.094	11.812
		-	140	-	1.250	11.500
		-	160	-	1.406	11.188
16	16.00	-	-	5S	.165	15.670
		-	-	10S	.188	15.624
		-	10	-	.250	15.500
		-	20	-	.312	15.376
		STD	30	-	.375	15.250
		XS	40	-	.500	15.000
		-	60	-	.656	14.688
		-	80	-	.844	14.312
		-	100	-	1.031	13.938
		-	120	-	1.219	13.562
		-	140	-	1.438	13.124
		-	160	-	1.594	12.812
18	18.00	-	-	5S	.165	17.670
		-	-	10S	.188	17.624
		-	10	-	.250	17.500
		-	20	-	.312	17.376
		STD	-	-	.375	17.250
		-	30	-	.438	17.124
		XS	-	-	.500	17.000
		-	40	-	.562	16.876
		-	60	-	.750	16.500
		-	80	-	.938	16.124
		-	100	-	1.156	15.688
		-	120	-	1.375	15.250
		-	140	-	1.562	14.876
		-	160	-	1.781	14.438

Chart 6.2 Continued

20	20.00	-	-	5S	.188	19.624
		-	-	10S	.218	19.564
		-	10	-	.250	19.500
		STD	20	-	.375	19.250
		XS	30	-	.500	19.000
		-	40	-	.594	18.812
		-	60	-	.812	18.376
		-	80	-	1.031	17.938
		-	100	-	1.281	17.438
		-	120	-	1.500	17.000
		-	140	-	1.750	16.500
		-	160	-	1.969	16.062
22	22.00	-	-	5S	.188	21.624
		-	-	10S	.218	21.564
		-	10	-	.250	21.500
		STD	20	-	.375	21.250
		XS	30	-	.500	21.000
		-	60	-	.875	20.250
		-	80	-	1.125	19.75
		-	100	-	1.375	19.25
		-	120	-	1.625	18.75
		-	140	-	1.875	18.25
		-	160	-	2.125	17.75
24	24.00	-	-	5S	.218	23.564
		-	10	10S	.250	23.500
		STD	20	-	.375	23.250
		XS	-	-	.500	23.000
		-	30	-	.562	22.876
		-	40	-	.688	22.624
		-	60	-	.969	22.062
		-	80	-	1.219	21.562
		-	100	-	1.531	20.938
		-	120	-	1.812	20.376
		-	140	-	2.062	19.876
		-	160	-	2.344	19.312
26	26.00	-	10	-	.312	25.376
		STD	-	-	.375	25.250
		XS	20	-	.500	25.000

Chart 6.2 Continued

28	28.00	-	10	-	.312	27.376
		STD	-	-	.375	27.250
		XS	20	-	.500	27.000
		-	30	-	.625	26.750
30	30.00	-	-	5S	.250	29.500
		-	10	10S	.312	29.376
		STD	-	-	.375	29.250
		XS	20	-	.500	29.000
		-	30	-	.625	28.750
32	32.00	-	10	-	.312	31.376
		STD	-	-	.375	31.250
		XS	20	-	.500	31.000
		-	30	-	.625	30.750
		-	40	-	.688	30.624
34	34.00	-	10	-	.344	33.312
		STD	-	-	.375	33.250
		XS	20	-	.500	33.000
		-	30	-	.625	32.750
		-	40	-	.688	32.624
36	36.00	-	10	-	.312	35.376
		STD	-	-	.375	35.250
		XS	20	-	.500	35.000
		-	30	-	.625	34.750
		-	40	-	.750	34.500
42	42.00	STD	-	-	.375	41.250
		XS	20	-	.500	41.000
		-	30	-	.625	40.720
		-	40	-	.750	40.500

the same outside diameter regardless of schedule number. As the schedule number increases, the wall thickness increases, and the actual bore is reduced.

An older system called Iron Pipe Size (IPS) was used to designate the pipe sizes in Standard Wall (STD), Extra Strong (XS), and Double Extra Strong (XXS). The IPS number is the same as the NPS number. STD is identical to SCH 40 for NPS 1/8 to NPS 10, inclusive. XS is equivalent to SCH 80 for NPS 1/8 to NPS 8, inclusive. XXS is generally thicker than schedule 160. Standard wall thickness will vary for sizes 1/8" to 10", but from sizes 12" and beyond, it is 3/8". Similarly, extra strong wall thickness will vary from 1/8" to 6", but from sizes 8" and beyond, it is the same 1/2" wall thickness. Likewise, the wall thickness for 10" and 12" pipe is the same 1-in. thickness.

Stainless steel piping schedules generally match with piping schedules for carbon steel piping, but are always identified with the suffix S from 1/8" to 12". Schedules 40S and 80S are the same as their corresponding schedules 40 and 80 in all sizes except 12" in schedule 40.

Chart 6.2 summarizes the size, schedule, thickness — ID and OD — of pipe sizes from 1/8" to 42". **Chart 6.3** shows pip sizes, diameters, wall thickness, and work pressures in SI units. These measures are based on ASME/ANSI B 36.10 Welded and Seamless Wrought Steel Pipe, and on ASME/ANSI B 36.19 Stainless Steel Pipe. **Chart 6.4** compares carbon steel piping standards from the United States, United Kingdom, Germany, and Sweden. Flange sizes and drillings may vary.

Chart 6.3
SI Units Pipe Sizes, Diameters, Wall Thickness, and Working Pressures Based on ASME/ANSI B 36.10 Welded and Seamless Wrought Steel Pipe and ASME/ANSI B 36.19 Stainless Steel Pipe

Diameter Nominal		Schedule		Outside Diameter - *D* - (mm)	Wall Thickness - *t* - (mm)	Inside Diameter - *d* - (mm)	Inside Area (cm²)	Pipe Weight (kg/m)	Water Weight (kg/m)
(inches)	*(mm)*								
1/8	3	10S		10.3	1.245	7.811	0.479	0.277	0.048
		Std	40		1.727	6.846	0.368	0.364	0.037
		XS	80		2.413	5.474	0.235	0.468	0.024
1/4	6	10S		13.7	1.651	10.398	0.846	0.489	0.085
		Std	40		2.235	9.23	0.669	0.630	0.067
		XS	80		3.023	7.654	0.460	0.794	0.046
3/8	10	10S		17.145	1.651	13.843	1.505	0.629	0.151
		Std	40		2.311	12.523	1.232	0.843	0.123
		XS	80		3.2	10.745	0.907	1.098	0.091
1/2	15	5S		21.336	1.651	18.034	2.554	0.799	0.255
		10S			2.108	17.12	2.302	0.997	0.230
		Std	40		2.769	15.798	1.960	1.265	0.190
		XS	80		3.734	13.868	1.510	1.617	0.151
			160		4.75	11.836	1.100	1.938	0.110
		XXS			7.468	6.4	0.322	2.247	0.032

Chart 6.3 Continued

3/4	20	5S		26.67	1.651	23.368	4.289	1.016	0.429
		10S			2.108	22.454	3.960	1.273	0.396
		Std	40		2.87	20.93	3.441	1.680	0.344
		XS	80		3.912	18.846	2.790	2.190	0.279
			160		5.537	15.596	1.910	2.878	0.191
		XXS			7.823	11.024	0.954	3.626	0.095
1	25	5S		33.401	1.651	30.099	7.115	1.289	0.712
		10S			2.769	27.863	6.097	2.086	0.610
		Std	40		3.378	26.645	5.576	2.494	0.558
		XS	80		4.547	24.307	4.640	3.227	0.464
			160		6.35	20.701	3.366	4.225	0.337
		XXS			9.093	15.215	1.818	5.436	0.182
1 1/4	32	5S		42.164	1.651	38.862	11.862	1.645	1.186
		10S			2.769	36.626	10.536	2.683	1.054
		Std	40		3.556	35.052	9.650	3.377	0.965
		XS	80		4.851	32.462	8.276	4.452	0.828
			160		6.35	29.464	6.818	5.594	0.682
		XXS			9.703	22.758	4.068	7.747	0.407
1 1/2	40	5S		48.26	1.651	44.958	15.875	1.893	1.587
		10S			2.769	42.722	14.335	3.098	1.433
		Std	40		3.683	40.894	13.134	4.038	1.313
		XS	80		5.08	38.1	11.401	5.395	1.140
			160		7.137	33.986	9.072	7.219	0.907
		XXS			10.16	27.94	6.131	9.521	0.613
					13.335	21.59	3.661	11.455	0.366
					15.875	16.51	2.141	12.645	0.214
2	50	5S		60.325	1.651	57.023	25.538	2.383	2.554
		10S			2.769	54.787	23.575	3.920	2.357
		Std	40		3.912	52.501	21.648	5.428	2.165
		XS	80		5.537	49.251	19.051	7.461	1.905
			160		8.712	42.901	14.455	11.059	1.446
		XXS			11.074	38.177	11.447	13.415	1.145
					14.275	31.775	7.930	16.168	0.793
					17.45	25.425	5.077	18.402	0.508

Chart 6.3 Continued

2 1/2	65	5S		73.025	2.108	68.809	37.186	3.677	3.719
		10S			3.048	66.929	35.182	5.246	3.518
		Std	40		5.156	62.713	30.889	8.607	3.089
		XS	80		7.01	59.005	27.344	11.382	2.734
			160		9.525	53.975	22.881	14.876	2.288
		XXS			14.021	44.983	15.892	20.348	1.589
					17.145	38.735	11.784	23.564	1.178
					20.32	32.385	8.237	26.341	0.824
3	80	5S		88.9	2.108	84.684	56.324	4.5	5.632
		10S			3.048	82.804	53.851	6.436	5.385
		Std	40		5.486	77.928	47.696	11.255	4.770
		XS	80		7.62	73.66	42.614	15.233	4.261
			160		11.1	66.7	34.942	21.240	3.494
		XXS			15.24	58.42	26.805	27.610	2.680
					18.415	52.07	21.294	31.925	2.129
					21.59	45.72	16.417	35.743	1.642
3 1/2	90	5S		101.6	2.108	97.384	74.485	5.158	7.448
		10S	40		3.048	95.504	71.636	7.388	7.164
		Std	80		5.74	90.12	63.787	13.533	6.379
		XS			8.077	85.446	57.342	18.579	5.734
		XXS			16.154	69.292	37.710	33.949	3.771
4	100	5S		114.3	2.108	110.084	95.179	5.817	9.518
		10S			3.048	108.204	91.955	8.340	9.196
					4.775	104.75	86.179	12.863	8.618
		Std	40		6.02	102.26	82.130	16.033	8.213
		XS	80		8.56	97.18	74.173	22.262	7.417
			120		11.1	92.1	66.621	28.175	6.662
					12.7	88.9	62.072	31.736	6.207
			160		13.487	87.326	59.893	33.442	5.989
		XXS			17.12	80.06	50.341	40.920	5.034
					20.32	73.66	42.614	46.970	4.261
					23.495	67.31	35.584	52.474	3.558

Chart 6.3 Continued

5	125	5S		141.3	2.769	135.762	144.76	9.435	14.476
		10S			3.404	134.492	142.06	11.545	14.206
		Std	40		6.553	128.194	129.07	21.718	12.907
		XS	80		9.525	122.25	117.38	30.871	11.738
			120		12.7	115.9	105.50	40.170	10.550
			160		15.875	109.55	94.254	48.973	9.426
		XXS			19.05	103.2	83.647	57.280	8.365
					22.225	96.85	73.670	65.091	7.367
					25.4	90.5	64.326	72.406	6.433
6	150	5S		168.275	2.769	162.737	208.00	11.272	20.800
		10S			3.404	161.467	204.77	13.804	20.477
					5.563	157.149	193.96	22.263	19.396
		Std	40		7.112	154.051	186.39	28.191	18.639
		XS	80		10.973	146.329	168.17	42.454	16.817
			120		14.275	139.725	153.33	54.070	15.333
			160		18.237	131.801	136.44	67.300	13.644
		XXS			21.946	124.383	121.51	78.985	12.151
					25.4	117.475	108.39	89.258	10.839
					28.575	111.125	96.987	98.184	9.699
8	200	5S		219.075	2.769	213.537	358.13	14.732	35.813
		10S			3.759	211.557	351.52	19.907	35.152
					5.563	207.949	339.63	29.217	33.963
			20		6.35	206.375	334.51	33.224	33.451
			30		7.036	205.003	330.07	36.694	33.007
		Std	40		8.179	202.717	322.75	42.425	32.275
			60		10.312	198.451	309.31	52.949	30.931
		XS	80		12.7	193.675	294.60	64.464	29.460
			100		15.062	188.951	280.41	75.578	28.041
			120		18.237	182.601	261.88	90.086	26.188
			140		20.625	177.825	248.36	100.671	24.836
			160		23.012	173.051	235.20	110.970	23.520
					25.4	168.275	222.40	120.994	22.240
					28.575	161.925	205.93	133.887	20.593

Chart 6.4
Comparison of Carbon Steel Piping Standards from the United States, United Kingdom, Germany, and Sweden

United States	United Kingdom	Germany	Sweden
ASTM A 53	BS3601	DIN 1629	
Grade A SMLS	HFS 22 & CDS 22	St 35	SIS 1233-05
Grade B SMLS	HFS 27 & CDS 27	St 45	SIS 1434-05
ASTM A 53	BS3601	DIN 1626	
Grade A ERW	ERW 22	Blatt 3 St 34-2 ERW	
Grade B ERW	ERW 27	Blatt 3 St 37-2 ERW	
ASTM A 53	BS3601	DIN 1626	
FBW	BW 22	Blatt 3 St 34-2 FBW	
ASTM A 106	BS3602	DIN 17175 [2)]	
Grade A	HFS 23	St 35-8	SIS 1233-06
Grade B	HFS 27	St 45-8	SIS 1435-05
Grade C	HFS 35		
ASTM A 134	BS3601	DIN 1626	
	EFW	Blatt 2 EFW	
ASTM A 135	BS3601	DIN 1626	
Grade A	ERW 22	Blatt 3 St 34-2 ERW	SIS 1233-06
Grade B	ERW 27	Blatt 3 St 37-2 ERW	SIS 1434-06
ASTM A 139	BS3601	DIN 1626	
Grade A	EFW 22	Blatt 2 St 37	
Grade B	EFW 27	Blatt 2 St 42	
ASTM A 155 Class 2	BS3602	DIN 1626, Blatt 3 with certification C	
C45		St 34-2	
C50		St 37-2	
C55		St 42-2	
KC55	EFW 28	St 42-2 [2)]	

Chart 6.4 Continued

KC60		St 42-2 [2]	
KC65	EFW 28S	St 52-2	
KC70		St 52-2	
API 5L	BS3601	DIN 1629	
Grade A SMLS	HFS 22 & CDS 22	St 35	SIS 1233-05
Grade B SMLS	HFS 27 & CDS 27	St 45	SIS 1434-06 *
API 5L	BS3601	DIN 1625	
Grade A ERW	ERW 22	Blatt 3 St 34-2 ERW	
Grade B ERW	ERW 27 [1]	Blatt 4 St 37-2 ERW	
API 5L	BS3601 Double welded	DIN 1626	
Grade A EFW	ERW 22	Blatt 3 St 34-2 FW	
Grade B EFW	ERW 27 [1]	Blatt 4 St 37-2 FW	
API 5L	BS3601	DIN 1626	
FBW	BW 22	Blatt 3 St 34-2 FBW	

[1] Specify API 5L Grade B testing procedures for these steels, [2] Specify SI-killed

Chart 6.5 summarizes some common valve and fitting standards from the American Society of Mechanical Engineers (ASME)

Chart 6.5
Common Valve and Fitting Standards (ASME)

- → **ASME A105/105M** Standard for carbon steel forgings for piping applications
- → **ASME A181/181M** Standard for carbon steel forgings for general purpose piping
- → **ASME A182/182M** Standard specification for forged or rolled alloy-steel pipe flanges, forged fittings and valves, and parts for high-temperature service
- → **ASME A727/727M** Standard specification for carbon steel forgings for piping components with inherent notch toughness
- → **ASME A961** Standard specification for common requirements for steel flanges, forged fittings, valves, and parts for piping applications
- → **ASME B16.10** Face-to-face and end-to-end dimensions of valves

Chart 6.5 Continued

- **ASME B16.34** Valves — flanged, threaded, and welding end
- **ASME B462** Standard specification for forged or rolled UNS N08020, UNS N08024, UNS N08026, UNS N08367, and UNS R20033 alloy pipe flanges, forged fittings, and valves and parts for corrosive high-temperature service
- **ASME B834** Standard specification for pressure consolidated powder metallurgy iron-nickel-chromium-molybdenum (UNS N08367) and nickel-chromium-molybdenum-columbium (Nb) (UNS N06625) alloy pipe flanges, fittings, valves, and parts
- **ASME D5500** Standard test method for vehicle evaluation of unleaded automotive spark-ignition engine fuel for intake valve deposit formation
- **ASME F885** Standard specification for envelope dimensions for bronze globe valves NPS 1/4 to 2 El-1996 R (1996)
- **ASME F992** Standard specification for valve label plates El-1997 R (1997)
- **ASME F993** Standard specification for valve locking devices El-1997 R (1997)
- **ASME F1020** Standard specification for line-blind valves for marine applications
- **ASME F1098** Standard specification for envelope dimensions for butterfly valves — NPS 2 to 24 EI-1993 R (1993)
- **ASME F1271** Standard specification for spill valves for use in marine tank liquid overpressure protections applications EI-1995 R (1995)
- **ASME F1370** Standard for pressure reducing valves for water systems, shipboard
- **ASME F1508** Standard specification for angle style, pressure relief valves for steam, gas, and liquid services
- **ASME F1565** Standard specification for pressure-reducing valves for steam service
- **ASME F1792** Standard specification for special requirements for valves used in gaseous oxygen service
- **ASME F1793** Standard specification for automatic shut-off valves (also known as excess flow valves, EFV) for air or nitrogen service
- **ASME F1794** Standard specification for hand operated, globe-style valves for gas (except oxygen gas), and hydraulic systems
- **ASME F1795** Standard for pressure-reducing valves for air or nitrogen systems
- **ASME A230** Standard specification for steel wire oil-tempered carbon valve spring quality
- **ASME A232** Standard for chromium-vanadium alloy steel valve spring quality
- **ASME A350** Standard specification for forged or rolled carbon and alloy steel flanges forged fittings and valves and parts for low-temperature service
- **ASME A338** Standard specification for ultrasonic examination of heavy steel forgings
- **ASME A694** Standard specification for forgings carbon and alloy steel for pipe flanges fittings valves and parts for high-pressure transmission service
- **ASME A404** Standards specification for forged or rolled alloy-steel pipe flanges forged fittings and valves and parts specially heat treated for high temperature service
- **ASME A522** Forged or rolled 8% and 9% nickel alloy steel flanges fittings valves and parts for low-temperature service.

TUBES

The nominal dimensions of tubes are based on the outside diameter. The inside diameter of a tube depends on the thickness of the tube, often specified as a gage. The tolerances are stricter with tubes compared to pipes. Wall thickness of pipe is designated in mm or decimal of an inch, whereas the wall thickness of a tube is given as a gage number, the most common being Birmingham Wire Gage (BWG). **Chart 6.6** shows gage numbers and their corresponding sizes.

Chart 6.6
Gage Numbers and Corresponding Sizes

Gage	Birmingham Wire Gage B.W.G.(INCHES)		Gage	Birmingham Wire Gage B.W.G.(INCHES)
00000 (5/0)	0.5		17	0.058
0000 (4/0)	0.454		18	0.049
000 (3/0)	0.425		19	0.042
00 (2/0)	0.38		20	0.035
0	0.34		21	0.032
1	0.3		22	0.028
2	0.284		23	0.025
3	0.259		24	0.022
4	0.238		25	0.02
5	0.22		26	0.018
6	0.203		27	0.016
7	0.18		28	0.014
8	0.165		29	0.013
9	0.148		30	0.012
10	0.134		31	0.01
11	0.12		32	0.009
12	0.109		33	0.008
13	0.095		34	0.007
14	0.083		35	0.005
15	0.072		36	0.004
16	0.065			

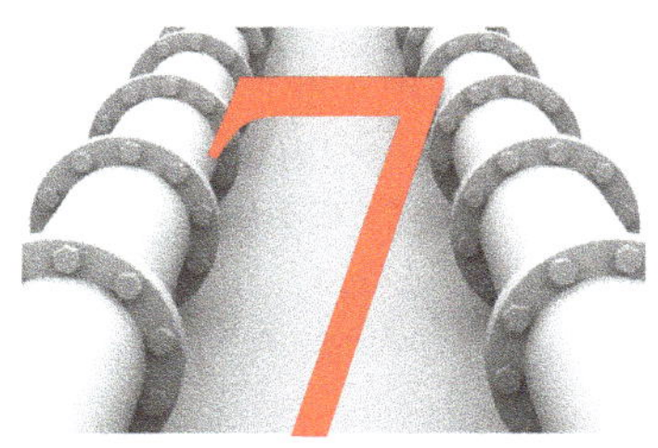

Flange Types and Materials

Flanges are used to connect a piping length to another piping length or pipes to equipment like pumps or compressors. To connect and disconnect piping, two mating plates are inserted over the pipes and fastened with bolts. A gasket used in between these flanges gets compressed by the bolting and seals the surfaces. Flanges are made in various materials, thicknesses, and designs suitable to various pressure and temperature ratings.

Forged steel flanges are manufactured in seven primary classes based on pressure/temperature relationship. These ratings are 150#, 300#, 400#, 600#, 900#, 1500#, and 2500#. The ratings and certification marks are usually stamped on the circumference of the flange. It should be understood that actual working pressure of a flange is dependent on the temperature within its ultimate limits. As the operating temperature goes up, the pressure subjected on the flange should be reduced. Class 150 flange of A105 grade can be operated at 230 PSI at 347°F/175°C whereas it can be operated only at 80 PSI at 800°F/427°C. The case for all flanges is similar in all grades that define the safe operating pressure and temperature.

Needless to say, the flange materials should be compatible with piping materials and both should be to flow medium. Different types of flanges in the same class have the same outside diameter, bolt circle diameter, thickness, number of bolts, bolt hole size, and bolt size. But the length of the flange may vary with each type of flange in the same rating and size.

Flanges are classified

1. By the construction — flanges are either slip on or weld neck
2. By the way the face of the flange is made; say raised face or flat face, tongue and groove, octagonal ring joint, or lens ring joint
3. By the standard into which the flanges are made and holes are drilled, the class of flange
4. By the material of construction

FLANGES BY CONSTRUCTION

SLIP-ON FLANGES

Slip on flanges, as the name suggests, are slipped onto the pipe. They are industry favorites because the pipe line fitters find it easy to align them compared to weld neck flanges. The bore of the slip-on flanges of a given size matches the OD of the pipe. The flange is then fillet welded up around on the inside and outside of the pipe and the flange, needing two weld joints (see Figure 7.1A and B). These flanges are used in lower class and general class applications; they are not recommended in high pressure lines. ASME B16.5 Code limits their use in the 1500#–2500# (lbs.) weight classes. The joint is very weak under fatigue conditions.

Figure 7.1A

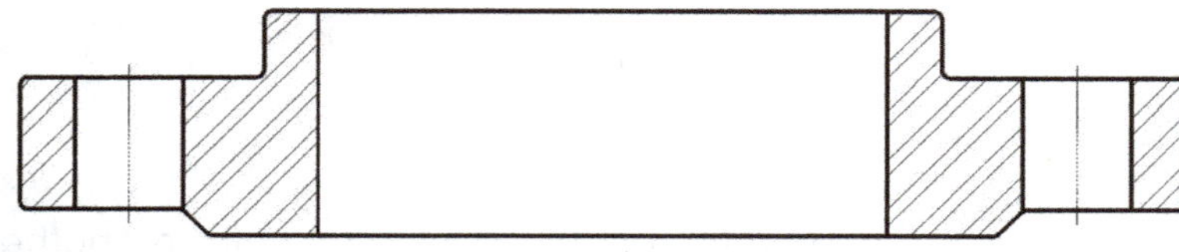

Figure 7.1B

Slip on Flanges

WELDING NECK FLANGES

Welding neck flanges have a hub on the backside of the flange taper matching the pipe to be welded (see Figures 7.2A and 7.2B). The inside diameter of the flange is bored to match the mating pipe. This arrangement avoids turbulence in flow and prevents erosion. Stress is distributed well and there is no restriction of flow. Welding neck flanges are an industry favorite for severe services such as high pressure and high temperatures, or sub-zero temperatures with wide fluctuations, where there may be certain bending of the flanges. Sturdy tapered hub and butt welding resists dishing. The flange and pipe are prepared and then butt welded. Weld face flanges have V-beveled ends so that they can be attached to the adjoining pipe with only one weld joint. It is a longer flange with its tapered hub that should be considered while laying lines. Unlike other flanges, this type of flange can be radiographed for NDT.

Figure 7.2A

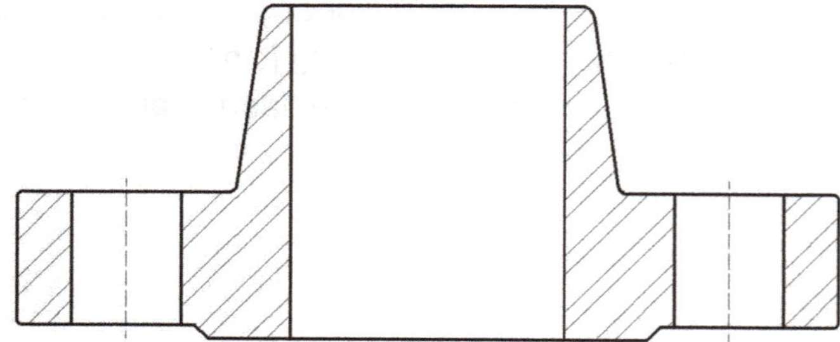

Figure 7.2B

Welding Neck Flanges

LAP JOINT FLANGES

Lap joint flanges are similar to slip-on flanges, but they are not welded to the pipe (see Figure 7.3A and B). A stub end, along with its flange face, is actually inserted into the flange and the lap joint stub end is welded to the pipe (see Figure 7.4A and B). The bore has a curved radius in the bore at the face in order to hold the pipe in place. The flange is a backing flange which is free to rotate on the pipe; it will be easy to match the bolts of the opposite flange. The stub end has the gasket face — these flanges do not have raised faces. The stub ends are normally 3" long, but some that are 6" long are also available. These are used in the pipelines where they are required to be opened frequently for cleaning or unplugging the choked lines. They are also used sometimes to cut the costs where the piping is stainless steel, but the flange material could be lower grade such as carbon steel. Many of the plastic flanges, particularly HDPE, are made like this.

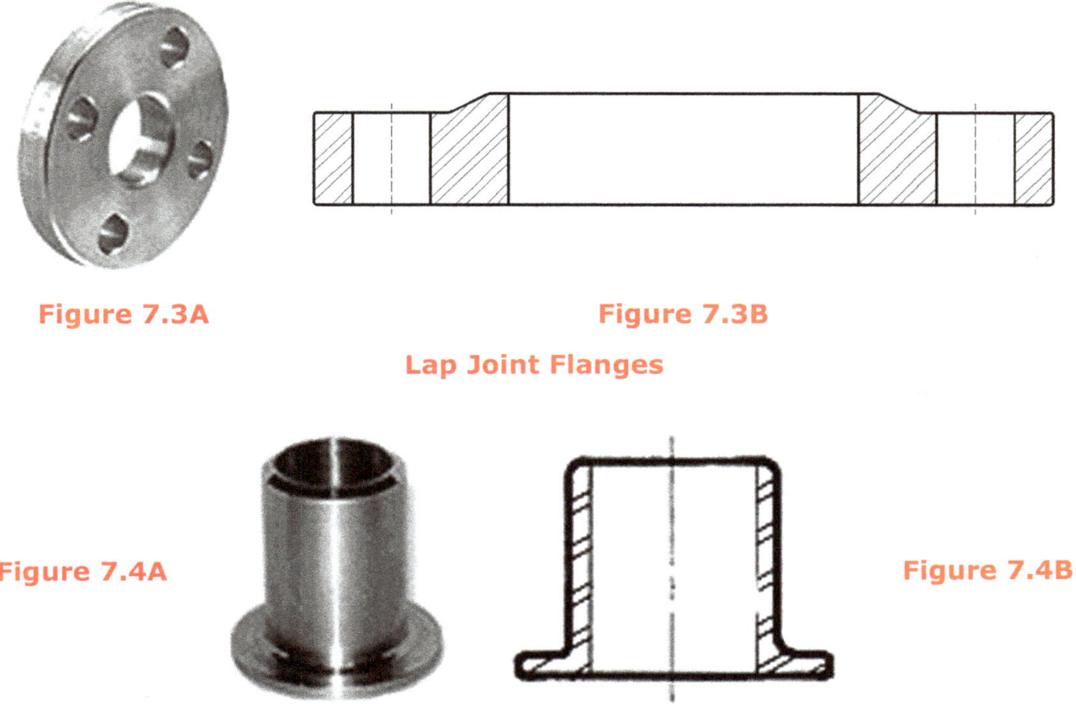

Figure 7.3A Figure 7.3B

Lap Joint Flanges

Figure 7.4A Figure 7.4B

Lap Joint Stub Ends

SOCKET WELDING FLANGES

Socket welding flanges are similar to slip-on flanges. They have a counter bore diameter on the hub end equal to the OD of the pipe on which they are used, and a bore diameter equal to the bore of the inside diameter of the pipe (see Figure 7.5). This arrangement makes up a socket into which the pipe is inserted. Hence the pipe schedule must be specified for ordering these flanges. The pipe is fillet welded to the flange on the hub end. As the pipe diameter on the inside and the flange inside diameter are equal, these flanges offer an unrestricted flow of fluid in the pipeline.

The pipe should not be inserted into the flange fully, but a certain gap or clearance should be allowed for expansion of the pipe. The ASME Boiler and Pressure Vessel (B & PV) Code Sec. III requires 1/16 inch gap between the pipe and fitting in the socket weld. They are now generally used with low pressure and temperature.

Figure 7.5
Socket Welding Flange

THREADED FLANGES

In threaded flanges, threads are made in the inside of the flange matching to the threads on the pipe (see Figures 7.6A and 7.6B). Pipes are screwed into the flanges. Sometimes the joints are seal welded to make them stronger and safer. They are not suitable for higher diameters because of the difficulty in machining proper threads and tightening the joint. They are used instead for smaller size pipes in low pressure services and also where welding is not allowed.

Figure 7.6A

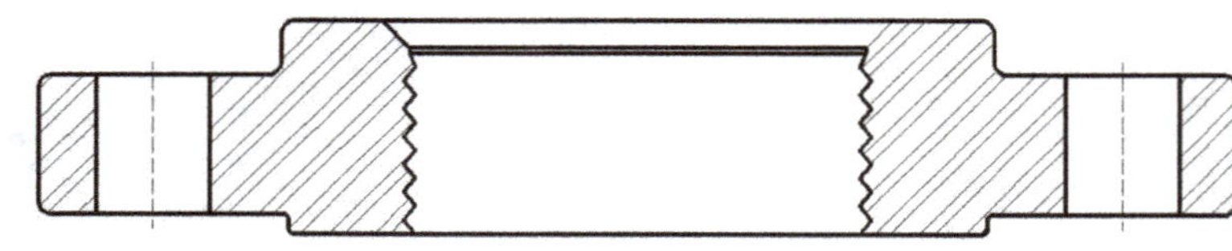

Figure 7.6B

Threaded Flanges

BLIND FLANGES

Blind flanges are just that. They are without a bore and are used to close or seal the end of the pipe lines (see Figures 7.7A and 7.7B). They are made in all classes and types, with or without a hub. They are deliberately fixed on the end of some lines which are required to be opened frequently for cleaning or unplugging. These are also used as covers for pressure vessels normally known as man ways or manholes. It must be understood that blind flanges are subjected to more stresses than any other type of flange. Blind flanges sometimes have a provision for threaded tappings for instrument connections or sampling take-offs.

Figure 7.7A

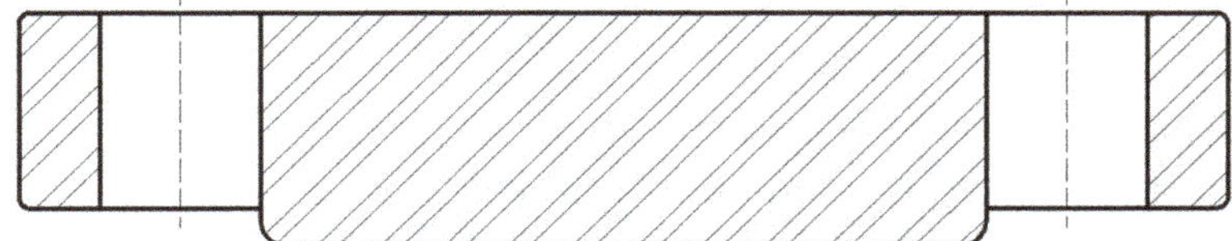

Figure 7.7B

Blind Flanges

ORIFICE FLANGES

Orifice flanges are used in conjunction with flow metering systems. An orifice plate is made as per individual flow requirements and is inserted between two of these flanges. These flanges are provided with tapped holes on the periphery of the flange to make connections for measuring the pressure drop in the flow through the orifice plate (see Figure 7.8). These tappings are accurately made in the flanges with reference to the orifice plate. These flanges are generally available as weld neck in raised face or in RTJ in all classes and in all materials. They also have jack screws on the flanges which allow separation of the flanges to change or inspect the orifice plate. They are usually supplied as a combination of both flanges and used as such. More often than not, these flanges are welding neck type. Nevertheless, inside welds should be ground smooth to avoid turbulence and false readings.

Figure 7.8
Orifice Flange

REDUCING FLANGES

Reducing flanges are standard flanges with a bore less than that of the main flange rating. They are used for reducing the line size of the piping. For instance, a 4" by 3" Class 150 flange will have flange dimensions and drillings of a 4" flange and bore size of 3" pipe. These flanges are also used in jacketed or gutted piping. They are available in all classes, in all materials and types.

FLANGES BY FACINGS

RAISED FACE FLANGES

The most commonly used flange is a raised face flange. In this flange, the gasket surfaces are raised above the rest of the face by about .060" or .250" depending on the pressure class of the flange (see Figures 7.9A and 7.9B). The pressure rating of the flange standardizes the height of the raised face. The finish of these flanges may be either phonographic serrated or concentric serrated. The most popular and standard finish is phonographic. Raised face flanges use a gasket that fits inside the bolts. The raised face of these flanges concentrates more pressure on a smaller gasket area and thereby adopts a higher gasket load on the joint. It is now advantageous to use a ring gasket, which encircles the inside of the bolt circle diameter. Traditionally, asbestos gaskets are used, but the industry is fast changing over to environmentally friendly materials. Spiral wound gaskets are the norm. *More details of these gaskets can be found in Chapter 12, Gaskets*. Raised face flanges are available in all materials except for cast flanges. In the case of cast flanges, the entire face of the flange has the same surface, i.e., flat.

Figure 7.9A

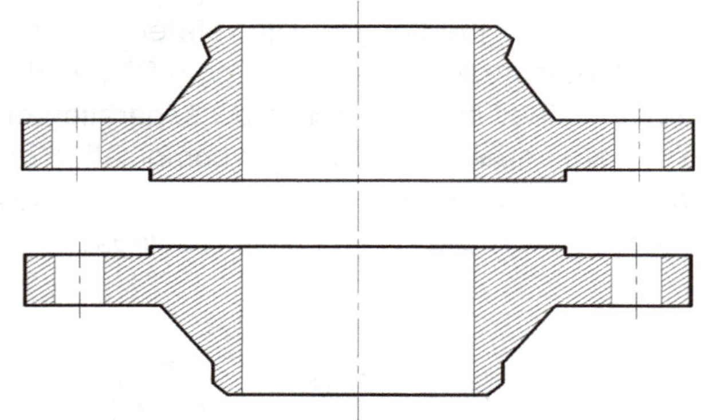

Figure 7.9B

Raised Face Flanges

FLAT FACE FLANGES

The flat face or full face flange has no machined step on the gasket surface and is generally made without any serrations. Flat face flanges are found in the cast flanges like cast iron or bronze, mostly on the pump flanges and cast iron pipelines.

But the problem arises when the flanges are to be mated with a raised face flange, which is an industry norm for steel piping. If the flanges are joined as they are now, the cast flange may break because it tends to bend into the gap caused by the raised face. Differences in height along the flange face can exert undue stress on the cast flanges and can even break the castings when bolted up. To avoid this problem in such a necessity, the raised face must be machined out flat before joining and a full-face gasket should be used. A gasket should cover the entire face of the flange. But connecting a raised face flange to flat face that is either steel or stainless steel should not pose any problem because materials of equal strength are used.

RING TYPE JOINT (RTJ) FLANGE

Ring type flanges and joints were originally developed to meet the demands of the petroleum and petrochemical industries. They offer a very good metal-to-metal seal in high temperature and pressure applications. The gasket in these flanges is a metal ring with a suitable profile. These flanges have a matching groove cut on their flange faces as per standard dimensions (see Figure 7.10). When the bolts are tightened, the metal gasket compresses into the mating grooves of both the flanges and makes a tight metal-to-metal joint. Joints are robust, and assembly and disassembly are easy.

Figure 7.10
Ring Type Joint (RTJ) Flange

API ring joint gaskets are generally available in two basic designs: oval and octagonal cross section. Flange grooves are made correspondingly. Octagonal cross sections are preferred because of their higher sealing efficiency. Flange grooves must be finished to 63 micro inches and must be cleaned of all burrs, chatter marks, and foreign material before insertion of the ring. They act as a seal due to the wedging forces and the initial line contact. Needless to say, those gaskets should be a bit softer than the companion flanges. ASME B16.20, API 6A, and ASME/ANSI B 16.5 standardize these flanges. A round bottom groove can accept only an oval cross section, whereas the recent flat bottom groove will accept either the oval or the octagonal gaskets.

These joints are standardized as styles R, RX, SRX, BX, and SBX. The efficiency of the seal increases as the internal pressure of the system increases because the designs make use of pressure energized effect. Some styles have a pressure passage hole to equalize the pressure on the sealing faces. Some flange groves can accept other profiles, but not all are interchangeable.

TONGUE AND GROOVE

In the case of tongue and groove, both flanges are used in conjunction with each other. One flange has a raised ring machined onto the surface whereas the mating flange has a matching groove. A gasket must be inserted in the groove and the flanges must mate or sit properly. There are large groove (see Figure 7.11) and small groove (see Figure 7.12) varieties in this type, but these are fast becoming obsolete.

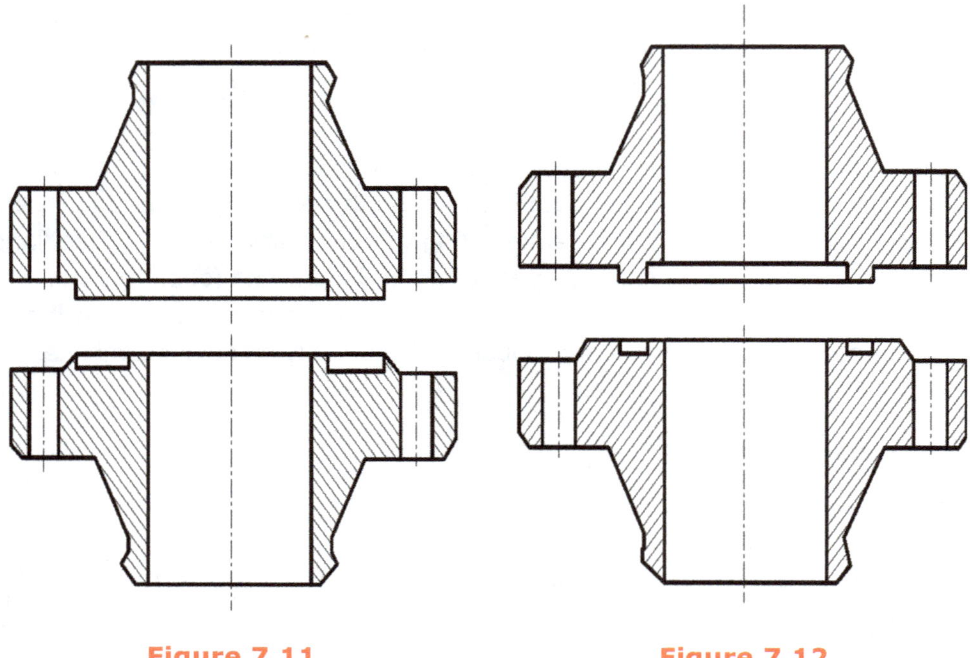

Figure 7.11 **Figure 7.12**

Tongue and Groove Joints

MALE AND FEMALE

Male and female flanges are used in conjunction with each other; both the flanges must be matched. There is a kind of thicker raised face on one flange and matching groove on the companion flange (see Figure 7.13). This combination is also getting obsolete — flanges of this type are made for replacement requirements.

Do not ever try to match dissimilar flanges!

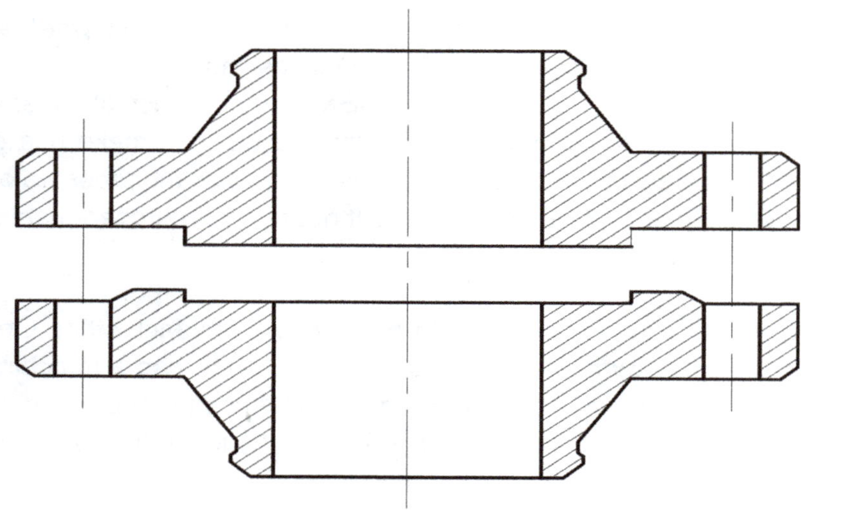

Figure 7.13
Male and Female Joint

LENS TYPE

A lens type joint is used in high pressure piping systems and in pressure vessel closures. The flanges, which have a spherical gasket surface, seal with a line contact (see Figure 7.14). The bolt loading can be smaller and overloading is inconsequential.

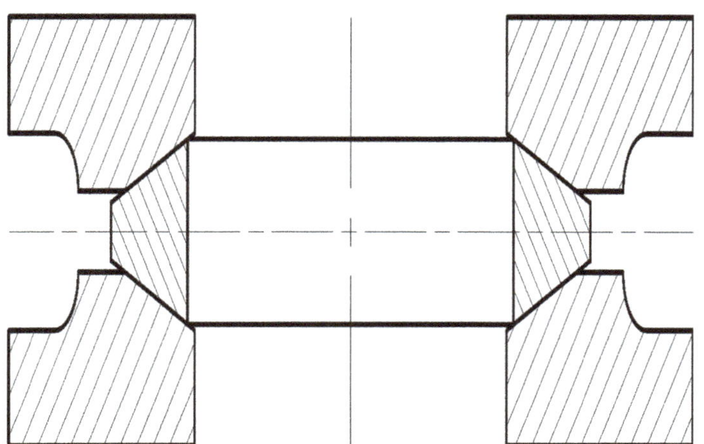

Figure 7.14
Lens Type Joint

TRICLOVER CLAMPS

Triclover clamps are used widely in the food, dairy, drug, beverage, and cosmetic industries. They do not require special tools and can be easily adopted for different applications. Some of these pipe lines carry viscous or semi-viscous fluids which may often tend to choke the lines. These lines may have to be opened and closed frequently with least trouble and no contamination. Triclover clamps make use of a clamp, two welded ferrules on both sides, a gasket, and a single fastener (see Figure 7.15).

Figure 7.15
Triclover Joint

OTHER DESIGNS

There are other designs to seal under extreme pressures, like Delta and Bridgeman. Most of them are used on the covers of pressure vessels or valve bonnet joints. *More details of these gaskets can be found in Chapter 12, Gaskets.*

FLANGE FACE FINISH

The flange face is the most important area as it is where the gasket sits. This face must be protected at all times during transport, storage, installation, operation, and maintenance. Maintenance persons knew that it will be very difficult to stop leaks from bruised flange faces. Flange faces are machined to standard finishes.

Arithmetic Average Roughness Height (AARH) or Root Mean Square (RMS) are different methods of calculation to define the quality of finish required on flange facings. It is essentially visual comparison shown as Roughness Average (Ra) in micro-inches (µin) or micrometers (µm). The normal Ra is 125 to 250 µin (125 to 250 AARH) or 3.2 to 6.3 µm (3.2 to 6.3AARH).

FLANGES BY STANDARDS

As already stated, many different flange standards are abounding worldwide. Popular standards include ASA/ANSI (United States), PN/DIN (Europe), BS10 (Britian, Australia), and JIS/KIS (Japan, Korea). More often than not these standards are not interchangeable (e.g., an ANSI flange will not mate against a DIN flange). *Do not try to mate by short cut methods such as enlarging bolt holes or using smaller size bolts. Occasions do arise where different standard flanges need to be mated, for example in the case that a ship has a JIS flange and shore connections are ANSI. Then fabricate a suitable spool piece of the right materials but with different flanges on both sides and use the spool piece. Store it properly. The occasion will come again!*

The most widely used standards are ANSI and DIN.

ASTM A105/A105M standards specify forgings made out of carbon steel for piping components such as flanges, fittings, valves, and similar parts limited to a maximum weight of 10,000 lb [4540 kg] ambient and higher temperature service in pressure systems.

ASTM A181/A181M standards specify forgings made out of carbon steel for piping components such as flanges, fittings, valves, and similar parts limited to a maximum weight of 10,000 lb [4540 kg] for general service.

A182/A182M standards specify forgings or rolled alloy-steel pipe flanges, forged fittings, and valves and parts for high temperature service.

ASME B16.5 — 1996 pipe flanges and flange fittings standard covers pressure-temperature ratings, materials, dimensions, tolerances, marking, testing, and methods of designating openings for pipe flanges and flanged fittings.

Chart 7.1 shows dimensions and tolerances for flanges in pipe sizes from 1/2" through 24", in pressure classes from 150 lb through 2500 lb. **Chart 7.2** shows the maximum allowable non shock pressure (psig) and temperature ratings for steel pipe flanges and flanged fittings according to ANSI B16.5 — 1988. The flange standards API 6A and ASME/ANSI B16.5 are similar dimensionally. However, the API 6A flanges are rated for higher pressures, as shown in **Chart 7.3**. **Chart 7.4** shows the pressure and temperature ratings for ASTM A105B, 304SS, 316SS, Alloy 20, and Hastalloy C276 in Class 150, 300, and 600 ratings. **Chart 7.5** compares ANSI and DIN flanges. As you'll see, flange dimensions and drillings do vary.

Chart 7.1
Dimensions and Tolerances for Flanges

Nominal Pipe Size NPS (inches)	150 psi			
	Diameter of Flange (inches)	No. of Bolts	Diameter of Bolts (inches)	Bolt Circle (inches)
1/4	3-3/8	4	1/2	2-1/4
1/2	3-1/2	4	1/2	2-3/8
3/4	3-7/8	4	1/2	2-3/4
1	4-1/4	4	1/2	3-1/8
1-1/4	4-5/8	4	1/2	3-1/2
1-1/2	5	4	1/2	3-7/8
2	6	4	5/8	4-3/4
2-1/2	7	4	5/8	5-1/2
3	7-1/2	4	5/8	6
3-1/2	8-1/2	8	5/8	7
4	9	8	5/8	7-1/2
5	10	8	3/4	8-1/2
6	11	8	3/4	9-1/2
8	13-1/2	8	3/4	11-3/4
10	16	12	7/8	14-1/4
12	19	12	7/8	17
14	21	12	1	18-3/4
16	23-1/2	16	1	21-1/4
18	25	16	1-1/8	22-3/4
20	27-1/2	20	1-1/8	25
24	32	20	1-1/4	29-1/2

Nominal Pipe Size NPS (inches)	300 psi			
	Diameter of Flange (inches)	No. of Bolts	Diameter of Bolts (inches)	Bolt Circle (inches)
1/4	3-3/8	4	1/2	2-1/4
1/2	3-3/4	4	1/2	2-5/8
3/4	4-5/8	4	5/8	3-1/4
1	4-7/8	4	5/8	3-1/2
1-1/4	5-1/4	4	5/8	3-7/8
1-1/2	6-1/8	4	3/4	4-1/2
2	6-1/2	8	5/8	5
2-1/2	7-1/2	8	3/4	5-7/8

Chart 7.1 Continued

3	8-1/4	8	3/4	6-5/8
3-1/2	9	8	3/4	7-1/4
4	10	8	3/4	7-7/8
5	11	8	3/4	9-1/4
6	12-1/2	12	3/4	10-5/8
8	15	12	7/8	13
10	17-1/2	16	1	15-1/4
12	20-1/2	16	1-1/8	17-3/4
14	23	20	1-1/8	20-1/4
16	25-1/2	20	1-1/4	22-1/2
18	28	24	1-1/4	24-3/4
20	30-1/2	24	1-1/4	27
24	36	24	1-1/2	32

Nominal Pipe Size NPS (inches)	**400 psi**			
	Diameter of Flange (inches)	No. of Bolts	Diameter of Bolts (inches)	Bolt Circle (inches)
1/4	3-3/8	4	1/2	2-1/4
1/2	3-3/4	4	1/2	2-5/8
3/4	4-5/8	4	5/8	3-1/4
1	4-7/8	4	5/8	3-1/2
1-1/4	5-1/4	4	5/8	3-7/8
1-1/2	6-1/8	4	3/4	4-1/2
2	6-1/2	8	5/8	5
2-1/2	7-1/2	8	3/4	5-7/8
3	8-1/4	8	3/4	6-5/8
3-1/2	9	8	7/8	7-1/4
4	10	8	7/8	7-7/8
5	11	8	7/8	9-1/4
6	12-1/2	12	7/8	10-5/8
8	15	12	1	13
10	17-1/2	16	1-1/8	15-1/4
12	20-1/2	16	1-1/4	17-3/4
14	23	20	1-1/4	20-1/4
16	25-1/2	20	1-3/8	22-1/2
18	28	24	1-3/8	24-3/4
20	30-1/2	24	1-1/2	27
24	36	24	1-3/4	32

Chart 7.1 Continued

Nominal Pipe Size NPS (inches)	600 psi			
	Diameter of Flange (inches)	No. of Bolts	Diameter of Bolts (inches)	Bolt Circle (inches)
1/4	3-3/8	4	1/2	2-1/4
1/2	3-3/4	4	1/2	2-5/8
3/4	4-5/8	4	5/8	3-1/4
1	4-7/8	4	5/8	3-1/2
1-1/4	5-1/4	4	5/8	3-7/8
1-1/2	6-1/8	4	3/4	4-1/2
2	6-1/2	8	5/8	5
2-1/2	7-1/2	8	3/4	5-7/8
3	8-1/4	8	3/4	6-5/8
3-1/2	9	8	7/8	7-1/4
4	10-3/4	8	7/8	8-1/2
5	13	8	1	10-1/2
6	14	12	1	11-1/2
8	16-1/2	12	1-1/8	13-3/4
10	20	16	1-1/4	17
12	22	20	1-1/4	19-1/4
14	23-3/4	20	1-3/8	20-3/4
16	27	20	1-1/2	23-3/4
18	29-1/4	20	1-5/8	25-3/4
20	32	24	1-5/8	28-1/2
24	37	24	1-7/8	33

Nominal Pipe Size NPS (inches)	900 psi			
	Diameter of Flange (inches)	No. of Bolts	Diameter of Bolts (inches)	Bolt Circle (inches)
1/2	4-3/4	4	3/4	3-1/4
3/4	5-1/8	4	3/4	3-1/2
1	5-7/8	4	7/8	4
1-1/4	6-1/4	4	7/8	4-3/8
1-1/2	7	4	1	4-7/8
2	8-1/2	8	7/8	6-1/2
2-1/2	9-5/8	8	1	7-1/2
3	9-1/2	8	7/8	7-1/2
4	11-1/2	8	1-1/8	9-1/4
5	13-3/4	8	1-1/4	11

Chart 7.1 Continued

6	15	12	1-1/8	12-1/2
8	18-1/2	12	1-3/8	15-1/2
10	21-1/2	16	1-3/8	18-1/2
12	24	20	1-3/8	21
14	25-1/4	20	1-1/2	22
16	27-3/4	20	1-5/8	24-1/2
18	31	20	1-7/8	27
20	33-3/4	20	2	29-1/2
24	41	20	2-1/2	35-1/2

Nominal Pipe Size NPS (inches)	**1500 psi**			
	Diameter of Flange (inches)	No. of Bolts	Diameter of Bolts (inches)	Bolt Circle (inches)
1/2	4-3/4	4	3/4	3-1/4
3/4	5-1/8	4	3/4	3-1/2
1	5-7/8	4	7/8	4
1-1/4	6-1/4	4	7/8	4-3/8
1-1/2	7	4	1	4-7/8
2	8-1/2	8	7/8	6-1/2
2-1/2	9-5/8	8	1	7-1/2
3	10-1/2	8	1-1/8	8
4	12-1/4	8	1-1/4	9-1/2
5	14-3/4	8	1-1/2	11-1/2
6	15-1/2	12	1-3/8	12-1/2
8	19	12	1-5/8	15-1/2
10	23	12	1-7/8	19
12	26-1/2	16	2	22-1/2
14	29-1/2	16	2-1/4	25
16	32-1/2	16	2-1/2	27-3/4
18	36	16	2-3/4	30-1/2
20	38-3/4	16	3	32-3/4
24	46	16	3-1/2	39

Nominal Pipe Size NPS (inches)	**2500 psi**			
	Diameter of Flange (inches)	No. of Bolts	Diameter of Bolts (inches)	Bolt Circle (inches)
1/2	5-1/4	4	3/4	3-1/2
3/4	5-1/2	4	3/4	3-3/4

Chart 7.1 Continued

1	6-1/4	4	7/8	4-1/4
1-1/4	7-1/4	4	1	5-1/8
1-1/2	8	4	1-1/8	5-3/4
2	9-1/4	8	1	6-3/4
2-1/2	10-1/2	8	1-1/8	7-3/4
3	12	8	1-1/4	9
4	14	8	1-1/2	10-3/4
5	16-1/2	8	1-3/4	12-3/4
6	19	8	2	14-1/2
8	21-3/4	12	2	17-1/4
10	26-1/2	12	2-1/2	21-1/4
12	30	12	2-3/4	24-3/8

Chart 7.2
Maximum Allowable Non-Shock Pressure (psig) and Temperature Ratings for Steel Pipe Flanges and Flanged Fittings According to ANSI B16.5 — 1988

Maximum Allowable Non-Shock Pressure (psig)							
Temperature (°F)	Pressure Class (lb.)						
	150	**300**	**400**	**600**	**900**	**1500**	**2500**
	Hydrostatic Test Pressure (psig)						
	450	1125	1500	2225	3350	5575	9275
−20 to 100	285	740	990	1480	2220	3705	6170
200	260	675	900	1350	2025	3375	5625
300	230	655	875	1315	1970	3280	5470
400	200	635	845	1270	1900	3170	5280
500	170	600	800	1200	1795	2995	4990
600	140	550	730	1095	1640	2735	4560
650	125	535	715	1075	1610	2685	4475
700	110	535	710	1065	1600	2665	4440
750	95	505	670	1010	1510	2520	4200
800	80	410	550	825	1235	2060	3430
850	65	270	355	535	805	1340	2230
900	50	170	230	345	515	860	1430
950	35	105	140	205	310	515	860
1000	20	50	70	105	155	260	430

Chart 7.3
Flange Standards API 6A and ASME/ANSI B16.5

API vs. ASME/ANSI Flanges				
Flange	Pressure Class Rating (psi)		Nominal Size Range (inches)	
	ASME/ANSI B16.5	API 6A	ASME/ANSI B16.5	API 6A[1)]
Weld neck	600	2000	1/2 – 24	1 13/16 – 11
	900	3000		
	1500	5000		
Blind and Threaded	600	2000		1 13/16 – 21 ¼
	900	3000		1 13/16 – 20 ¾
	1500	5000		1 13/16 – 11

[1)] In the old API standard, flanges ranged from 1 1/2 to 10(20) inches.

Chart 7.4
Pressure/Temperature ratings for ASTM A105B, 304SS, 316SS, Alloy 20 and Hastalloy C276 in Class 150, 300 and 600 ratings

CLASS 150 FLANGE PRESSURE RATINGS (PSI)
PER ANSI B16.5-1988

	A105 C. STEEL	304 SS	316 SS	304L SS 316L SS	ALLOY 20	625 C276
-20 to 100 °F	285	275	275	230	230	290
200 °F	260	235	240	195	215	260
300 °F	230	205	215	175	200	230
400 °F	200	180	195	160	185	200
500 °F	170	170	170	145	170	170
600 °F	140	140	140	140	140	140
650 °F	125	125	125	125	125	125
700 °F	110	110	110	110	110	110
750 °F	95	95	95	95	95	95
800 °F	80	80	80	80	80	80
850 °F	65	65	65	65		65
900 °F	50	50	50			50
950 °F	35	35	35			35
1000 °F	20	20	20			20

CLASS 300 FLANGE PRESSURE RATINGS (PSI)
PER ANSI B16.5-1988

	A105 C. STEEL	304 SS	316 SS	304L SS 316L SS	ALLOY 20	625 C276
-20 to 100 °F	740	720	720	600	600	750
200 °F	675	600	620	505	555	750
300 °F	655	530	560	455	525	730
400 °F	635	470	515	415	480	705
500 °F	600	435	480	380	470	665

Chart 7.4 Continued

600 °F	550	415	450	360	455	605
650 °F	535	410	445	350	450	590
700 °F	535	405	430	345	445	570
750 °F	505	400	425	335	440	530
800 °F	410	395	415	330	430	510
850 °F	270	390	405	320		485
900 °F	170	385	395			450
950 °F	105	375	385			385
1000 °F	50	325	365			365

CLASS 600 FLANGE PRESSURE RATINGS (PSI)
PER ANSI B16.5-1988

	A105 C. STEEL	304 SS	316 SS	304L SS 316L SS	ALLOY 20	625 C276
-20 to 100 °F	1480	1440	1440	1200	1200	1500
200 °F	1350	1200	1240	1015	1115	1500
300 °F	1315	1065	1120	910	1045	1455
400 °F	1270	940	1030	825	960	1410
500 °F	1200	875	955	765	935	1330
600 °F	1095	830	905	720	910	1210
650 °F	1075	815	890	700	900	1175
700 °F	1065	805	865	685	890	1135
750 °F	1010	795	845	670	880	1065
800 °F	825	790	830	660	865	1015
850 °F	535	780	810	645		975
900 °F	345	770	790			900
950 °F	205	750	775			775
1000 °F	105	645	725			725

Chart 7.5
ANSI and DIN Flanges

ANSI	**Flange Class**	150	300	600	900	1500	2500
DIN	**Flange Pressure Number, PN**	20	50	100	150	250	420

Pipe Fittings

Pipe fittings are used to join lengths of pipe or to turn at various angles, to reduce or increase the size of pipe ends, to divert or divide flow, and combine or return it. The weakest links in piping are the connections; in any piping line, we have a host of them. Hence it is necessary that we get the most suitable fitting for the given service and install it in the best way to get a most reliable service. We have here some of the fittings and end connections generally used in piping. The list is not complete, though exhaustive. Fittings are available in different metallurgies as per standards and mechanical design. Fittings should preferably be seamless whereas sometimes larger fittings are fabricated.

Generally most of the fittings fall into the following broad categories. End connections are screwed, welded, or flanged. Few of the flanges are again screwed — most of them are welded to the individual pipes. Specific code requirements dictate the joint method, thread, or welding. Normally, socket-welded fittings are used in 2" size and smaller, and butt-welding is used in all sizes, particularly in 2" size and larger.

The description below of various fittings is generally for the screwed fittings, whereas flanged or butt welded fittings are available in almost all sizes and invariably for high pressures. *Materials of construction for fittings are the same as valves, and the fittings come in corresponding pressure classes. They are made with the same end connections as valves and in all standard pipe sizes.* But, they may not be available in certain metallurgies and sizes.

TYPES OF FITTINGS

SCREWED ENDS

Valves and fittings with screwed ends are the most widely used in the industry with metallurgies including brass, iron, steel, and other alloys. They are made suitable to all pressures, except high pressures. Screwed connections are confined to smaller pipe sizes because it is normally difficult to make and use screwed connections for large sizes. The most popular thread is American national pipe thread, known as NPT, whereas all other standard threads are also used in the industry. It is easier to align screwed fittings and there is no problem of weld metal intrusion into the pipe. Threads on the pipe and screwed joints can easily be made at site. Screwed joints can be used when welding is not permitted due to process hazards.

However, a screwed joint is weak because there is a certain reduction of metal at the joint due to threads. If the threads and fit up are not proper, the joint may leak. These joints are not recommended when shock, vibration, erosion, and crevice corrosion are anticipated. The threaded joint is sometimes welded to get a leak proof joint, but ASME does not concede that it adds to the strength of the pipe.

WELD ENDS

Welded fittings and valves are available only in steel and its alloys. They are used for all pressures, invariably for higher pressure and high temperature, and for sub zero temperature services. They are not recommended for lines requiring frequent dismantling. Butt and socket welding fittings are two types of welding end materials. Butt-welding valves and fittings come in all sizes; socket welded ends are usually limited to smaller sizes. There are standard codes of practice in welding these fittings.

Butt welded fittings are most common and a practical way of joining larger size piping, though commonly used for all sizes of piping. Butt welded fittings are invariably used for 2" and above NB pipe sizes. It is necessary to prepare the joint suitably for welding. It provides a reliable joint that can be radiographed. However, there is a possibility of weld metal intrusion into the pipe that may obstruct the flow or damage downstream fittings.

Socket welded are easier to prepare and easier to align. Chances of weld metal intrusion into the bore are minimal. But there is 1/16" recess where the pipe meets inside the socket; this recess can harbor foreign material or weaken the joint. Hence, these joints are not recommended by code where severe erosion or crevice corrosion can occur.

FLANGED ENDS

The more popular method of joining pipelines and fittings is by bolting flanges together. It is invariably the method of joining pipes and fittings in higher pressures and temperatures, and certain metallurgies like cast iron. Flanges are either slip on or weld neck (see Figure 8.1).

Figure 8.1
Various Flanges

Slip on flanges, as the name suggests, are slipped onto the pipe. The flange is then welded up around on the inside and outside of the pipe and the flange. In the case of a butt welded or weld neck flange, the flange and pipe are prepared and then butt welded. Screwed flanges are screwed to the pipes. Both flanges are bolted together with a gasket making tight seal. Flanged joints should be minimized to reduce the potential point of leakage.

Pipeline flanges are designated and available in a number of classes, say from 125 to 6600 or more. But then metallurgies dictate the temperature and pressure at which they can be operated.

Detailed discussions of flanges and gaskets can be found in Chapters 7 and 12 respectively.

VARIOUS PIPE FITTINGS

ELBOWS OR "ELLS"

An elbow, often called an *ell* like the letter *L*, is used to make an angular turn in piping. An angle — such as 90°, 45°, or 60° — designates elbows, the most common being the 90 degree or right-angle elbow (see Figure 8.2A). Figure 8.2B shows a 45° elbow. Reducing elbows are also available to join lines of different sizes. Socket welded elbows and flanged elbows are normally available; flanged elbows are the norm in higher sizes and certain metallurgies like cast iron. Screwed elbows are inside threaded, but occasionally you may find an elbow with external threads. The street elbow is one; it has one female end and one male end (see Figure 8.2 C).

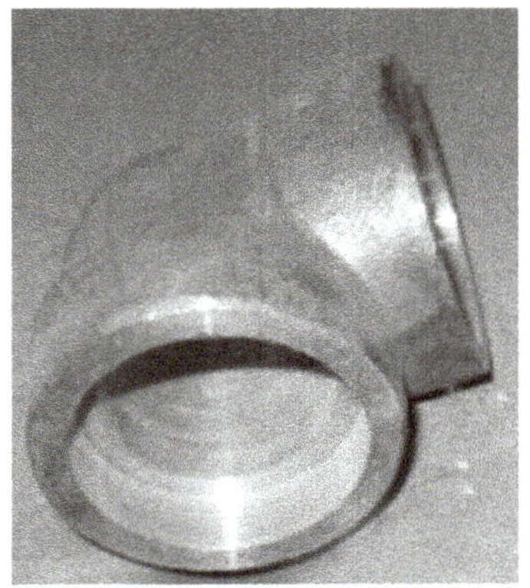

Figure 8.2A
90° Elbow

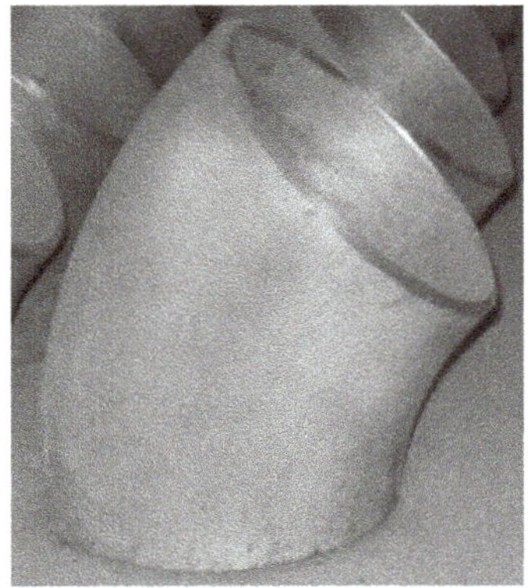

Figure 8.2B
45° Elbow

Figure 8.2C
Street Elbow

Figure 8.3A shows a long radius butt welded elbow. Long radius elbows are used where reduced pressure drop across the elbow is required. In turn, Figure 8.3B shows a 180° return elbow which is used to take a U-turn of the pipe line, like in the heat exchanger coils. Most general return bends are in butt weld configuration.

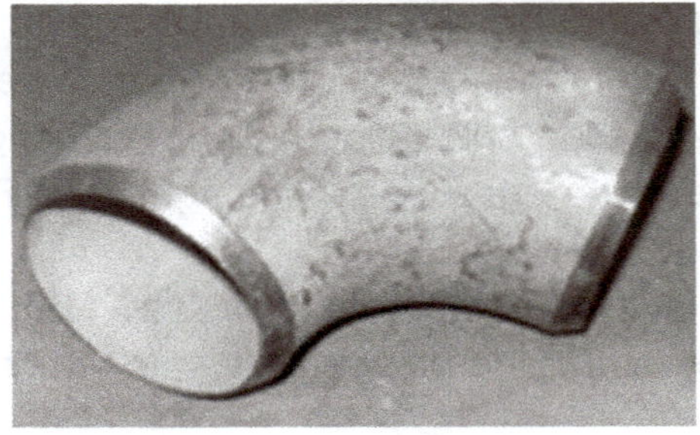

Figure 8.3A
Long Radius Elbow

Figure 8.3B
180° or Return Elbow

TEES

As the name implies, a tee looks like the letter *T* and is a 3-way fitting (see Figure 8.4). It is used to take a branch at right angles from a pipeline. Tees are also made with reducing ends to permit joining lines of different sizes. Socket welded tees and flanged tees are normally available. Screwed tees are normally inside threaded and are confined to smaller sizes.

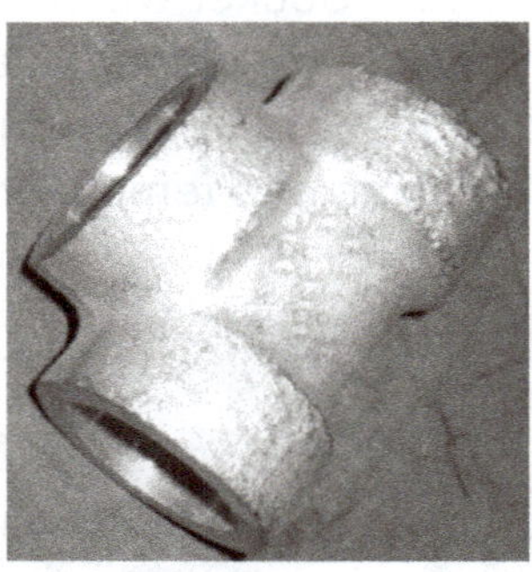

Figure 8.4
Socket Welded Tee

CROSS

A cross is a four-way fitting (see Figure 8.5). Occasionally you may find a cross with reducing ends. Flanged, screwed, and welded fittings are common.

Figure 8.5
Cross

LATERAL

Laterals are used when the line needs to be branched off at a 45° angle (see Figure 8.6).

Figure 8.6
Lateral

COUPLING

A coupling is used to connect two pieces of pipe of the same size in a straight line (see Figure 8.7). Also known as sockets, they are screwed or socket welded. Butt welded sockets are also available. Flanges are invariably used for joining higher sizes or where the lines need to be disconnected.

Figure 8.7
Coupling

NIPPLES

Nipples are short lengths of pipe male threaded on both ends (see Figure 8.8A). A close nipple is threaded its entire length and is used to make close connections. Nipples are also available with only one end threaded, the other end used for a socket welded into another pipe. Hex nipples have hexagonal flanks on the middle for using with a spanner instead of pipe wrench. Straight nipples without threads are used as connectors between lines for welding. They are also available as reducing nipples or swage nipples with one end threaded to a smaller size (see Figure 8.8B).

Figure 8.8A
Straight Nipple

Figure 8.8B
Swage Nipple

UNIONS

A union is essentially an easily demountable pipe coupling. With the help of a union, the piping can be opened or closed in position without the complication of right and left threads and without disturbing the entire piping. It is a coupling with an additional screwed joint in the middle (see Figure 8.9). This has male and female faces brought tight by an external nut. Hence it is a three-piece construction. By opening or closing the nut, the line can be opened or closed without bothering the rest of the piping. It will be virtually difficult to open or close pipelines without unions at strategic locations, unless you want to open lines end to end.

Figure 8.9
Screwed Union

REDUCERS

As the name suggests, a reducer is used to connect the line to a smaller size line or vice versa. There are two kinds of reducers: concentric reducer (see Figure 8.10A) and eccentric reducer (see Figure 8.10B). Concentric reducers share the same central axis for higher and smaller diameter of the fitting. In the eccentric reducer, these are offset. Specific uses and lay out demands the use of any one kind of reducer or the other. For instance, eccentric reducers are most often used at the pump suctions.

Figure 8.10A
Concentric Reducer

Figure 8.10B
Eccentric Reducer

BUSHINGS

Bushings are fittings threaded externally for higher diameter and internally for smaller diameter. They are essentially reducers that are screwed into the end of fittings or valves to reduce the size of the end opening. They are available in different thread sizes and have advantages in reducing lines to smaller length. The most widely used bushings are in stainless steel (see Figure 8.11).

Figure 8.11
Bushing

CAPS AND PLUGS

Caps (see Figure 8.12) are used for shutting off male threaded ends of pipe. Plugs (see Figure 8.13) are used to close female threaded end openings. They are available in screwed and welded construction. Blind flanges are used for closing the flanged pipe lines.

Figure 8.12
Screwed Cap

Figure 8.13
Screwed Plug

PIPE TAKE-OFFS

These fittings are essentially take-off fittings in a pipe line, when a smaller size tapping is taken from a large pipe. *They must be used in high pressure, high temperature, and critical service lines, instead of making a hole in the larger line and welding a smaller line.*

Thread-o-let (see Figure 8.14A) is used when a threaded tapping is required, say for a pressure gage. Weld-o-let (see Figure 8.14B) is used for welding a small line tapping from a large line. The fittings are contoured to suit both the pipes, the higher size and the lower size. Although this description is cited as an example, almost all combinations of fittings are manufactured to meet the piping demands. The names also may vary depending on the manufacturer or country.

Figure 8.14A
Thread-O-Let

Figure 8.14B
Weld-O-Let

Variations of almost all the pipe fittings described are sometimes made especially to suit individual piping requirements. Fittings are the limbs of any piping system and unfortunately the weakest. Wisdom tells not to compromise on them.

Hand Tools in Piping

The very first hand tools used by man must be the stones used for killing animals and for making fire. But now the modern world has so many tools around for every conceivable human activity that we have became over-familiar with them. Familiarity breeds contempt; we often use them with little attention. We often ignore the fact that there is a technique bordering on art in using them. Just as a small human mind contains within it the entire universe, even the biggest machinery is built by small hand tools. The care in proper use and caution against misuse should form a part of the industrial training program. In any process industry, more pipes and fittings are closed or opened than any other part on any day. The right selection, use, and maintenance of hand tools will not only reduce the risk of unwanted injuries, but also improve the efficiency of the worker. Presence of mind is an important qualification one must always possess.

A hammer or a pipe wrench is so common that every engineer must have seen some time or the other; the pipe wrench used as a hammer and a slide wrench as a lever. Although more people get injured by their misuse, it is a pleasure watching some people using them to the best advantage. More injuries happen because workers are tempted to use a wrong tool just because the right tool is not readily on hand. They will not consider walking to their tool box just a few minutes away. Skilled workers are ones who make sure that they plan and take all the right tools required for a given job. They also take good care of them because tools are faithful servants. They will be careful to have sufficient tools for the job on hand, to select the proper tools for each stage of the job, to use them correctly, to maintain them in good condition, to promptly replace them when beyond repair, and to store them properly.

This chapter gives a brief account of the tools used in piping. Some of the tools — like pipe wrenches — are made exclusively for work on piping whereas other tools are also used in every other activity of maintenance. The descriptions hold good for all.

TOOLS REQUIRED IN PIPING WORK

GENERAL TOOLS

1. Hammers
2. Open End Wrenches
3. Box End Wrench or Ring Spanner
4. Combination Wrench
5. Ratchet Wrench and Socket Set
6. Torque Wrench
7. Spud Wrench
8. Striking Wrench
9. Slide Wrench or Adjustable Spanner
10. Chisel
11. Hand Hacksaw
12. Bolt Cutter
13. Wire Brush
14. C-clamp

PIPING TOOLS

15. Pipe Wrench
16. Strap Wrench
17. Chain Wrench
18. Other Piping Wrenches
19. Pipe Cutter
20. Pipe Die
21. Pipe Tap
22. Thread Extractor
23. Flange Splitter or Flange Spreader
24. Drift Pin
25. Pry Bar

MEASURING TOOLS

26. Tape Measure
27. Steel Rule
28. Plumb Bob
29. Steel Square
30. Bevel Protractor
31. Combination Square
32. Spirit Level

VISES

33. Bench Vise
34. Pipe Vise
35. Portable Yoke Vise
36. Chain Vise

GENERAL TOOLS

HAMMERS

The hammer is one of the most extensively used and misused tools. It is one of the earliest tools used by humans and the most widely used in whatever form. Although a variety of hammers have been developed to meet the demands of various engineering trades, the present discussion is limited to two kinds of hammers: ball pein and sledge hammers.

PARTS OF A HAMMER The most widely used and really handy tool is the ball pein hammer (see Figure 9.1A). The rounded part of the hammer is known as the pein. The bottom portion, which actually hammers, is the face. There is an eye in the center into which the wooden handle is driven. The part between the eye and the face is known as the post. The whole unit other than the handle is called the hammerhead.

Sledge hammers are heavier hammers in 4lb, 8lb, 16lb, etc., sizes (see Figure 9.1B). They do not have a pein and the construction is towards heavier blows. Both the faces can be used for

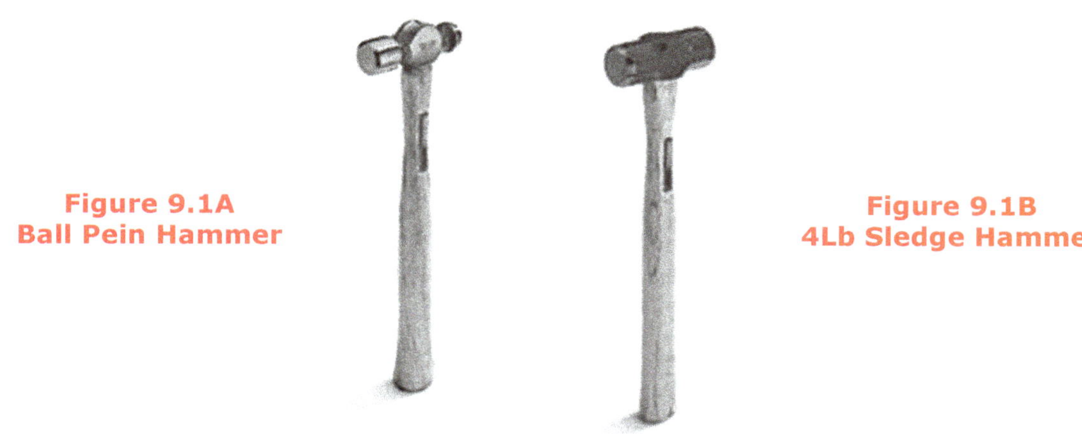

Figure 9.1A
Ball Pein Hammer

Figure 9.1B
4Lb Sledge Hammer

hammering. The present trend is towards use of Fiberglass handles in place of wooden handles, because they are impervious to weather conditions. There are techniques for fixing and holding the handle.

FITTING THE HANDLE The eye of the hammer is made to be smaller in the middle than at the ends. The wooden handle is attached into this tapered eye. Drive the wooden handle tight into position in the eye, cut off the extra portion of the handle sticking out of the eye with a hacksaw. Then hammer a steel wedge into the wood in the eye. This forces the wooden handle further tight against the walls of the eye. The handle should be a little less than 90° with the face. This helps in keeping the hand lower than the work, which results in more force applied to the face. *The handle should be comfortable on the hand, easy and firm to grip, and properly balanced.*

TECHNIQUE OF THE HAMMER *When using the hammer, the face of it should be at a plane parallel with the object being hammered.* Off-center blows can dislodge metal from the struck piece and can throw dangerous splinters into the air. Raise the arm straight away from the object to be struck. Bring it down with a sharp, swift motion. If it is a light blow, raise your arm just enough for the required length and weight for the blow. If it is heavy blow, the hammer should be grasped firmly, add a little more length and energy. *Always avoid over and under blows.*

The handle should be grasped about 1/4 to 1/3 of the way from the free end. *Do not choke the hammer by holding near the head. The chances are that your fingers will hit the object. If held too close to the free end of the handle, the chances of slipping are more. Be careful both ways.*

SAFETY WITH A HAMMER

1. Take a good grip on the handle and strike the object carefully after pointing at it properly.
2. Take care so that the best part of the face of the hammer comes in contact with the object to be struck.
3. When striking on another tool, such as a chisel, punch, or wedge, the face of the hammer should be at least 3/8" larger than the face of the tool.

4. Attention while hammering is an extreme necessity.
5. Do not use the handle as a prying tool.
6. Even a small trace of oil or grease can spell disaster. Keep hammers dry and clean.
7. Many use the back of handle to drive sensitive jobs, but this can split the handle.
8. Always use a hammer with the proper size and weight depending on the job. It is easier and safer.
9. Never use on hammer to strike another hammer.
10. Do not use a hammer with a loose or damaged handle. Steel wedges are driven into the hammerhead to keep it tight; frequently check the condition for the tightness. Some kinds of wood shrink under certain conditions. Sometimes, soaking the hammer in water overnight restores the tightness. Do not use nails to keep the handles tight even as a makeshift method; instead, use wedges.
11. Do not use hammers with heads that are mushroomed, cracked, dented, or chipped, or that show signs of wear.
12. Do not weld, grind, or heat-treat hammers.

OPEN END WRENCHES

As the name suggests, these wrenches are open ended with two sides covering the nut (see Figure 9.2). U-shaped openings grip the two opposite sides of the bolt or nut. Open-end wrenches are widely used because they can make a job faster, particularly for initial tightening or final opening. They are made in the standard sizes with a clearance of .005–.015" so that they can be inserted and removed from the nut easily. Actual spanner ends are set at an angle so that the spanner can be turned over to get a better travel. These spanners are made conveniently with two different sizes on both ends to fit two different nuts. Hence, they are generally known as double ended or DE wrenches. Open-end wrenches are also available with only one end. They are available in inch and metric sizes made to standard sizes.

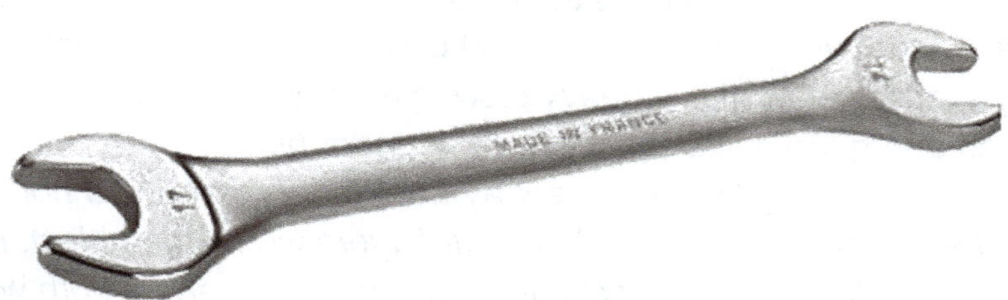

Figure 9.2
Open End Wrench

SAFETY WITH DE WRENCHES:

1. DE wrenches can easily slip. Hence, use them with care.
2. Watch if the wrench is worn out. *Worn-out wrenches slip more.*
3. The handle is short and does not offer much leverage. Do not use extensions on it as these wrenches can easily slip.

4. It is normal practice in the industry to insert the eye of a ring spanner into the other open end of this wrench and use it as an extension with a little more leverage. Be careful because it has great potential to slip.
5. Do not use hammers on these wrenches or exert too much pressure.
6. Keep oil and dirt away from these wrenches.
7. DE wrenches must always be pulled. Make it short and steady.

BOX END WRENCH OR RING SPANNER

The box end wrench or ring spanner is safer than other wrenches — use it whenever possible. It fits the nut better than an open-end wrench because it surrounds the entire nut (see Figure 9.3). Maneuverability with a ring spanner is better than with an open-end wrench. These wrenches can be used in confined areas with as little as 15° swings.

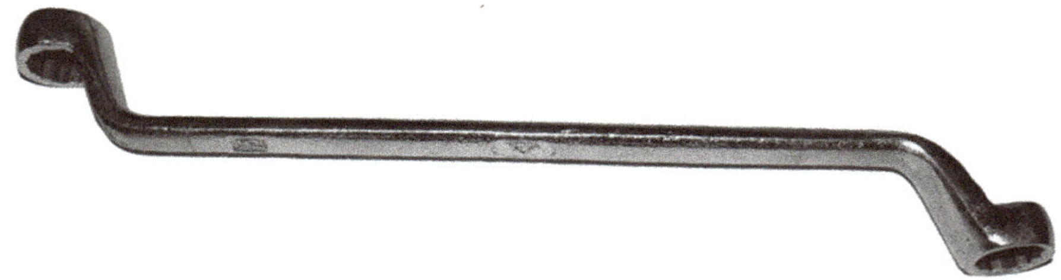

Figure 9.3
Box End Wrench

Box-end wrenches are made in standard sizes, with two different sizes on both ends to fit two different nuts. The lengths are sized as per the size of the nut so that proportionate torque can be exerted. Normally the internal opening has six-points or twelve-points for use with hex nuts or bolt heads. The twelve-point fits onto the fastener at twice as many angles, an advantage when the swing of the wrench is limited. The wrenches can be inserted and removed from the nut easily. The handle is set at an angle so that the spanner can be turned over for better travel. Handles are offset at an angle of 15° to give hand clearance, but flat ring spanners are also available. The wrenches are available in inch and metric sizes made to standard sizes.

SAFETY WITH RING SPANNERS

1. The box end wrench or ring spanner must always be pulled. Make the pull short and steady.
2. Keep your wrenches clean and in good shape, without any grease or oil.
3. Think of the clearance at the end of the pull as your fingers may get injured. *Many finger knuckles get injured because of the failure to watch this.*
4. Do not use extensions on these wrenches. It is normal practice to insert an open-end wrench or a ubiquitous valve persuader into the eye of this wrench and get a little leverage. *Be careful because it has great potential to slip.*
5. Similarly, do not hammer on the box end wrenches. If you have to use hammers, use them with a striking wrench instead.
6. The points become rounded over a period; replace such wrenches.

7. Points also develop unseen cracks across — inspect and replace them.
8. Do not substitute pliers for work a wrench should do.
9. Never put your face or head level with a wrench handle.
10. Use a box or socket wrench to free a tight or frozen nut.

COMBINATION WRENCH

As the name suggests, a combination wrench is a wrench with an open ended wrench on one end and a ring wrench on the other (see Figure 9.4). The sizes of both are generally the same and they fit the same size nut or bolt head. It has the advantage of getting good grip with the ring wrench and exerting force on the nut. Then when the nut is loose, the wrench can be turned over to do a fast job. These wrenches are available in inch and metric dimensions made to standard sizes.

Figure 9.4 Combination Wrenches

RATCHET WRENCH AND SOCKET SET

A socket wrench or a ratchet wrench uses different sizes of sockets that fit the nuts or bolt heads snugly (see Figure 9.5). Unlike other wrenches, this one has a ratchet mechanism which allows the nut to be tightened or loosened, without requiring that the wrench be removed and repositioned after each turn. The tightening and loosening mode can be selected by a small lever on the wrench. The ratchets have a square head or drive into which the sockets are fixed. These drives are standardized in 1/4 inch, 3/8 inch, 1/2 inch, and 3/4 inch, and so on.

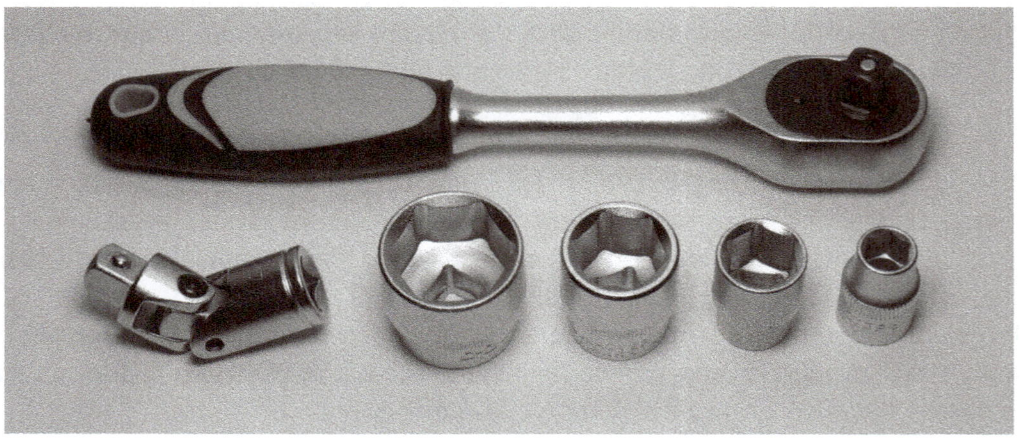

Figure 9.5
A Sockets Wrench Set:
Ratchet Wrench, Universal Joint, Sockets

Socket sets are generally known as box sockets. Like box end wrenches, they are also 6- or 12-pointed inside a socket. They are available in inch and metric sizes made to standard sizes. They have a square hole on top where a short lever or a ratchet handle is inserted. Sockets are made to suit nuts as per standards. The socket fits snugly over the nut and is turned with the short handle or a ratchet drive. Ratchet handles do a faster job because you do not need to remove the socket from the nut and refit it. Instead, just swing the ratchet handle back and forth. By changing the drive lever position, the nut is opened or tightened. These wrenches are safer because they hold the nut tightly. Universal joints, adopters, and extensions are available to add versatility to socket sets.

SAFETY WITH SOCKETS

1. Do not use extensions or cheater bars.
2. Pull socket handles and ratchets toward you. Do not push.
3. Maintain your balance.
4. Prevent slipping. Keep tools clean.
5. Misused sockets develop cracks. Please check them before use.

TORQUE WRENCH

A torque wrench is used to open or tighten a bolt or a nut with controlled torque (see Figure 9.6). Torque wrenches are available in many models, designs, and torque ranges. Manual torque wrenches are ideal for applications where a reasonable torque of about 600 ft-lbs or less is required. Enough space should be available to operate the long handle of this wrench.

Figure 9.6
Torque Wrench

The most common torque wrench is a click torque wrench. It has torque readings marked on a rotary dial, which can be used to adjust the required torque. The wrench is set to the required value before using and then put into use. When the set torque is reached, the torque wrench clicks. The escape mechanism makes the the drive slip so that no further torque can be applied.

Hydraulic torque wrenches are also available; they are normally used on high pressure heat exchangers and columns. They may be a little unwieldy for routine requirements due to the space and time in setting up hydraulic pumps, hoses, etc. But they are indispensable for specific applications.

SAFETY WITH TORQUE WRENCHES

1. Torque wrenches are precision tools. Use them with respect.
2. Do not push the wrench. Always pull the wrench towards you.
3. Do not hammer on the wrench or use it as a hammer, lever, or prying bar (even in anger!!)
4. Do not use extensions or "cheater bars" on torque wrenches.
5. Do not try to exceed the rated capacity by some means.

SPUD WRENCH

Spud wrenches are used for matching holes in flanges in the pipe line so that bolts may be placed in corresponding holes. They have a wrench on one side and a sharp pointed spud on the opposite end (see Figure 9.7). There are two types of spud wrenches: one with a twelve-point ring end and the other with an open end. Apart from aligning holes, these can double up as spanners because they are available in various spanner sizes. Adjustable spud wrenches are also now available in both inch and metric sizes.

Figure 9.7
Spud Wrench

Fix the tapered spud end of the wrench in a hole. Then maneuver this tapered end into the hole in the other flange by leveraging and prying. Push bolts into holes wherever they are matched, preferably adjacent to the spud. Place the spud wrench in another hole, leverage it to bring the flange holes close, and place the possible bolts. A few attempts may be required to align holes in a pair of flanges.

Excessive pulling on the flanges indicates that they are exerting strain on the pipeline, particularly if the flanges are close to a pump or compressor. This must be understood as misalignment of mating flanges — one of the major reasons of pump or equipment failure.

STRIKING WRENCH

As the name itself suggests, one can really hammer onto striking wrenches to open or tighten tough bolts and nuts (see Figure 9.8). These wrenches are also known as knocking wrenches, slugging wrenches, slogging wrenches, and hammering wrenches. They are made in 6-point or 12-point design and in almost all sizes metric or inch. Occasionally, square or octagonal designs also can be procured. Similarly wrenches with flat or offset striking handles are available. A close-fitting striking spanner, a big hammer, and strong hands are all that are needed for forcing an obstinate nut in or out. A striking wrench can often create sparks, which could be dangerous in certain chemical industries. Hence non-sparking tools have been developed and are available.

All safety precautions essential in using spanners and hammers must scrupulously be followed while using these tools, with presence of mind.

Figure 9.8
Striking Wrench

SLIDE WRENCH OR ADJUSTABLE SPANNER

The slide wrench is an open-end wrench with one adjustable jaw so that it can fit a range of different size nuts (see Figure 9.9). These spanners come in sizes to suit a number of larger or smaller size nuts. The angle of the opening to the handle is 22 1/2". They are basically made to fit odd-size nuts, but now it is a common practice of many technicians to use them as one spanner for many, adjusting the jaw to fit the nut snugly. *It is good practice not to use a slide wrench when a standard size spanner is available. The adjustable spanner is more likely to slip than an open-end spanner. Next to the hammer, this tool is the most widely used and misused tool.*

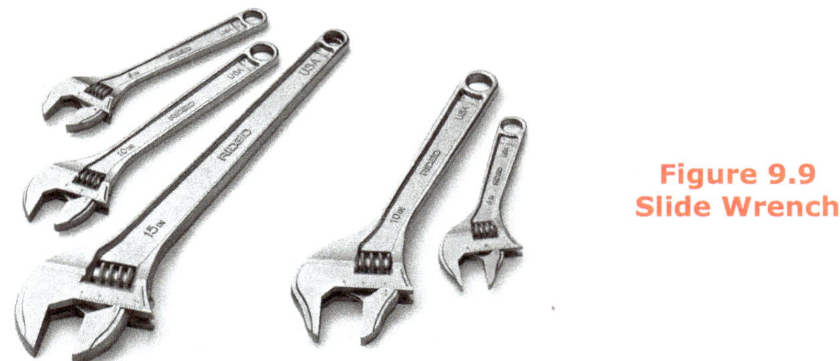

Figure 9.9
Slide Wrench

SAFETY WITH SLIDE WRENCH

1. Do not use pipes, extenders, or cheater bars. A cheater bar is a pipe piece slipped over the handle to increase leverage. This may result in breakage. Maintenance people tend

to keep and hide small pieces of pipe securely stashed under the hollows of columns. Be careful.

2. Pull the wrench toward you. *Do not push.*
3. Maintain your balance because the slide wrench can slip easily.
4. Do not use wrenches with worn out screws or jaws. *They will invariably slip because they cannot hold tight on the nut.*
5. Do not hammer on the slide wrench or use it as hammer.
6. Do not use it as a prying bar.

CHISEL

Chisels are occasionally used in piping work, sometimes to force a stubborn nut or for making an opening between the flanges. In both cases, better tools are available.

Chisels are made out of chisel steel in octagon shapes with a cutting edge at the end of the blade (see Figure 9.10). The cutting edge is hardened from the end to one inch back. They are tempered to remove any sharp line between the hardened section and the chisel body. Then they are annealed to make them tougher and stronger. Chisels will cut any metal that can be filed.

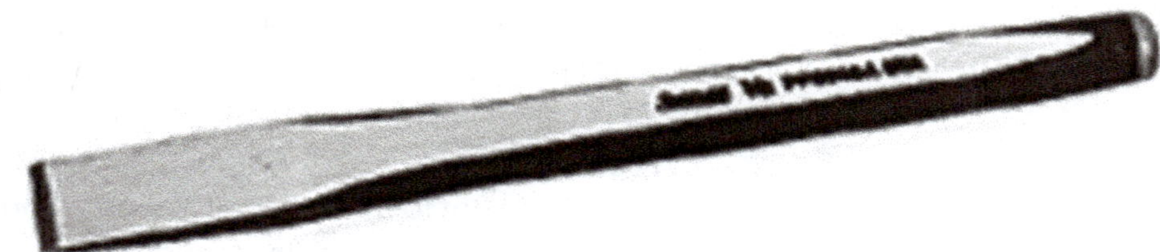

Figure 9.10 Chisel

As the name suggests, cold chisels are made to cut cold metal only and not for hot metal. When grinding chisels, care must be taken not to heat the chisel too much. The proper angle to grind a flat chisel is 70° total included angle.

SAFETY WITH CHISELS

1. Use safety goggles. Serious eye injury is possible by the flying chips.
2. Do not use a damaged chisel. Replace.
3. Select a chisel size that fits the job.
4. Chisels should be discarded if their edges show chip out or brittleness.
5. If these chisels become mushroomed, they should be dressed on a grinder.
6. Keep the cutting edge of a chisel away from people.
7. Do not use cold chisels on material harder than the cutting edge of the chisel.
8. Use of a punch or the chisel holder is a good practice to prevent hand injury.
9. Do not use cold chisels on stone or concrete.

HAND HACKSAW

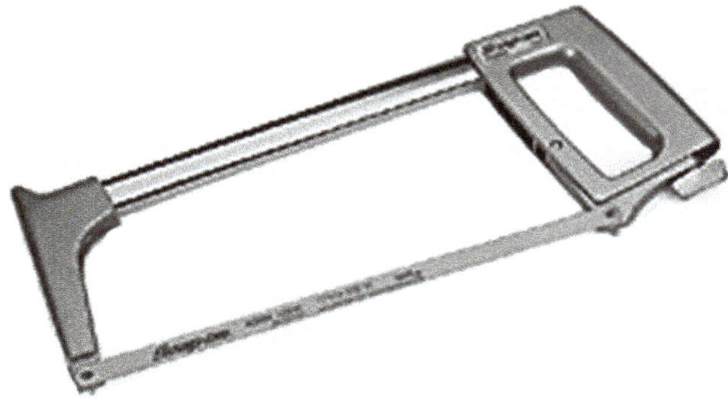

Figure 9.11
Hand Hacksaw

Hacksaws are used widely in piping work for cutting new pipes, removing old ones, and cutting stubborn bolts (see Figure 9.11).

Hacksaw blades come with 32, 24, 18, and 14 teeth to an inch. Blades with more number of teeth are used for thinner sections, whereas those with fewer teeth are used for thicker materials. Hacksaws are selected so that two to three teeth should always be in contact with the metal being cut. This is more important if a stroke should be started on a corner and on pipelines. If the metal gets between the teeth, it may break the teeth or the blade. Hacksaw blades with 18 teeth per inch are commonly used in piping work.

Use a good frame and adjust it in straight and taut condition. Insert the blade in the frame with the teeth facing forward. A loose blade will not cut straight; it may also twist and break. Too much tension will also twist and potentially break the blade. Give a little pressure on the forward stroke, enough to make the teeth cut. The blade should be released a little on the return stroke — the blade does not cut on the back stroke. On the other hand, dragging on the back stroke dulls the blade. Try to make thirty to forty cutting strokes per minute using the full length of blade.

Avoid cutting corners. (Pun is intentional!!) But if necessary, make light and steady strokes until a good cut has been made on the pipe and more teeth can be accommodated. Use a blade with coarse teeth on soft materials as these materials tends to clog into the teeth.

Another important point of caution: just before finishing the cut, the strokes should be made lighter in order to avoid dropping the hand against the work. Also hold the cut piece of pipe so that it does not fall on anybody. It is common to get injured for not being careful.

BOLT CUTTER

Bolt cutters are used to cut away obstinate bolts (see Figure 9.12). Cover the bolt to be cut with a rag or gunny cloth to restrain flying chips. Use the correct size cutter for the required bolt. Keep bolt cutter jaws at a right angle to the material being cut. Insert the bolt into the cutter jaws as far as possible. Apply slow, steady pressure and cut away the bolt.

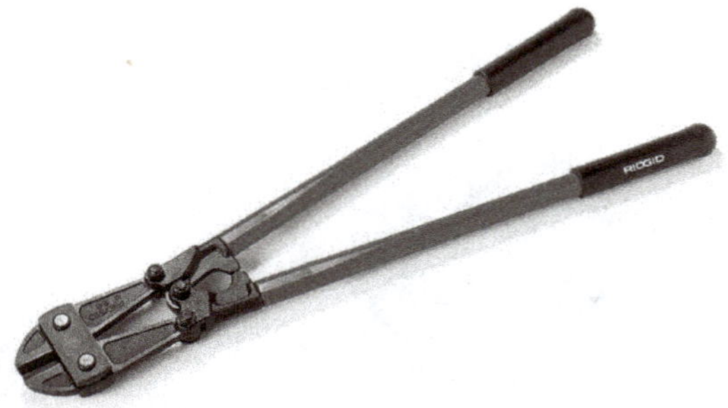

Figure 9.12
Bolt Cutter

SAFETY WITH BOLT CUTTERS

1. Do not use "cheaters" to extend the length of the bolt cutter handles to make the job apparently easier or for use over heavier bolts. This may end up in breakage of the bolt cutter handles or jaws.
2. Wear safety glasses at all times.
3. Do not use the bolt cutter on heated bolts. This could soften the jaws and spoil them.
4. Do not use the bolt cutters to bend the material being cut.
5. Do not hammer on the bolt cutter handles.

WIRE BRUSH

A wire brush is used to clean threads on bolts and pipe (see Figure 9.13). Always clean threads after they are made to remove any leftover dirt or metal pieces. Clean threads result in ease of work. Wire brushes are also used to clean the slag from the weld metal after welding is completed.

Figure 9.13
Wire Brush

Figure 9.14
C-clamp

C-CLAMP

You may wonder why a C-clamp has a place in piping tools! But it does, and yes, it is one of the most wanted tools. A C-clamp is one tool that comes in handy when an emergency repair is on hand with a threat to shut down the plant. They come just in time to seal a pinhole leak or a small crack in the pipe line. A *more detailed discussion* is available in Chapter 32, Emergency Repairs. Suffice it to say that C-clamps are available in a number of sizes which can be selected depending on the pipe size. It has a pressure pad fixed on a ball joint, which can be moved in or out by the screw (see Figure 9.14). *C-clamps are also used with an angle iron welded to it for use in pipe line up.*

C-clamps require certain precautions with C-clamps because they are used while the pipe line is active.

SAFETY WITH C-CLAMPS

1. Always use a perpendicular clamping force. C-clamps can slip easily, particularly when used on round pipe lines.
2. Engage the pressure pad fully; otherwise the clamp can slip.
3. Do not use the C-clamp to pull, lift, or carry something just because it looks handy for that.
4. Use safety goggles because C-clamps are used in piping on filled or partially filled lines. *Serious eye injury is possible*.
5. Do not use C-clamps without pressure pads. *They will slip all the more*.
6. Do not use C-clamps with screws that are bent or otherwise damaged.
7. A C-clamp on pipe lines is generally used for emergency jobs. So it is extremely necessary that proper physical and mental balances are maintained.

PIPING TOOLS

PIPE WRENCH

Although most tools are used by technicians and engineers for pipe work and many other purposes, some wrenches are made exclusively made piping work: pipe wrenches, strap wrenches, and chain wrenches.

Pipe wrenches are available in sizes of 8", 10", 14", 18", 24", 36", and 48" and are indispensable in piping work (see Figure 9.15A).

Figure 9.15A
Pipe Wrench

Figure 9.15B
Proper Use of Pipe Wrench

A pipe wrench is similar to an adjustable wrench but with teeth on the jaws to get a grip on the pipe. The teeth are arranged so that pressure to the wrench is always applied toward the open end of the jaws. The teeth on the movable jaw slant towards the closed end, whereas the teeth on the solid jaw slant towards the open end of the wrench. The more pressure that is applied towards the open end of the wrench, the tighter that the wrench grips the pipe. Pipe wrenches do not take grip in the opposite direction.

Turn the thumbwheel to get the proper size of the pipe and fix it on the pipe. Place the pipe wrench in the direction you want, either of opening or closing. Take a good bite on the pipe and then start turning the wrench in the direction you want.

While using any pipe wrench, maintain a gap of approximately 1/2" (12mm) between the shank of the hook jaw and the pipe itself (see Figure 9.15B). This facilitates proper gripping action of the wrench with the pressure of the two gripping points (the heel jaw and the teeth of the hook jaw). Otherwise, the wrench may slip, damaging the pipe and the jaws and resulting in an injury.

Do not try to operate a pipe wrench in reverse. It won't work!

SAFETY WITH PIPE WRENCHES

1. The teeth should be cleaned periodically. Remove dirt or metal particles that clog the teeth. *Failure to do so will result in slippage.*
2. If the teeth are worn out, renew the pipe wrench.
3. Similarly, if there is a lot of slippage on the adjusting nut, the wrench should be replaced.
4. Never use an extension or a cheater on the pipe wrench; use a higher size pipe wrench. Pipes are inserted into the handles of pipe wrenches and used to open obstinate lines (see Figure 9.15C). It is a common practice for the maintenance technicians to keep aside small pieces of pipes to be used as leverages. Though extensively used, this practice can be dangerous.

Figure 9.15C Do Not Use Extension on Pipe Wrench

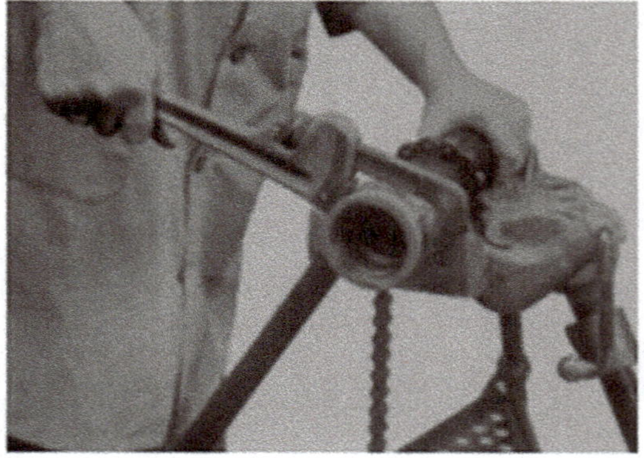

Figure 9.15D Use of Incorrect Size Wrench (Smaller Here)

5. Likewise, do not hammer on a pipe wrench or use it as a hammer.
6. Do not use a pipe wrench to open an ordinary nut. Unfortunately, this practice is often done to save on the number of tools carried or to save time. A pipe wrench is for pipes and not for nuts! (Yes!! No pun intended!!!)
7. Do not use a wrench with a bent or twisted handle. *A bent handle indicates the wrench has been overloaded and probably has internal damage. Do not try to straighten it.*
8. Do not use an incorrect size wrench (see Figure 9.15D). A wrench nut should not be located at the very end of the hook jaw. *Using a wrench beyond its capacity may result in wrench failure. Using a larger wrench may spoil the work piece*. Never use an extension or a cheater on the pipe wrench; instead, use a higher size pipe wrench. Pipes are inserted into the handles of pipe wrenches and used to open obstinate lines. It is a common practice for maintenance technicians to keep aside small pieces of pipes to be used as leverages. *Though extensively used, this practice can be dangerous.*
9. Do not apply a side load to the handle. It is not designed for it.
10. Do not use a pipe wrench as a lever or a lifting device, or to bend tubing. Any of these practices is dangerous.
11. Do not use a pipe wrench near a flame or on heated parts, for example, trying to open heated nuts with a pipe wrench. Do not be tempted just because the pipe wrench is the only tool possible because the flats of the nut would have rounded off by this time anyway. Teeth will soften on a pipe wrench, or the chain on a chain wrench could also become soft.
12. Lubricate the non-painted portion, and store in a dry place.

Daniel C. Stillson, a steamboat fireman, invented the pipe wrench. Until then, serrated blacksmith tongs had been used for screwing pipes. The owner of a heating and piping firm, James Walworth, challenged Stillson to make a prototype which "either twist off the pipe or break the wrench." His design was then patented on September 13, 1870, and Walworth manufactured the wrench. Stillson was paid about $80,000 in royalties during his lifetime.

STRAP WRENCH

A strap wrench is a friction type wrench that fits pipes from 1/2" to 10". It is meant for turning shafts or piston rods or for polished pipes without damaging the finish of the job. It does not have much grip and is not meant for general piping. It has a serrated or woven plastic tape to get a good grip on the pipe (see Figure 9.16).

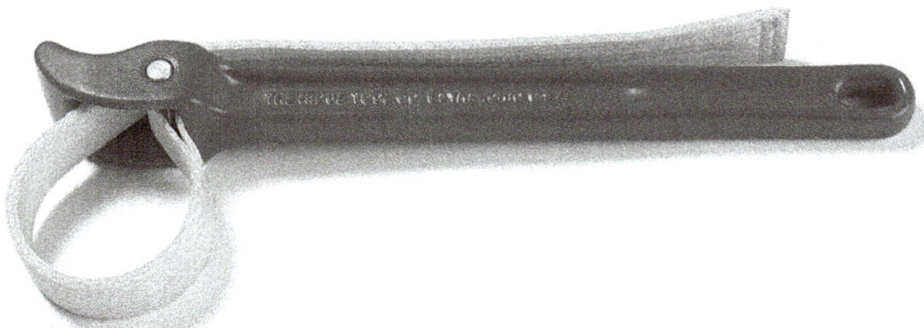

Figure 9.16
Strap Wrench

Inspect the strap and replace if it is frayed or cut. These problems reduce the strength of the strap, which may actually snap while on job.

Do not let the wrench slip due to grease, oil, or dirt that may have impregnated the strap or that got smeared on the job.

CHAIN WRENCH

Figure 9.17
Chain Wrench

Chain wrenches are generally used for large size pipes; say 3" to 8" pipes. They are made with a single jaw or double jaw (see Figure 9.17). A double jaw model gives fast, ratchet-like action in both directions. Some models are available with replaceable jaws. The jaw is placed over the pipe and the chain is slipped over tight. The chain grips the pipe tight. They are good for large size pipes where much leverage is required and where pipe wrenches can be cumbersome.

SAFETY FOR CHAIN WRENCHES

1. Clean the chain with a wire brush as the dirt may result in the wrench slipping.
2. Replace if the links are worn or damaged. Check for separation of the links.
3. Lubricate the non-painted portion and store in a dry place.

While on the subject, it is good to have a look on some other wrenches used on pipe work.

END PIPE WRENCH

End pipe wrenches used for fast, easy grip for close-to-wall and parallel work (see Figure 9.18).

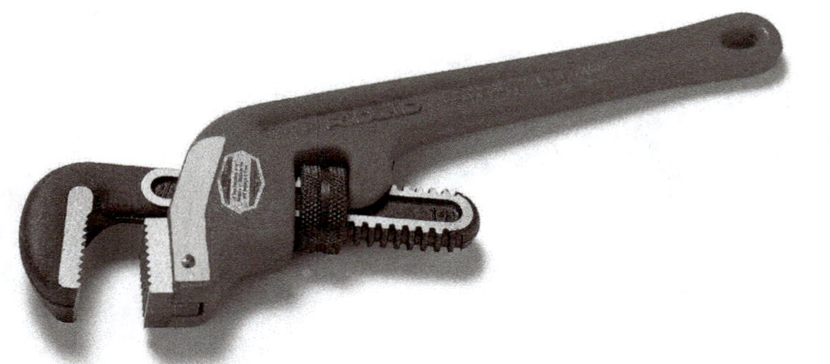

Figure 9.18
End Pipe Wrench

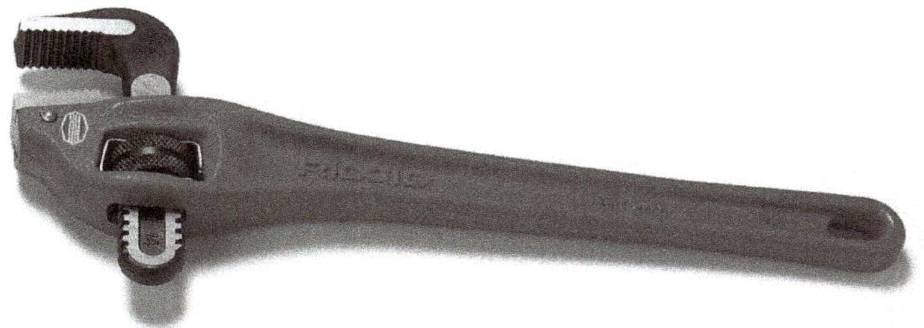

Figure 9.19
Offset Wrench

OFFSET WRENCH

Offset wrenches provide easy entry to tight spots and awkward angles (see Figure 9.19).

COMPOUND LEVERAGE WRENCH

Compound leverage wrenches multiply leverage. Hence, they are very useful for seized joints (see Figure 9.20).

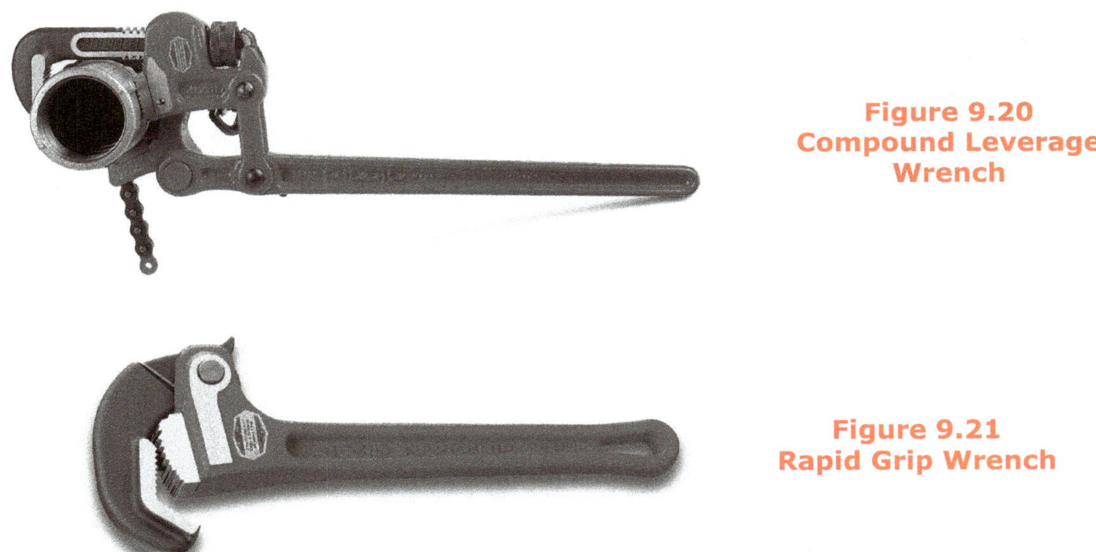

Figure 9.20
Compound Leverage Wrench

Figure 9.21
Rapid Grip Wrench

RAPID GRIP WRENCH

Rapid grip wrenches can be used for faster jobs. However, make sure to center the work piece in the "v" of the hook (see Figure 9.21). If not, the wrench may lose the gripping action, resulting in failure of the hook.

PIPE CUTTER

The best way to cut pipe is by using a pipe cutter because, by this method, the pipe is cut squarely, accurately, and quickly. Pipe cutters are available in different sizes and designs. The most commonly used pipe cutter with two rollers and one cutting wheel is easier to use, though there must be enough clearance to revolve it completely around the pipe (see Figure 9.22A). Pipe cutter wheels make grooves and they are worked deeper and deeper until the pipe is cut through. Actually no metal is removed; it is only displaced by the cutter.

Pipe should be held tight in a pipe vice or combination vice. Place the rotary pipe cutter in position on the pipe. Then tighten the handle just until the cutting wheel touches the pipe. Make a turn and check. If OK, tighten a little more and make a few turns around the pipe. Keep tightening such that the cutter bites more at every turn until the pipe is cut through. If the edge of the pipe is flared a little, smooth it out. Use a little cutting oil or other lubricant during cutting to reduce wear on the cutting wheel and also to allow easier operation of the pipe cutter.

Four-wheel pipe cutters (see Figure 9.22B) are used in areas where a complete turn is impossible. The model at the top has a short handle for use in confined areas. The model at the bottom shows an additional lever to operate the cutter by two workers.

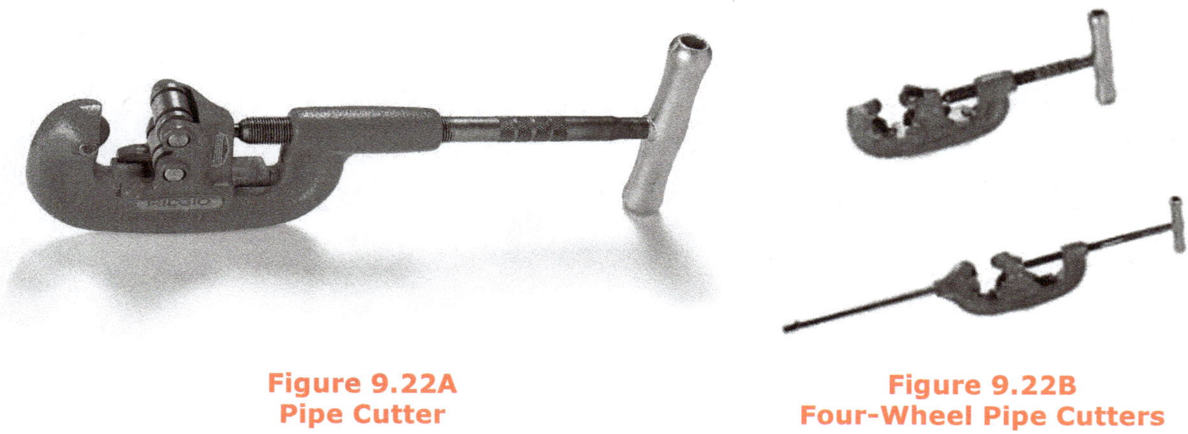

Figure 9.22A
Pipe Cutter

Figure 9.22B
Four-Wheel Pipe Cutters

SAFETY FOR PIPE CUTTERS

1. Select the correct model of pipe cutter for the job and material. Do not try to cut pipe with the wrong cutter or cutter wheel. *Serious injury and damage to the pipe could be the result.*
2. Always inspect the cutter wheel before use to see that it is not blunt, chipped off, or damaged in any way. Examine for proper tracking and cutter wheel sharpness.
3. Do not force the cutter wheel into the pipe just to finish the job fast. The life of the cutter wheel will be shortened and possibly it will get damaged.
4. Always clean the pipe cutter after each job and lubricate the feed screw, rollers, and cutter wheel.
5. Store them by hanging in a warm dry area.

PIPE DIE, PIPE TAP, AND THREAD EXTRACTOR

A detailed discussion about pipe dies, pipe taps, and thread extractors is given in Chapter 10, Cutting, Threading, and Welding Pipes.

FLANGE SPLITTER OR FLANGE SPREADER

Flange splitters or flange spreaders are just that — they split the flanges. Though used occasionally, they are very useful tools. In any process industry, there will necessarily be occasions to open the flanges and separate them for removing or fixing gaskets or for inspection. Ring-type-joint flanges used in refineries and petrochemical industries should invariably be split wide to access the gasket. In any industry, there will be a few flanges that will not budge. The higher the sizes, the more difficult it will be to separate them.

There is a better way to separate flanges than using steel wedges and hammering them into whatever little opening exists. Instead, use flange splitters. There are number of designs, however. Figure 9.23 shows a typical flange splitter. The pin is inserted in place of a convenient flange bolt. The bolt may be more convenient, but the pin this will not obstruct access to the gasket! The central wedge is placed into the flange gap and is screwed down. As the wedge goes down, the flange is opened up.

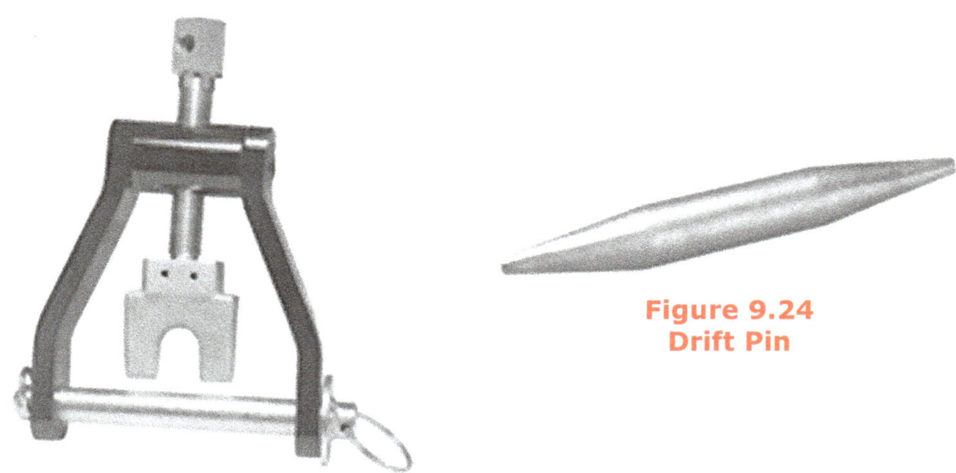

Figure 9.23 Flange Splitter

Figure 9.24 Drift Pin

DRIFT PIN

Normally the spud wrench should be able to align the holes in a pair of flanges. If, however, the flanges are badly misaligned, then drift pins or punches are used (see Figure 9.24). Drift pins are made in a variety of sizes to match bolt holes. One or both ends are tapered.

Take the correct size pin and place it in one hole in any one flange. Locate a hole in the other flange by manipulating or moving the taper end of the drift pin this side and that. Try to locate the pin into any nearest hole even a little. Then hammer the pin to bring the hole of the other flange in line. When the holes come in line, place a bolt in a hole adjacent to it or wherever the holes are matched. Remove the pin and put a bolt in that hole. If necessary use the pin in the rest of any holes that did not match and align them.

The need to use a drift pin indicates that the pipe or flange is not in proper alignment and that pulling the line into position will exert a certain strain on pipe and fittings. This must be understood as it is one of the major reasons of pump failure.

SAFETY WITH DRIFT PINS

1. Use safety goggles. *Serious injury is possible as chips may break out of drift pins and hammers.*
2. Do not use a damaged drift pin. Replace damaged ones.
3. Select a drift pin or punch that fits the job.
4. Drift pins should be discarded if pins show chip out or brittleness.
5. If these pins become mushroomed, they should be dressed on a grinder.
6. It is safer to hold the drift pin with a pin or punch holder or a chisel holder to prevent hand injury.

PRY BAR

Figure 9.25
Pry Bar

A pry bar is also a flange aligning tool (see Figure 9.25). Pry bars are used on heavier piping and where the misalignment is large.

Fix the tapered end of the pry bar in a hole and maneuver this end into the hole in the other flange by leveraging and prying. Push bolts into holes wherever they are matched, preferably the adjacent ones. Place the bar in another hole, and do the alignment again. A few attempts will be required to align holes in a pair of flanges.

The use of a pry bar indicates that the pipe or flanges are very badly misaligned. Bringing them back into position will exert excessive strain on the pipe, fittings, and associated equipment. This problem is one of the major reasons for rotating equipment failure. *Follow the procedure in Chapter 14, Bolting Up, to bring the flanges back into alignment.*

MEASURING TOOLS

TAPE MEASURE

Measuring tapes are available in 10 ft, 50 ft, 100 ft, and corresponding metric sizes (see Figure 9.26). Self-retracting tapes are spring loaded. Measuring tapes are handy and much easier than a steel rule because they can be kept in a pocket.

Normally the tape is pulled out by one hand, and measurements are taken. If the tape is let off, it goes back into the casing. Some tapes have a thumb lock which can be moved to lock or unlock the tape. Some tape measures have a small spirit level as well.

Figure 9.26
Tape Measure

Tape measures of either 100 ft or 30 meters are required in piping work; they are generally not self-retractable. The tape is pulled out by hand, but retracted by rotating a small lever.

SAFETY WITH TAPE MEASURE

Do not pull the tape beyond its rated length. You will be wasting the rest of your time trying to push it back. Similarly, tape edges can cut your fingers, particularly when it is retracting fast.

STEEL RULE

Steel rules are also used to make small accurate measurements. They are available in both inch and metric measurements, generally marked on both edges of the rule in a number of standard lengths. Figure 9.27A shows a foldable rule and Figure 9.27B shows the steel rule. They are also used in conjunction with plumb bobs to measure deviations in verticality, etc. Straight edges are used for checking straightness of lines, fittings, and flanges. Steel rules are accurate — do not use them for anything other than measurements. Without much difficulty, they can be bent and distorted!

Figure 9.27A
Foldable Rule

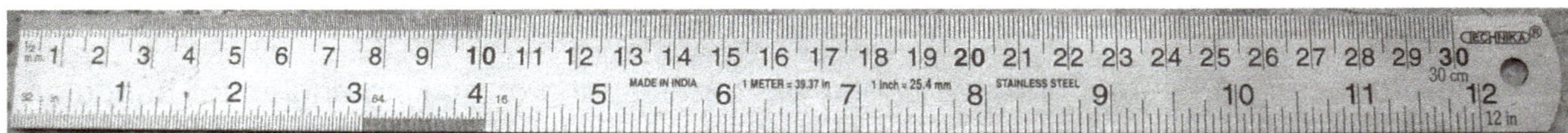

Figure 9.27B
Steel Rule

PLUMB BOB

Developed more than 5000 years ago, the plumb bob is one of the oldest and simplest tools used to measure the verticality of a pipeline or any structure. It is simply a small round weight or bob with a pointed tip attached to a strong thread (see Figure 9.28). It is suspended along the pipe line and checked for the verticality. It is used to transfer points from ceiling to floor and from floor to wall.

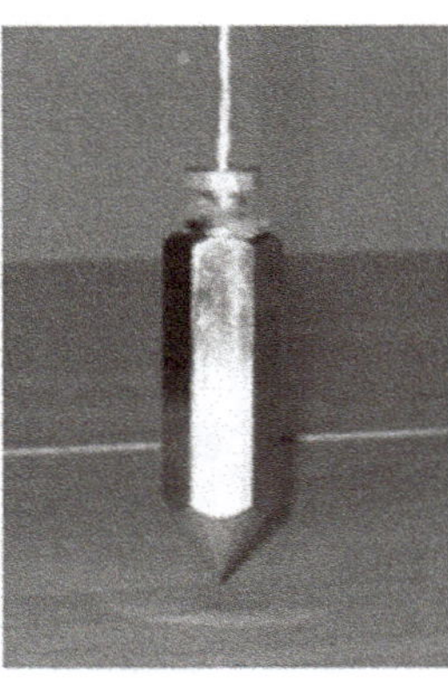

Figure 9.28
Plumb Bob

STEEL SQUARE

Steel squares used in piping work are 24" long on one leg and 16" long on the other leg (see Figure 9.29). They are used in pipeline fabrication for checking the square and straightness. Often misused, steel squares easily get out of shape. *All too often, you see pipe fitters tapping the just-weld tacked pipe fittings with the back of a square to bring it just into position required. This is a bad practice!*

Figure 9.29
Steel Square

BEVEL PROTRACTOR

The protractor and center finder is useful when measuring or making angles for bends and miters. The bevel protractor features a graduated scale and a reversible 180° double protractor (see Figure 9.30). The protractor can be moved accurately on the scale and rotated to get the measurements. It has also a small spirit level. One side of the protractor is flat, allowing for laying flat on the job. The protractor can also be used as a depth gauge. This is a precision tool. Hence, use it with due respect and care.

Figure 9.30
Bevel Protractor

COMBINATION SQUARE

A combination square is a versatile tool which is essentially a 12-inch ruler with a sliding head that has two precision flat surfaces; one at 90-degrees and the other at 45-degrees to the ruler (see Figures 9.31A and 9.31B). It also contains a spirit level. Some models have a protractor. The head can be slid anywhere along the ruler by loosening a knob and be locked in place with it. Some combination squares have a scriber housed in the head for marking and some have center punches for center marking. This is a precision tool. Hence use it with due respect and care.

Figure 9.31A
Combination Square

Figure 9.31B
Various Parts of Combination Square

SPIRIT LEVEL

Pipe lines must run vertically, horizontally or, occasionally, at predetermined angles for proper flow and outlook. The flanges must be placed either vertically or horizontally. This simplifies later maintenance work on the lines. In addition, the work looks neat and pipe strains are less.

The spirit level is used to make sure that the pipe is vertical or horizontal when installing pipe (see Figures 9.32A and 9.3B). In piping work, the level has a frame consisting of one or more transparent tubes, filled with alcohol, ether, or similar fluid. It is not fully filled, and thus has a small air bubble. The position of the bubble within the tube indicates whether the instrument is horizontal or not. The level is normally provided with two bubbles; one for plumb or vertical, the other one for level or horizontal. Some levels have a third bubble meant for 45°. The longer the level, the more accurate the work will be. The surface is curved to facilitate use on pipe lines.

Figure 9.32A

Figure 9.32B

Spirit Levels

VISES

BENCH VISE AND PIPE VISE

Bench vises and pipe vises are the most common types of vises. They are used for firmly holding a piece of work or pipe so that some mechanical work can be carried out. Pipe vises should be used for piping work and bench vice for other normal jobs. There is also a combination vise that can accommodate piping and other works. It has square upper jaws for holding normal work and gripper jaws for pipe. See Figure 9.33 for a bench vise and Figure 9.34 for a pipe vise.

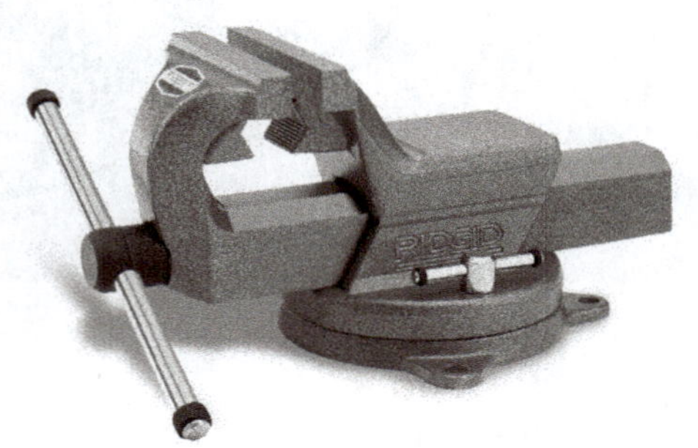

Figure 9.33
Bench Vise

Figure 9.34
Pipe Vise

To avoid squeezing the component parts, always fix the pipe, not the fitting or the valve, in the vise. When repairing a valve, fix it with strips of soft metal like aluminum, copper, or brass to avoid damage to finished parts. It should be fixed in the vise with valve ends — not valve sides — to prevent damage to the valve body.

PORTABLE YOKE VISE

As the name suggests, a portable yoke vise is portable (see Figure 9.35). It can be easily attached to a work table, plank, or similar thing to work on the pipe.

Figure 9.35
Portable Yoke Vise

CHAIN VISE

Chain vises are used mostly on the field, particularly for heavier jobs. They can be fixed on a tripod and carried up and down various places. They permit holding of irregular shapes (see Figure 9.36).

Figure 9.36
Chain Vise

A SAFE JOB IS A BETTER JOB

Every day avoidable accidents occur because of using the wrong tool for the right job or the right tool being used in the wrong way and with short cut methods. Hammering on spanners, pipe extensions on pipe wrenches, chisel with chipped ends, wrenches used as levers, and screwdrivers as chisels are only some of the examples. The use of hand tools is a matter of experience and not subject for a treatise. Tools are made for specific use. Small points, though often repeated, will keep injuries at bay.

- Plan your job. There used to be a placard on every table in our office, "People do not plan their jobs to fail; they fail to plan their job."
- Select the correct tools for the job. Do not return often to the tool box for proper tools. This delays the job and prompts you to take short-cut methods. *Use hand tools as they are intended to be used.* For instance, do not use your screw driver as pry bar. It will surely damage the tool and cause serious injury as a bonus.
- Take enough number of tools. *Lack of enough tools prompts use of a wrong tool.*
- Use them correctly. Learn the proper technique and adopt it. *Wrong practice spells danger.*
- Maintain your tools in good condition. Clean them. Replace them when damaged. Check for make-shift repairs, loose or cracked handles, mushroomed heads, dull cutting surfaces, spread out jaws, and worn working surfaces. Some tools — like hammers and chisels — may with time develop some defects. Rectify the defects or replace the tools.
- Store your tools carefully. When all the tools are easy to locate, the technician will almost always use the correct tool for the job. *Tools placed out of the way are a hazard themselves.*
- Keep them clean. A tool free of grease, dirt, and rust is always a safer tool. In particular, keep non-sparking tools free from ferrous or other contaminants. They may impair the non-sparking properties. Manufacturers also suggest that they must not be kept in the direct contact of acetylene, which may create explosive mixtures in moist conditions.
- Always select a spanner which has a close and tight fit on the nut or bolt head. This is particularly important with non-sparking tools because their hardness is less than that of steel tools; thus, loose fitting spanners can slip easily and also wear out quickly.
- Use safety goggles wherever flying chips or chemical injuries are possible, even while repairing tools like chisels and hammers.
- Check for end clearances while using spanners in confined places.
- Take a body position suitable for the body and tools. It should be a safe position to give a fast escape route in case of a fault. *Wrong body position while using tools can result in permanent problems in human body.*
- Buy the best tool you can pay for. *Cheaper tools get spoiled easily and injuries are much costlier.*

Concentration and care matters, no matter how trivial the task is.

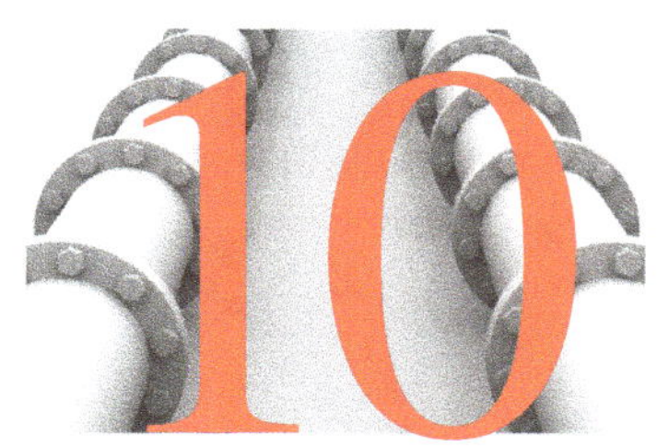

Cutting, Threading, and Welding Pipes

Three normal activities are central in pipe joining whether it is pipe to pipe or pipe to a fitting. They are cutting, threading, and joining. Pipe joining can be screwed, welded or flanged. This chapter briefly describes these regular activities in the piping practice. Note that pipe joining by flanges is taken up separately in Chapter 14, Bolting Up. *Safety is of paramount importance. Tools are unforgiving friends.*

CUTTING PIPE

A number of methods can be used to cut the given length of pipe or pipes. Demands of production, volume, cost, and time ultimately dictate the method of cutting. For most maintenance work, the most common method for cutting pipes is hand saw cutting.

HAND HACKSAW

A hacksaw is used to cut the pipe in the field. *Make sure that the cut is straight throughout. Hacksaws can cut metal, but can also cut fingers!* A more detailed discussion of hacksaws is available in Chapter 9, Hand Tools in Piping.

MACHINE SAW

When large size pipes or pipes with heavier schedules are to be cut, they will be difficult to cut by hand. Thus, machine saws are used. They are part of the machine shop equipment used routinely for cutting rods and other materials. Blades are very much like hand hacksaw blades, but are heavier, stronger, and longer. Various sizes are available. Arrangements for feeding the blade and cooling it are normally available on these machines.

PIPE CUTTER

Pipe cutters can also be used to make a straight cut. A cutter with two rollers and one cutting wheel is easier to use, though there must be enough clearance to revolve it completely around the pipe (Figure 10.1). Three-wheeled cutters obviate this problem, but they require more skill. Use a little cutting oil or other lubricant during cutting to reduce wear on the cutting wheel and also allow easier operation of the pipe cutter. A discussion on pipe cutters is available in Chapter 9, Hand Tools in Piping.

Figure 10.1
Pipe Cutter

COLD SAWING

A circular rotating blade cuts through the pipe (see Figure 10.2) while cutting fluid is applied. The process does not produce a HAZ (Heat Affected Zone). The cuts are precise and square with minimum burrs. The process is much faster than with a hand hacksaw.

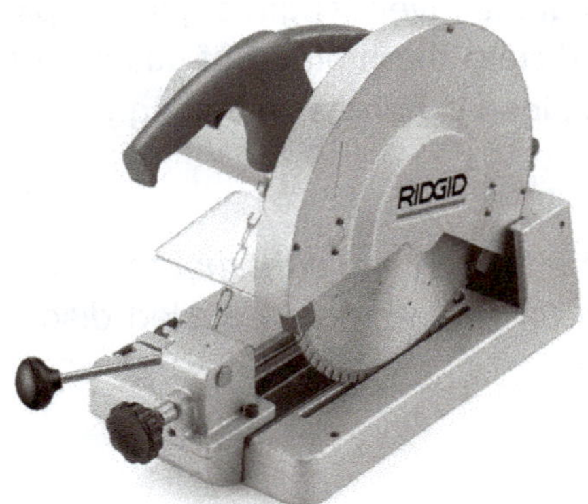

Figure 10.2
Cold Saw

BAND SAW CUTTING

Band saw cutting is good for high volume cutting of rod, bar, pipe, and tubing. The process can be fully automatic and blades are available in various tooth configurations. The blade is a continuous band of metal fixed either vertically or horizontally, rotating between drive and driven wheels. If the blades snaps or breaks, there is a provision on the machine itself for spot welding the joint.

ABRASIVE CUTTING

Abrasive sawing is similar to circular saw cutting. It is manual but can be fast if a small number of pipe pieces are needed urgently. It operates with a circular abrasive blade that grinds through the pipe. Abrasive sawing is an easy method but leaves a burr, which needs to be cleaned before further activity is taken up. It also creates a HAZ due to the heat in the grinding action.

LATHE CUTTING

Lathe cutting is good for production of small quantities of emergency requirements. Lathe cutting is suitable for cutting large diameter and a heavier schedule of pipes. Essentially a cold cutting method, lathes can make square cuts, chamfer, and also deburr. When a large number of short lengths or nipples are required, lathe cutting is often used, especially when the required numbers of nipples are not readily available in the plant warehouse.

LASER CUTTING

Laser cutting is the latest method; it can be fully automatic to meet the demand for a large number of pipes. A number of advantages can be developed into the system such as cutting, deburring, inspecting, and even packing. Laser cutting produces narrow kerf, tight tolerances, and minimum HAZ (Heat Affected Zone.) It is good for cutting stainless steel alloys, hard materials like nickel alloys, and titanium. Presently it suffers from high capital cost and, hence, may not be suitable for maintenance applications.

GAS CUTTING

The most common method of cutting pipe is by oxy-gas flame cutting (simply, gas cutting), followed by plasma-arc cutting at the shop floor. Air carbon-arc cutting is rarely practiced in piping except in some sacrificial cases. The oxy-acetylene flame cutting process makes use of a cutting torch. The torch mixes the acetylene and the oxygen in proper proportions to produce a preheating flame of about 1400–1600°F. Then by pressing a lever on the torch itself, a jet of oxygen is directed towards the cutting area. This jet quickly oxidizes the iron and blows out the reactants from the joint. The narrow slit formed in the metal is called the kerf.

Most often, the gases are supplied by the cylinders. An ordinary oxygen cylinder is made from a seamless steel shell which stores it at 2200 psi at 70°F. The acetylene cylinders are different in construction from oxygen cylinders. Acetylene cylinders are packed with a porous material, the fine pores being filled with acetone, which can absorb many times its own volume of acetylene. Acetylene is drawn off through a valve at the top of the cylinder. Pure acetylene is self-explosive and becomes extremely dangerous if used above 15 pounds pressure. If the cylinder is not upright, liquid acetone can be drawn from the cylinder. Therefore, do not keep acetylene cylinders on their side. Cylinders have caps that screw on the neck rings to protect the valve from damage during shipment and handling.

Oxygen valves are double seated to prevent leakage around the valve stem when the valve is fully opened. The oxygen valve is especially designed to operate at high pressures. When in use, the valve should be opened with the key handle as far as it will go. To prevent cylinder rupture, a safety device is generally incorporated in the oxygen valve that will burst if pressures get too high. The acetylene valves are similar in construction; they do not need to withstand as much high pressure as oxygen valves. A key handle operates the valve and should be left in place during use of the cylinder. Open the valve only one-half turn or, at the most, one and one-half turns. Opening the valve more will result in liquid acetone being drawn from the cylinder.

Most regulators have two gages: one indicates the available cylinder pressure when the valve is opened and the other indicates the pressure of the gas coming out. Regulators have at least two relief devices to protect against the release of high-pressure gas. All regulator gages have a provision of blowout backs which release the pressure from the back before the gage glass explodes. Outlets on the oxygen regulators have right hand threads whereas those of acetylene are left hand to avoid accidental wrong connections.

The hoses that make connections between the torch and the regulators should be strong, yet light in weight, flexible, and nonporous, rated at 100 psig. Oxygen hoses are green and acetylene hoses are red. Oxygen connection nuts have *right-hand* threads and acetylene connection nuts have *left-hand* threads and a grooving around the nut which readily identifies it. Never interchange oxygen and acetylene hoses.

BACKFIRE AND FLASHBACK

Back fire can be caused when the tip touches the work, it is overheated, or the torch is operated with wrong gas pressure settings. The flame goes out with a loud snap or pop. *Immediately close the torch valves, and check all connections, loose tip, or dirt.* If this happens repeatedly, train the worker on proper operating techniques.

When the flame burns back inside the torch, more often than not with a hissing noise, a flashback occurs. *Immediately close the oxygen valve at the torch cylinder —both gases must be cut off.* You should close the torch oxygen valve that controls the flame to stop the flashback at once. Let the torch cool before relighting it. Remove accumulated soot in the oxygen hose by blow through, clean the tip, and check the nozzles.

CUTTING AND BEVELING PIPE

Pipe cutting and beveling with a cutting torch is the most common practice in piping fabrication. It requires a steady hand and a trained technique to obtain an even and exact bevel, which is essential for a good weld joint.

Cut the pipe square and remove all the slag. The best cutting results are achieved by positioning the torch tip very close to the work at the appropriate angle. Do not cut a bevel in one go, particularly on a heavy pipe. Point the torch toward the center line of the pipe. Begin at the top and proceed down the side. Then start at the top again and go down the other side, finishing at the bottom of the pipe. Make a coarse cut and trim it suitably.

When making a T- or Y-joint, a template or pattern is developed and the pipe is marked accordingly. It is preferable to punch mark it and cut accordingly. Do not leave the punch marks on the pipe however.

HANDLING AND STORAGE OF GAS CYLINDERS

1. Never fill your own cylinders.
2. Never alter or fix the safety devices on a cylinder.

3. Never store cylinders near a heat source, in direct sunlight, or in a closed or unventilated space.
4. Store the cylinders in a cool, dry place. Chain the cylinders in an upright position.
5. Protect the cylinder valves and safety devices. Always replace the cylinder cap.
6. Never mix empty cylinders with full cylinders. Mark the cylinder "Empty" or "MT." *Do not mix cylinders that contain different gases.*
7. Never use slings or magnets to carry cylinders.

PLASMA ARC CUTTING (PAC)

Nonferrous metals and stainless steels cannot be cut by the oxy-gas cutting method. In normal steels, rapid oxidation ensures the cutting process. But stainless steels do not respond well due to higher oxidizing temperatures of alloying elements. Hence, the plasma arc cutting (PAC) method was developed. An arc is formed between the electrode and the work piece. It is constricted by a fine-bore copper nozzle, which increases the temperature and velocity of the plasma emanating from the nozzle. The temperature of the plasma can go in excess of 20,000°C and the velocity can reach the speed of sound. A narrow kerf is made while cutting through.

The process can cut any metal, and it can cut carbon steel ten times faster with equal, if not better, results. However the initial cost of the equipment is high.

PIPE THREADING

The screwed joint is by far the most common method of joining pipe. American Pipe Thread Standards are the most common standards in the piping industry, though piping with other standards is also seen in the industry. NPT thread, as is commonly known, has a taper of 3/4" per foot length for all sizes (see **Chart 10.1**).

Chart 10.1
Specifications for National Pipe Threads

Pipe Size Inches	Threads per inch	O.D. Inches	N.P.Thread length
1/8	27	13/32	3/8
1/4	18	35/64	9/16
3/8	18	43/64	9/19
1/2	14	27/32	3/4
3/4	14	1-3/64	3/4
1	$11^{1/2}$	1-5/16	15/16
1-1/4	$11^{1/2}$	1-21/32	15/16
1-1/2	$11^{1/2}$	1-29/32	1
2	$11^{1/2}$	2-3/8	1-11/32

NOTE: Thread length tolerance = ± one thread

NPT pipe dies from 1/8" to 2" nominal pipe size are commonly available. These dies are used to make male thread on the pipe. They are housed in ratchet type handles. Each die head consists of four die pieces which can be inserted into the head. The entire head is fixed into the ratchet handle and used (see Figures 10.3A, 10.3B, and 10.3C).

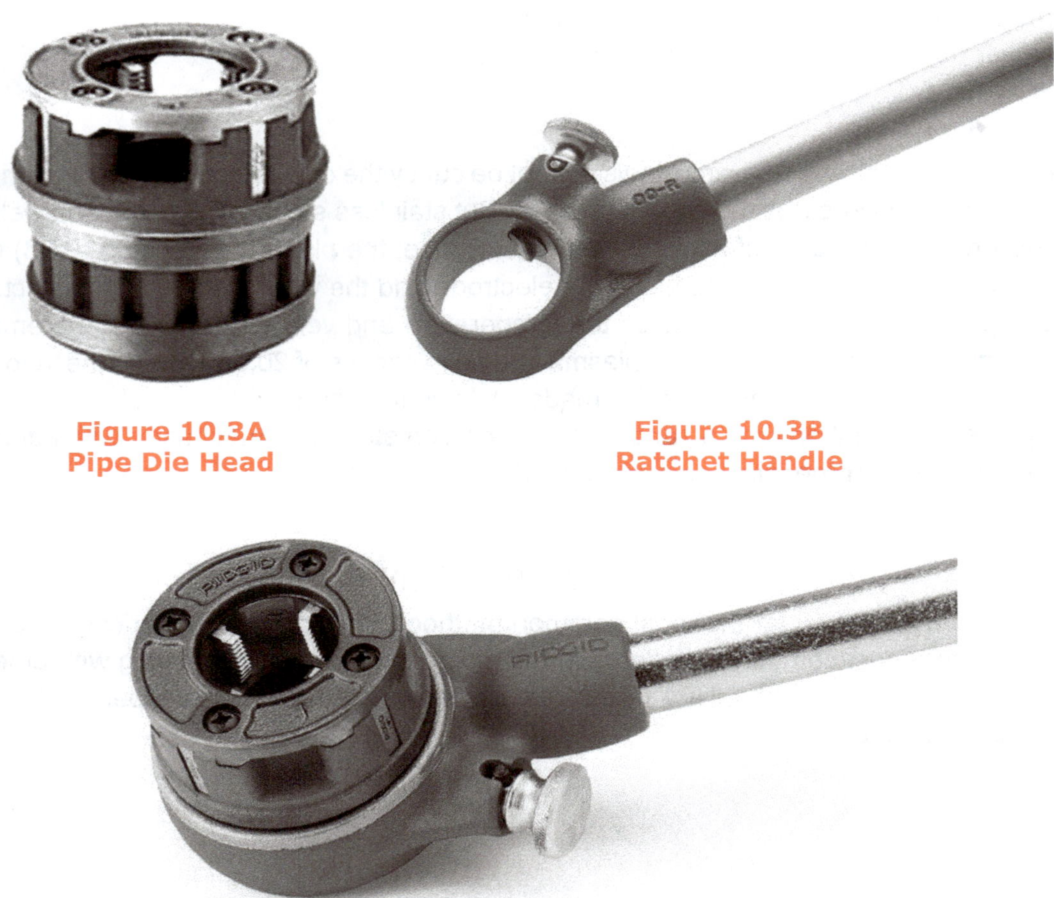

Figure 10.3A
Pipe Die Head

Figure 10.3B
Ratchet Handle

Figure 10.3C
Pipe Die with Ratchet Handle

Take the required size die with its ratchet handle and fix it over the pipe at its end to be threaded (see Figure 10.4). There is a small lever on the handle (Note: it is in Figure 10.3B), which makes the die to turn in one direction and slip on the other direction. This lever is engaged for making the thread and disengaged to remove the die from the finished piece. Engage the lever. Then push the pipe die on to the pipe, engage it on the pipe, press a little hard on the pipe, and hold it square. Turn it to make a bite. Apply a little oil to make the action smooth and remove the heat. Turn it clockwise. After a few turns, back off and check if the formed thread is good. If it is OK, turn the pipe die until all of it is over the pipe and full threads are made. Turn over the direction lever and remove the pipe die from the pipe. Clean the threads, wipe off any foreign material, burrs, etc. Please note that pipe threads can be sharp and can cut your fingers. See Figure 10.4 for proper positioning of the pipe in a pipe vise.

Figure 10.4
Proper Position of Pipe in the Pipe Vise

The thread is tapered to make a tighter joint. Without the use of cutting oil while making pipe threads, the dies and taps get damaged and bad thread will be the result.

The previous discussion was on making pipe threads manually, but a number of models in pipe threading machines are available to cater to the industry where a large number of pieces are needed for use or for sale. On the other hand, machine tools like lathes are made good use of for making threading on pipes for maintenance works, when a good number of threaded pieces are required or in an emergency. By and large, hand die are widely used all over.

PIPE TAP

Pipe taps are used to make, clean, or chase female threads. The operation is similar to pipe dies. However, instead of ratchet handles, pipe taps have separate handles. These handles are the same as the ones used for other hand tapping of threads. A typical pipe tap is shown in Figure 10.5.

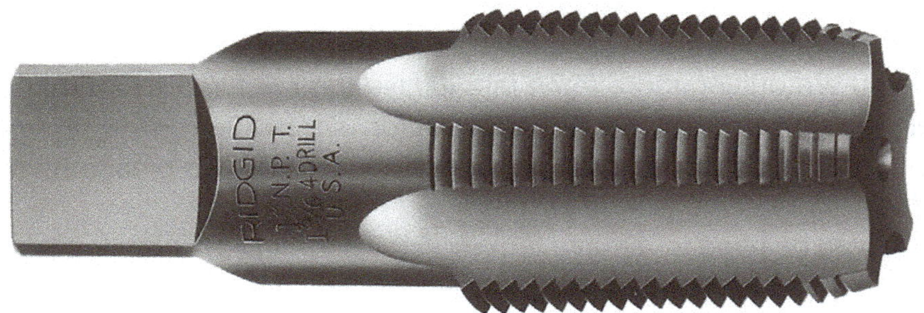

Figure 10.5
Pipe Tap

Fix the tap inside of the pipe squarely and firmly. Press the tap tight and square. Fix the handle on the square end and turn it. Apply a little oil to make the action smooth and remove the heat. Turn the tap clockwise. After a few turns, back off and check if the thread formed is good. If it is OK, turn the pipe tap until full threads are made. Remove the tap. Clean the threads and wipe off any foreign material, burrs, etc. *Please note that pipe threads can be sharp and can cut your fingers.*

HOW TO MAKE UP A SCREWED JOINT

After the threads are made on the pipe, clean them. If it is a bought-out nipple or fitting, check the threads and clean them. Normally, a wire brush is enough to clean them, but if they are even a little bit damaged, run the suitable die over the threads to correct them. Run a tap to correct internal threads.

Next, apply a small amount of good thread lubricant or a pipe thread compound on the male threads. Use only a small amount, as excess of this will surely settle on valve seats. Lubricant is good — but not when it is inside a pipeline. Pipe dope or shellac, even hemp rope, can be used to make a leak proof joint. However, present practice is to use Teflon tape over the threads. Follow the direction of threads while wrapping the tape or it may become loose when the joint is tightened. Teflon tape helps in sealing the threads, but does not compensate for bad threads.

Engage the threads and feel for the right contact. If it is right, tighten up the fitting as far as it will go. If the threading job is up to standard and made properly, the fitting will go up a good length. Further turns with a wrench will finish the job. All the male threads need not be driven into the joint. Due to the imperfections of the die, one or two could be left exposed. But do not try to force and push all the threads into a fitting, particularly a valve, as it will surely damage it.

HOW TO EXTRACT A BROKEN NIPPLE

THREAD EXTRACTOR

It is not out of the way here to dwell a little on pipe extractors. You may find yourself in a difficult situation where a piece of nipple is broken inside the pipe or fittings. We have a tool for that — a thread extractor — which is actually an extractor for broken pipe pieces. It is a very hard reverse-thread bit. Thread extractors are available in different sizes. Select the one that matches the inside diameter of the broken piece of nipple. Gently tap the extractor inside the hole, and turn as if you are opening the nipple. The reverse flutes on the extractor make it to bite harder into the metal as you are putting more force into it. Use the tap handle to turn it evenly on both sides. Any other wrench will surely break the extractor. *Thread extractors are made of very hard metal; hence, they are very brittle and can easily break inside. So please be careful, as more often than not the broken piece comes off easily.*

A thread extractor can also be used to remove broken screws. Select the proper size extractor and drill a hole into the broken screw. Generally extractors are marked with an appropriate drill required to be used with it (see Figure 10.6). Please be very careful when drilling the hole in the screw, so as not to drill into the material surrounding the screw. It is preferable to drill a small pilot-hole first. Use a punch to mark the spot before you start drilling. Drill slowly and stop often to check your progress. Gently tap the extractor inside the hole, and turn as if you are opening as suggested above.

Figure 10.6
Various Sizes of Thread Extractors

PIPE WELDING

Pipe welding is a skill by itself and is the simplest method of joining pipes and fittings. It is the only method for higher sizes and heavier schedules of piping, and for high pressure, high temperature, and sub-zero temperature service. Even though welding of pipes is similar to welding other materials, there are certain skills, techniques, hazards, and demands in pipe welding which makes it a distinctive profession.

Welding on pipes and fittings must be performed by persons with proper credentials. Welding on pressure piping must be carried out by applying qualified welding procedures. There are certain ASME (American Society for Mechanical Engineers) and API (American Petroleum Institute) codes, and AWS (American Welding Society) standard welding procedures for welding pipe lines which must be followed.

We have umpteen numbers of welding procedures like shielded metal arc welding (SMAW), gas tungsten arc welding (GTAW, formerly known as tungsten inert gas welding or TIG), and plasma arc welding. Electrodes are available for all the procedures and in almost all metallurgies. We have flux coated electrodes (normal day-to-day weld rod or stick electrode), bare electrodes (filler wire for GTAW welding), and tubular electrodes. A treatise on welding would be needed to cover all that. But....

Manual metal arc welding is carried out by AC or DC. AC welding is carried out by a transformer. DC welding is carried out by a rotary generator coupled to an AC motor or with an AC transformer and rectifier arrangement. That includes state-of–the-art electronics. In the field rotary, DC generators are driven by suitable petrol or diesel engines. Of late, inverter-based welding machines are becoming popular. All these machines are equipped with a current control device with a long cable which can be adjusted by the welders themselves.

The bulk of the pipe welding is carried out by either by SMAW or by GTAW. The most common method is SMAW, routine welding method. An arc welding electrode or stick electrode is a suitable metal wire with necessary flux coating. Flux coating carries out a number of important functions. Fundamentally flux excludes air from the weld metal area, thus saving it from oxidation. It stabilizes the arc and improves weld metal penetration. The weld metal should be compatible with the base metal to be welded.

Stick electrodes or welding electrodes are manufactured for use with AC, DC, or both. In DC welding, the polarity of the electrode is specified. DC straight polarity is when the electrode is connected to the negative and DC reverse polarity is when the electrode is positive. Electrodes are manufactured for straight, reverse, or both polarities. DC welding offers a smoother arc, easier overhead welding, better looking welds, easier starts, less spatter, and fewer arc drop outs. DC reverse polarity gives about 10 percent more penetration whereas DC straight polarity welds thinner metals better for a given amperage than AC. However, if the part is magnetized, it is worthwhile to weld with AC because AC alternates between polarities.

Normally, a machine with 225–300 ampere capacity would suffice for pipe line work. For heavy schedule pipes, multiple passes are made. Duty cycle is the number of minutes out of a 10-minute cycle that a welder can operate. The higher that the current is, the less time there is for arc and the more time for cooling time in between.

As per the American Welding Society's coding system, E indicates Electrode, generally for SMAW, or submerged arc welding. The first two numbers of the electrode are the tensile strength of the weld metal of the electrode. The third number indicates the position that the electrode can be used in — whether it is all positions, horizontal, or vertical. 1 is an all-position electrode; 2 is for horizontal fillet and flat welding; 3 is no longer used; and 4 is for flat, horizontal, overhead and vertical down. The last number indicates the flux of the welding electrode or the current and polarity (AC, DC STRAIGHT POLARITY OR REVERSE POLARITY) that can be used on this particular electrode. Certain electrodes can be used only on DC, some on AC, and some on both. **Chart 10.2** summarizes the last number designation.

Chart 10.2
Last Number Designation of Electrode Code

Fourth Digit	Welding Current
0	DC Electrode Positive
1	AC, DC Electrode Positive
2	AC, DC Electrode Negaitive
3	AC, DC Electrode Positive or Negative
4	AC, DC Electrode Positive or Negative
5	DC Electrode Positive
6	AC, DC Electrode Positive
7	AC, DC Electrode Negaitive
8	AC, DC Electrode Positive
9	AC, DC Electrode Positive or Negative

Consider the E7018 electrode — the most commonly used stick electrode in the carbon steel piping lines. Here, 70 indicates the tensile strength (70,000 psi in this case) of the weld bead. The third number, 1, indicates the position of the welding. Thus, number 1 indicates an all position welding. Finally, 8 indicates an all current electrode AC, DC, and both polarities.

For specific applications, gas tungsten arc welding is employed which makes use of a filler wire instead of stick electrode. It is a favored procedure especially for the thinner sections of stainless steel, aluminum, and copper alloys. Tungsten wire, which is ground to a tapered tip, is used as the electrode. Arc is formed between this non-consumable tungsten electrode and the metal being welded. Argon gas normally serves the purpose of gas shielding. Inert gas is fed into the torch, out on to the weldment for shielding. Most often, filler wire of suitable material and size is used and added to the weld pool separately.

As a general practice, all important welding jobs are carried out with GTAW root pass followed by welding with stick electrode. In some critical jobs, complete welding is carried out by GTAW. Though difficult to master, the joints with GTAW welding are stronger and are of better quality. Specific electrode and procedure must be properly selected and adhered to for an effective job. This is all the more true when welding alloy steels or exotic metals.

PIPE WELDING POSITIONS

Five basic positions are used in pipe welding:

1. Horizontal Rolled Position (1G)
2. Vertical Position (2G)
3. Horizontal Fixed Position (5G)
4. Pipe Inclined Fixed (6G)
5. Pipe at 45° Angle Fixed (6GR)

WELDING PROCEDURE

Following is a brief procedure for welding pipe lines. There are two major classes of weld: fillet and butt.

Fillet welds are carried out between two surfaces not in the same plane. Welding is roughly triangular in cross section and done along the sides of the parts being joined. Butt weld is carried out between two surfaces in the same plane.

Butt welding is the most common procedure for joining pipe to pipe or pipe to fitting, and also for welding a neck flange. In a butt joint, both the ends of pipe or fitting are brought together, aligned, and welded. Edges are prepared to V or U shape as per the engineering requirements.

Most pipe welds take more than one pass (layer) to make a complete joint unless the pipes are very thin (say schedule 5 or 10). Runs (popular name for the weld passes) are made by actually weaving in an oscillating motion. Slag from each layer should be scrupulously removed; each layer should be wire brushed before the next pass is made.

Fillet welds are used for welding slip-on flanges, or socket weld fittings, or for seal welding threaded fittings to pipe. Fillets should fill both the inside and the outside of the flange. Sometimes after threading the pipe into the fitting, the joint is seal welded. Seal welding is used to obstain a leak free joint in certain applications. ASME does not accept the theory that seal welding aids the strength of the joint.

Electrodes must be compatible with the materials of the fittings. Flanges need to be aligned vertically and horizontally, and should be square. Care should be taken to position the flanges so that the bolt holes will match them on the companion flanges.

JOINT PREPARATION AND FIT-UP

Pipe joints must be prepared for getting good results. Cleaning is most important. Rust, slag, paint, oil, any foreign material, and scales must be removed. Miscellaneous oils applied on the surface to protect pipes from rusting during storage are one of the dangerous culprits for a bad joint.

When making a butt joint, edges are beveled by machining or grinding, or by using a gas-cutting torch. Edges must be accurate, especially when the bevels are made in the field by gas cutting. Pipes and fittings must be properly aligned and maintained as such until the welding is completed. Generally a piece of an angle or channel is tacked to the pipe as a jig for leveraging the pipe into proper alignment during fitment and kept as such until a firm joint is made. This temporary jig must necessarily be of the same metal as the piping, particularly so in case of stainless steels; this procedure is not even allowed in such cases. However, there are now better tools and jigs to carry out this operation efficiently. Adequate root clearance or opening is required for proper fusion and penetration of the weld.

After positioning, the pipe joint is tack welded and adjustments are made by slightly tapping or hammering to bring both the pieces into proper alignment. Depending on the size and thickness of the pipe, two or more tacks with adequate thickness and strength are essential to keep the joint in position during welding. Tack welds should fuse well and should not ride over the pipe. Tack welding should be made with the same electrode as the entire weld.

Electrodes must be selected for the proper metallurgy, position, and type of welding. Similarly, proper current must be used. To achieve acceptable root passes, stainless steel and higher alloy piping is continuously purged during the root pass. If both pieces are different in metallurgy — for example, SS304 to 316 or steel to stainless steel — please consult the welding engineer for an appropriate electrode. Often it may be adequate to use a higher class of electrode.

Clean the tack thoroughly before going ahead. If the tacking is good, and pipes are in proper alignment, go ahead for the first pass or the root pass, with the proper electrode.

After the root pass is completed and cleaned, it is subjected to a non destructive examination (NDE). Although the dye penetrant check is the most common, radiography is used to assess the weld quality. After the tests are passed, further welding is continued. A number of passes are required to complete the joint. Either some or all the passes are subjected to NDE, depending in engineering requirements. If NDE fails, the joint should be ground and re-welded. Unless the defect is excessive, only the unacceptable portion is removed and fresh welding is carried out.

After the root pass, the weld area would be increasingly wider. It can be filled with a larger electrode, but it is both a better and a normal practice to weave the weld to fill the gap with the same size electrode. It is a technique of lateral movement of welding across the weld puddle.

Alloy steels and stainless steels may require specific preheat, welding, and post weld heat treatment. Appropriate procedures should be strictly adhered to or a failed joint may end up. An unsuccessful joint now may be a safer bet than a failed joint later. It will be difficult to get a good joint when it is moist, in rain or snow, or in high winds.

TECHNIQUES OF WELDING

Good welds are easy to come by. Correct amperage setting, right arc length, accurate rod selection, proper angle to the work, correct travel speed, and proper storage of welding rods are the key words.

CURRENT SETTING

Too little current makes a weak arc. It will be hard to strike and the electrode may stick to the work. Too much current creates excessive spatter and a large crater. The electrode box usually has labels that indicate current ranges, polarity, etc. Amperage should be based on the material's thickness, welding position, and the finished weld. Overhead welds require about 15 percent less heat than a flat weld.

LENGTH OF ARC

Too short an arc will make the rod stick. Too long an arc make pools of molten metal and will result in poor penetration, undercuts, and porosity. However the correct arc length varies with each electrode and application. As a guide line, arc length should be about the diameter of the metal core of the electrode.

ANGLE OF TRAVEL

The electrode angle affects the penetration. Hold the rod perpendicular to the joint and tilt the top of the electrode in the direction of travel, approximately 5-to-15 degrees. Too perpendicular a rod can cause slag to get trapped in the weld whereas too flat a rod reduces the penetration.

SPEED OF TRAVEL

Speed controls the amount of rod deposited and the uniformity of the bead. Too fast a travel results in a thin bead with shallow penetration. Too slow a travel lets the bead build up with edges that overlap the base metal with poor penetration and may even make holes in thin pipes. Adjust travel speed so that the arc stays within the leading one-third of the weld pool.

STORAGE

Electrodes pick up moisture from the atmosphere; electrodes with too much moisture cause cracking and create porous welds. They must be dry. Low hydrogen electrodes pick up more moisture.

WELDING STAINLESS STEEL

In general, most of the stainless steels can be welded using more or less the same techniques as carbon steel. However, stainless steels pose different problems because of their lower melting temperature and coefficient of thermal conductivity, their higher coefficient of thermal expansion, and electrical resistance. Proper electrode and procedure must be selected depending on each individual job. If the service is critical, repair is on old piping, or service temperatures are excessive, then special care must be exercised.

All of the austenitic stainless steels of 300 series are normally weldable except Type 303 and Type 303S (which contain high sulfur and selenium respectively to improve machinability). Because of their higher coefficient of expansion, higher electrical resistance, and lower thermal conductivity than mild-carbon steels, high travel speed welding is recommended. This will reduce heat input and carbide precipitation, and minimize distortion. The welding current can be a little lower due to lower melting temperature and lower thermal conductivity. A bit more care should be taken to prevent distortion.

Ferritic stainless steels are not hardenable by heat treatment and are magnetic. Almost all the ferritic types can be welded by the normal processes (though not as well as austenitics) except for the free machining grade 430, due to its high sulfur content. The coefficient of thermal expansion is almost equal to mild steel and lower than the austenitic types. These steels may need post-weld heat treatment. Filler rods can be of similar metallurgy, but these steels are often welded with austenitic filler rods to increase the toughness.

Martensitic stainless steels are hardenable by heat treatment and are magnetic. The low-carbon type can be welded without special precautions. Pre-heat and post-heat weld treatment may be required for those with more than 0.15% carbon. The weldment tends to air harden and may result in cracks. These steels are often welded with austenitic filler rods to increase the ductility.

Duplex stainless steels also have good weldability, though not as good as austenitics. Duplex stainless steels have a comparatively low coefficient of thermal expansion, closely matching that of carbon steels.

With some extra care, welding together of different metals can be carried out. For welding mild steel to stainless steel, electrodes E 309 or E 310 can be successfully used.

AVOIDING DEFECTIVE WELDS

We have listed out here some of the defects normally encountered in pipe welding and early recognition can save expensive downtime and loss at a later date.

UNDERCUT

Cracks aside, undercut is usually considered as the worst defect. It is a sharp narrow groove along the edge where the arc removes the metal, but is not filled. It reduces the strength but also makes a stress point which may initiate a crack. The wrong weaving technique, wrong electrode angles, higher current, and too fast a travel speed are some of the reasons for undercut.

LACK OF FUSION

Lack of fusion is the next bad defect, where the weld metal does not really fuse into the parent metal thereby making a cohesive joint. Incorrect electrode selection and manipulation, too low a current, and too slow travels are some of the causes.

SLAG

Inadequate removal of slag from the weld area — particularly in the case of multi-pass weld — can leave a weaker weld joint with improperly fused or penetrated weld. Although scrupulous cleaning is important before, during, and after welding bigger electrode, low current can cause slag inclusions. Slag inclusions reduce the strength, and may initiate cracks.

INCORRECT PREPARATION, FIT UP, AND PROFILE

These weaknesses will result in poor joint strength, bad appearance, and poor penetration. Proper alignment of the piping components to be welded, fit up, correct size, and type of electrode with correct current and electrode manipulation will save us from these defects.

INCOMPLETE PENETRATION

A weld electrode that does not really melt into the parent metal, thereby fusing them together, results in this incomplete penetration. The weld joint will not have enough strength. It may hold together by the brute force of the multiple passes. Bad preparation and fit up with insufficient root gap, too great a land added with a bigger electrode, low current, and incorrect angle of welding all contribute their mite to this defect.

CRACKS

The welding electrode must be compatible with the base metal. Electrodes must be dry and should not pick up any moisture, particularly low hydrogen varieties. Improper technique and travel, too high or too low a current, bad preparation, and incorrect support during welding can result in weld cracks. Cracked tack welds, when over-welded, can end up in cracked final welds.

POROSITY

Porous welds occur when flux coating breaks down due to excessive current, excessive moisture pickup by the electrode (particularly low hydrogen types), and impurities absorbed from the parent metal. Welding electrodes must be kept in the oven and should be heated for at least an hour at 230°F/110°C for general purpose types and 480°F/250°C for low hydrogen types. However, it is observed that the 6010 electrode operates more efficiently when not placed in the oven.

HEAT AND DISTORTION

Welding is a hot work where expansion and contraction occur, which can eventually result in certain distortion. If not restrained or held down properly, pipe joints can warp and distort. It will be difficult to bring them back without resorting to evasive techniques. The pipe joints must be properly supported by jigs and fixtures during cutting, welding, and pre- and post-heat treatment. To minimize unwanted distortion or stresses:

- Do not use excessive heat or more welding than necessary. Make intermittent welds rather than a continuous weld. Avoid excessive local heat buildup. Weld on both sides of the joint in a sequence by proper balancing.
- Sufficiently clamp the piping components during welding and cooling.
- Use the right joint preparation and fit up and avoid large weld grooves.

CARE AND SAFETY IN WELDING

1. The polarity switch which is provided for changing the electrode lead from positive (reverse polarity) to negative (straight polarity) should never be operated while under the load of a welding current.
2. Operating the rotary switch should be performed while the machine is idling and the welding circuit is open.
3. Every power circuit is grounded to prevent accidental shock by stray currents. Be sure that the welding machine is properly grounded.
4. Don't leave the electrode holder on the table top or in contact with a grounded metallic surface.
5. Don't permit cables to operate hot.
6. Don't operate with worn or poorly connected cables.
7. Don't operate electrode holders with defective jaws and with loose cable connections.
8. Don't use cracked or detective helmets or shields.
9. Don't under any circumstances look at an electric arc with the naked eye.
10. Don't use cracked, ill-fitting, or defective welding plates.
11. Don't use welding color plates without a protecting cover glass
12. Don't use poor, inadequate or worn-out work clothing.
13. Don't under any circumstances strike an arc on compressed gas cylinders.
14. Don't weld in the vicinity of inflammable or combustible materials.
15. Don't weld in confined spaces without adequate ventilation.
16. Don't pick up hot objects.
17. Don't do any chipping or grinding without goggles.

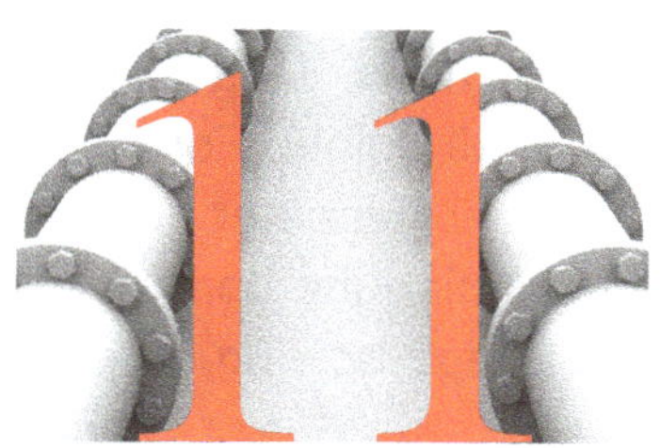

Bolts and Studs

Bolts and studs are extensively used as fasteners connecting pipes by flanges and for assembling system components such as valves. There are made up of various sizes and materials conforming to different standards.

BOLTS, STUDS, OR STUD BOLTS

A bolt has threads on one end only; the other end is formed as a head. Those bolts made with threads right up to the bolt head are known as full thread bolts whereas those with threads only until a certain length are known as standard thread bolt. In American standard, on the bolt head, the width across the flat is equal to one and one-half times (1 1/2) the bolt diameter plus 1/8". That gives the spanner size.

Stud bolts have threads on both ends or throughout the length, made to standard diameters with nuts on both ends. Bolts have heads, but stud bolts do not have heads (no pun). Studs rods are also made in simple carbon steel in lengths of one meter or two; please be careful of them. Some studs are manufactured with threads on both ends up to a certain length, with an unthreaded portion in between.

Effectively, a bolt is a threaded fastener to be used with a nut. A stud is a threaded fastener to be used with two nuts. The size and material of the fastener, and the length of engaged threads are some of the factors for selecting the fasteners. Welded joints or riveted joints can only be dismantled and fixed back after resorting to invasive methods. The sole advantage with the bolted joints is that they can be disassembled and assembled back. Studs are used for temperature-pressure piping. Continuous-thread studs are threaded from end to end and are used with two nuts made as per ANSI B16.5. On the other hand, tap-end studs have a short thread on one end with a chamfered point, called the tap end. This end is for screwing into a tapped hole. The other end is meant for the nut, which may have either a chamfered or round point. Double-end studs have equal-length threads on each end to accommodate a nut. Both ends have chamfered points. The length of the stud is measured overall.

Normally bolts and studs exert compressive pressure on the flanges and through them on to the gasket. Fasteners exhibit stress relaxation behavior dependent upon their material of construction. This will have a marked effect on the load they are able to generate on the flange / gasket assembly under operating conditions. In most flange assemblies, the distribution of pressure on the gasket is usually not the same at all points. For example, two large diameter bolts might provide the same overall load as 12 smaller diameter ones, but the distribution of the load would be very different. The gasket area surrounding the bolts or studs undergoes greater compression than midway between the bolts, due to flange bowing. Therefore, in order to distribute the gasket pressure as equally as possible, a higher number of properly-spaced bolts or studs should be used. Bolt spacing is calculated by the pressure exerted by them midway between the bolts. An insufficient bolting load can result in a leak from this point. But if the clamping force is excessive, the gasket may crush or creep.

Although it is impractical to tighten all the fasteners simultaneously, a proper tightening sequence should be followed to get a uniform loading of the gasket. Although experience plays a very vital role in tightening the flanges, the torque control method is an effective solution to obtain uniform and effective distribution of bolt load.

First and foremost, it is first necessary to seat the gasket properly under atmospheric conditions. This factor, known as **y** (yield), is the stress required to deform the gasket into the irregularities of the joint surface. It is dictated by the compressibility of gasket material. Then there must be sufficient clamp force to overcome the internal hydraulic thrust generated while it actually in service at its operating conditions. It is this **m** (maintenance or multiple) factor that imposes the clamp force of the fasteners to maintain sealing of the gasket at its operating conditions. It is a multiple of the internal pressure — the ratio of gasket contact pressure to process pressure. Values of the m factor for various gasket materials are standardized.

METALLURGIES

Bolts and studs are made as per various standards, ASTM, British or DIN. Care must be taken while procuring new bolts or renewing them in unknown service for equivalence in metallurgy and manufacture. Usually the grade of the bolt is indicated on the bolt head by raised or depressed standardized markings. Do not use cut pieces from so called stud rod or threaded rod just for any piping service. People have often landed in trouble and lost time while trying to fix mismatched bolts and nuts wittingly or unwittingly.

The following fastener alloys are the ones most commonly used for piping; many others are available that will meet very specialized applications. The alloy selected for a fastener application must not only maintain strength, etc., at the temperatures to which it will be heated, but also it must resist the environment in which it is used. Thus corrosion resistance of the metal selected must be considered.

CARBON STEEL

Carbon steels are most widely used for general applications. They are used on flanges in the pressure rating up to 150 class and a temperature rating up to 445°F/230°C. They should not be used on pressure code vessels. There are high tensile bolts and studs, some of which are as follows:

B-7 (CR-MO) These can be used for service temperatures up to 840°F/450°C and in the flange classes beyond 150 ASA. Even in the case of 150 class flanges, B7 studs are used if the bolt size is more than 1" diameter.

ALLOY B-14 (CR-MO-VA) Used like B7 studs, but more severe temperatures over 840°F/450°C.

L-7 These bolts are used in low temperature service.

STAINLESS STEEL

Stainless steels are used where certain corrosion resistance is required, such as in acidic service. There are a large number of stainless steel alloys, such as SS304, 316, 317, CD4MCu, etc., used for specific services. Use only the grade of material as required. They can be used in low temperature applications.

NON-FERROUS ALLOYS

These studs are used specifically for corrosion resistance, the most important service being salt water. They should not be used in place of steel bolts or on coded vessels. Brass or bronze studs are used in oil service at less than 345°F/175°C, for certain chemicals up to 445°F/230°C. However, Monel can be used for oils, chemicals, and utilities up to 550°F/288°C.

ASTM A193 — Specifies alloy steel and stainless steel bolting materials for high temperature service intended for use in pressure vessels, valves, flanges, and fittings. Threads are specified at 8 threads per inch (tpi) for diameters above one inch.

B7 — Alloy steel, AISI 4140/4142 quenched and tempered.
B8 Class 1 — 304 stainless steel, carbide solution treated.
B8 Class 2 — 304 stainless steel, carbide solution treated, strain hardened.
B8M Class 1 — 316 stainless steel, carbide solution treated.
B8M Class 2 — 316 stainless steel, carbide solution treated, strain hardened.

ASTM A194 — Specifies carbon and alloy nuts for bolts for high pressure and high temperature service.

2H — Quenched and tempered carbon steel heavy hex nuts.
4 — Quenched and tempered carbon-molybdenum heavy hex nuts.
7 — Quenched and tempered alloy steel heavy hex nuts.
8 — Stainless AISI 304 heavy hex nuts.
8M — Stainless AISI 316 heavy hex nuts.

ASTM A320 — Alloy steel and stainless steel bolting materials for low temperature.

L7 — Alloy steel, AISI 4140/4142 quenched and tempered.
L43 — Alloy steel, AISI 4340 quenched and tempered.

B8 Class 1 — 304 stainless steel, carbide solution treated.
B8 Class 2 — 304 stainless steel, carbide solution treated, strain hardened.
B8M Class 1 — 316 stainless steel, carbide solution treated.
B8M Class 2 — 316 stainless steel, carbide solution treated, strain hardened.
Obviously B8 series of ASTM A193 is repeated here in ASTM A 320.

ASTM A307 — Specifies carbon steel bolts and studs, 60 ksi tensile strength.
A — Bolts and studs intended for general applications.
B — Heavy hex bolts and studs intended for cast iron flanged joints.
That brings us to recommended fastener working temperatures as shown in **Chart 11.1**.

Chart 11.1
Fastener Working Temperatures

	TEMPERATURE				
MATERIAL		MINIMUM		MAXIMUM	
		°C	°F	°C	°F
Carbon steel		-20	-4	300	572
B7, L7		-100	-148	400	752
B6		0	32	500	932
B8		-250	-418	575	1067
B16		0	32	520	968
B17B		-250	-418	650	1202
B80A		-250	-418	750	1382

PROBLEMS AND SOLUTIONS

A number of methods are available for tightening the bolts, like torque control, angle control, and yield control, bolt stretch method, heat tightening, and use of tension indicating methods. Except in a few special occasions, only the torque method is used and generally without the use of a torque wrench for all sundry joints.

One of the more annoying problems with bolted joints is the self loosening of the fasteners. This occurs with all materials, more so with stainless steels. Stress relaxation, creep, and vibrations are some of the causes. Fasteners are tightened such that a certain preload is imparted on the joint. Then the joint is stressed due to operating conditions. When this stress is relieved over time, the bolt is actually relaxed. Then the nut self-loosens. Materials of the fasteners may creep under high temperature. By the selection of fasteners with proper metallurgy, these problems can be avoided. As a matter of fact when fasteners are properly selected and used carefully, they serve faithfully for years.

However, upon subsequent opening or closing of the joint, problems do develop if the materials used are not of the same quality and/or the method is not proper. We find situations occurring in the industry when just a single bolt or two are replaced. The bolt is replaced by another one that is just readily available. The replacement may be the same size, but too often it is not the same material. These bolts give up and yield and other bolts follow the suit. *Therefore, never replace bolts and studs with just any.* Confirm and use the right material, particularly when you are in hurry.

In most of the piping flanges, bolting is not done using washers, although in some cases the use of washers is recommended. By using washers, the load is more uniformly distributed. But washers tend to turn, rotating with the nuts which disturb the torque pattern. Conditions like pressure, temperature, flow parameters, and vibrations continuously vary in any service, resulting in variations on the flange loading, and relaxation of the fasteners, gaskets, etc. To compensate for this, Belleville washers (metal disc springs) are used.

Sometimes spring washers are used to prevent this self loosening. Self locking nuts offer certain resistance to self-loosening. A Nylock nut has a nylon insert in the nut which not only resists loosening but also prevents the ingress of water, oils, and ambient corrosive products.

Various other techniques are used to prevent self loosening. Though rarely used on pipe lines, one of them uses a check nut or double nut. Historically a thin nut is used over the standard nut. However, the currently accepted practice is to fix the thin nut on first, tighten it to about 30% of the full torque. Then fix the thick nut on top and tighten it to the full torque value. *Care must be taken so that the thin nut does not rotate when you are tightening the thick nut, by holding it with another spanner.* An open end spanner would come in handy for this purpose.

One of the annoying reasons for loosening fasteners on a pipe line is the vibrations. Line vibrations are induced by flow or mechanically by the vibrating rotary equipment. Result would be that the bolts may work out loose and fall or may break out. They may also breakout when inadequate number of them are used or of totally wrong and incompatible material. Many maintenance personnel have an inclination to use only alternate bolts. Do not fall into that habit. *Never use fewer fasteners than the number designed for the flange.*

Occasionally, galling occurs, where the mating parts are literally cold-welded. Materials are heavily loaded after intimately being brought together. Clumps of metal from any one part stick to the other at random, resulting in a seizure. The problem is compounded when the lubricant dries out, particularly when high temperatures are encountered. This happens more often with stainless steel, high alloy, titanium, and even zinc-coated bolts and nuts. Oxide films of the fasteners may break under severe loads, and the bolt and nut may lock together at the high points at the interface, eventually leading to a seizure. This can be prevented by using materials less prone to galling; lubricant on the mating parts, i.e., bolts and nuts; coarse threads instead of fine threads; or different metallurgies of bolts and nuts within the accepted limits. Anti-seize compounds are applied which prevent galling. In effect they improve corrosion resistance, providing certain ease in dismantling parts at a later date.

Always use a boundary lubricant such as Molykote or Neverseez while fixing bolts and nuts. This makes for easier assembly and removal. This is particularly important in high temperature service or where bolts are threaded into the body of equipment. Without this application, it will be extremely difficult to open studs after they are put in high temperature service. Gas cutting may have to be resorted to cut away the bolts and that may not be permitted always. Moreover, a non-lubricated bolt has an efficiency of 50% of a well-lubricated bolt.

Anti-friction coatings are boundary lubricants which contain suspended solid lubricants, such as graphite, PTFE, or molybdenum disulphide. The coatings are applied on the bolt or stud before bolting up, thus making up a lubricating film at mating point and avoiding metal-to-metal contact.

It is a cardinal rule in most industries that at least one or two threads of the bolt or stud must protrude beyond the nut. Normally the first two threads are not properly formed because of the chamfer. But if a greater number of threads protrude out, they may not add to any more mechanical strength, but can actually create a problem. Due to atmospheric corrosion, these threads may corrode and it will be extremely difficult to open them. Nuts should have at least 20% greater ultimate tensile strength than the fasteners.

Hence

- Select fasteners with sufficient yield strength.
- Select fasteners with the same modulus of elasticity.
- Ensure there is no corrosion of the fasteners.
- Tighten them to specified torque.
- Use thread lubricant.
- Use all the bolts in a given flange.
- Use compatible nuts.
- Use the same standard bolts and nuts.

One often comes across in three types of bolts made to BSW (British Standard Whitworth), Metric, and UNC (Unified National Corse).

BSW (British Standard Whitworth) is more or less obsolete. However one can find BSW nuts and bolts at some places. Old nuts die hard! Sizes are measured in inches and number of threads per inch as TPI. The thread angle is 55°.

American thread has two models: UNC (Unified National Corse) and UNF (Unified National Fine). UNF is seldom used in piping. The thread angle is 60°.

Metric threads are designed by The International Organization for Standardization or ISO, and are now used all over the world. Thread angle is 60°. Metric bolts are designated by three numbers followed by the letter M. The first number is the nominal major diameter of the thread, the second number is the pitch in millimeters and the third is the length. For example M10 x 1.0 indicates that the major diameter of the thread is 10mm and the pitch is 1.0mm. The thread form for Unified and Metric threads are identical. They are manufactured in different grades, such as 4.6, 8.8, and 10.9, which indicate the maximum torque that the bolt can withstand. The grade number is marked on the bolt head, and often on its nut.

It is generally believed that the screw thread was invented around 400 BC by Archytas of Tarentum (428 BC–350 BC), a contemporary of Plato. Known as the founder of mechanics, Archytas used screws for extraction of oils from olives and juice from grapes. Archimedes

(287 BC–212 BC) used screws to pump water, though these were used in Egypt much earlier. The screw was described in the first century AD in Mechanica of Heron of Alexandria.

For centuries, the making of screw thread depended on the skill of the craftsman. Around 1750, Antoine Thiout's innovations brought the screw drive onto the lathe, which would move the carriage semi automatically. In 1770 Jesse Ramsden made the first successful screw-cutting lathe.

Joseph Whitworth proposed standardization of threads in 1841; by 1860 they were practiced throughout Britain. In 1864 in America, William Sellers independently proposed what later became the U.S. Standard and, subsequently, the American Standard Coarse Thread (NC) and Fine Thread (NF). Metric thread standards were also developed around the same time in other parts of Europe.

Gaskets

In the modern world, gaskets are everywhere — on refrigerator doors to soft drink bottles to pressure cookers. A gasket is a stationery mechanical seal to prevent leakage between two mating surfaces, i.e., flanges in the case of piping. A gasket is used to create and retain a static seal between two flanges, a complete physical barrier against the fluid contained within the piping system. It must flow into any irregularities in the mating surfaces being sealed. At the same time, it must be sufficiently resilient to resist extrusion and creep under operating conditions. In other words, gaskets actually should compress into the gap and fill the machining irregularities. They should be flexible enough to withstand vagaries of process parameters, like temperatures, pressures, and line fluctuations. They must be capable of overcoming minor alignment and flange imperfections such as non-parallel flanges, distortion, troughs and grooves, surface waviness, surface scorings, and other imperfections.

The cost of a gasket is minuscule when the gasket fails, compared to the cost of production lost, cost of maintenance, loss of energy, and impact on the environment. All these could be saved if proper attention is given to:

- Proper design of the flanged joint
- Selection of the right gasket material required for the application
- Proper installation procedures

The pipe joint is subject to compressive pressure of the flanges, usually achieved by bolts. Under operating conditions, this pressure is opposed by hydrostatic end thrust, produced by internal fluid pressure, which tends to separate the flanges. Fluid pressure acts through the small clearances obtained by the gasket thickness and tries to extrude the gasket. Sufficient friction exerted by the soft gasket keeps it in place resisting extrusion.

SELECTION CRITERIA

Gaskets should be selected and used considering the following parameters:

- Compatibility with the operating medium
- Operating temperature and pressure

- Variations of operating conditions (fluctuating, start-stop)
- The type of joint involved

FLUID COMPATIBILITY

The gasket, along with the filler and the limiting rings, should invariably withstand the chemical attack of the fluid it seals. Acids, alkalis, solvents, and oils require gaskets made suitable especially for them. Also, the severity increases with temperature and time. Certain industrial applications require flame-retardant materials for gaskets. PTFE filler ring gaskets invariably require an inner ring.

FLANGE COMPATIBILITY

Basically the gasket is a renewable component in the joint system and not the flange. Therefore, it should be softer or more deformable than the mating surfaces. It must also be chemically compatible. For metal gaskets, this means that consideration must be given to galvanic corrosion. Galvanic effects can be minimized by selecting metals for gasket and flange which are close together in the galvanic series, or the gasket should be sacrificial, not the flanges.

TEMPERATURE

Temperature dictates most of the physical characteristics such as material state and shape, resilience, and burn out or oxidation point. Creep and chemical properties change with temperature. Hence, the gasket should be selected such that it stands operating conditions and the inevitable worst case scenarios. Higher temperatures or sub zero temperatures need special materials, like graphite or silicone. Corrosive applications may need Teflon, etc. Certain industries and applications require flame-retardant materials for gaskets. Chemicals behave more aggressively at higher temperatures.

PRESSURE

Pressure is the most important consideration — the gasket should never give away under pressure. Gasket material should be able to accept the compression required exerted by the bolts, or else it will tear off before being put in service. Hydrostatic force of the fluid in the line tends to separate the flanges. Flange clamping load or bolting load must always be better than this hydrostatic force and should overcome it. This pressure also tries to push the gasket out of the flange surfaces. Gasket and gasket loading must be able to cope with these forces and continue to make an effective seal during its life time. Figure 12.1 shows a diagram of forces acting on the gasket.

JOINT DESIGN

Clamping force on the gasket must be sufficient to seat the gasket intact in position and prevent flange separation due to confined fluid pressure. The flange must be sufficiently rigid to prevent excessive bending, which would cause localized unloading of the gasket. A similar condition

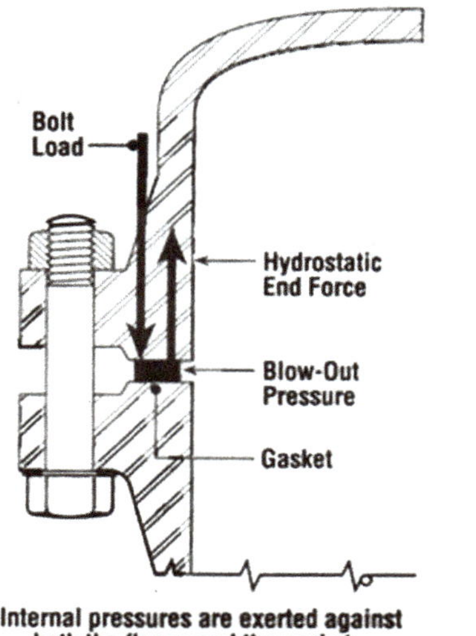

Figure 12.1

occurs when insufficient number of bolts are inserted and tightened on the flanges. Any flanges using full face gaskets must be thick enough to prevent bowing between adjacent bolts.

SURFACE FINISH

Each gasket type behaves best when the mating faces of the flanges have a specific and required surface finish. Surface finish requirements differ with gasket type and must be properly adopted. Elastomeric and PTFE gaskets need relatively rough gasket seating surface to prevent extrusion, whereas solid metal gaskets normally require a surface finish in the order of 63 micro inches. Spiral-wound gaskets fall between these two general types.

SEATING STRESS

Gasket sealing is accomplished by the flow of the gasket material into machine surfaces on the flange facings. The amount of force per unit of gasket area required to completely flow the gasket is known as yield or seating stress. The design includes a margin of safety of approximately 4:1.

FLANGE FACINGS

There are many types of flange facings in use. Flat face, raised face, ring type joint, tongue and groove, and male and female are some of these faces. Other special flange facings are those which use the Bridgeman closure and the lens ring.

GASKET THICKNESS

Ensure that the gasket is as thin as possible, which should be enough to compensate for the flange variations and surface finish, but should be strong and thick enough to resist extrusion. Thinner gaskets will be less prone to the attack of process chemicals and also will have higher gasket loading. In spite of all the equivalence charts and comparison of many materials, the performance of the same kind of gasket from different manufacturers may show marked differences, particularly in stringent applications. Hence, always use a tried and tested gasket rather than just any gasket. After all, the cost of a gasket is insignificant compared to the materials lost or discomfort caused. The gap between the flanges should be exact; a greater number of gaskets should not be used to cover the gaps in the flanges. Separating the flanges excessively to insert a gasket is also a bad practice.

GASKET MATERIALS

A wide variety of materials are used in the manufacture of gaskets; they can be broadly divided into:

- Elastomeric materials
- Fibrous materials
- Metallic materials
- Other materials

Gaskets can also be divided into three main categories: soft (non-metallic), semi-metallic, and metallic.

Soft gaskets (non-metallic) are normally limited to low-to-medium pressure applications over a range of corrosive and non corrosive applications. These are generally composite materials such as compressed asbestos fiber (CAF) and non-asbestos fiber materials, graphite, PTFE, etc.

Semi-metallic gaskets are suitable for both low and high temperature and pressure applications. They consist of metallic and non-metallic materials such as spiral wound gaskets, corrugated metallic, metal jacketed gaskets, and metal eyelets.

Metallic gaskets are fabricated from a single or combination of metals in special shapes specially designed for high temperature and pressure applications, such as Lens rings and ring type joints.

The metallurgy of metallic gaskets is not discussed here as you can find them in Chapter 3, Commonly Used Piping Materials.

ELASTOMERIC MATERIALS

Rubbers in wide variety are available; they can be formed to individual requirements, shapes, and sizes. Composition gaskets with rubber offer the advantage of rubber in chemical resistance.

Ethylene propylenediene (EPDM) has good resistance to a number of strong acids and alkalis, hot water, and steam, but is not advised for solvents and aromatic hydrocarbons as it is susceptible to oils. It is highly compressive and resists abrasion. It has a useful temperature range of –9°F to 300°F (–23°C to 149°C). It weighs less because of its low specific gravity — about 0.8 to 1.0.

Natural rubber shows good resistance to most inorganic salts, mild acids, and alkalis. However, it is not recommended for oils and solvents. Oxygen, ozone. and sunlight are detrimental to this rubber.

Neoprene (chloroprene, CR) is another excellent material for oils, petroleum solvents, and fuels, moderate acids, alkalis, salt solutions, and refrigeration applications. It has good weathering characteristics, but is not recommended for strong acids or hydrocarbons. Temperature range is –0.4°F to 180°F (–18°C to 82°C).

Buna "N" is commonly referred to as a nitrile rubber. It beats neoprene in chemical resistance and temperature properties with good abrasion and tear resistance. It offers good resistance to hydrocarbons and oils, but is not recommended for chlorinated hydrocarbons, esters, ketones, and strong oxidizing agents. It has a useful temperature range of –0.4°F to 180°F (–18°C to 82°C).

Viton (Vinylidene Fluoride-Hexafluoropropylene) is a fluorocarbon elastomer which was primarily intended for handling hydrocarbons such as solvents, gasoline, and jet fuels. It is not recommended to use this in services normally using Buna N, where it will swell itself. It has better chemical compatibility and temperature range of –0.4°F to 350°F (–18°C to 177°C).

Silicone has better temperature capabilities, and is not affected by ozone and sunlight (which are not good for many hydrocarbons and steam).

Teflon is no doubt a versatile material, withstanding most of the chemicals, and it has a temperature range better than most rubbers. Its highly compressive, resilient, and anti-stick characteristics make it an excellent material for gaskets. It exhibits good dielectric properties. It can be processed into variety of shapes and sizes. It can be mixed with a number of materials such as carbon, graphite, glass, and bronze to achieve desired results. This material is considered the industry standard for soft sealing components due to its almost universal chemical resistance and its low coefficient of friction.

Kalrez is a proprietary material of DuPont; it combines the best Viton and fluorocarbons in service conditions and chemical resistance. It is very costly and so it is sparingly used in specific services.

TFE (Tetrafluoroethylene/propylene dipolymer) is manufactured by 3M and has excellent chemical resistance. It can be used in temperatures to 204°C. It offers specific resistance to automobile fluids, battery acid, rocket fuels, etc. Valves made with this material are used in ozone water treatment systems.

Styrene-butadiene (SBR) is a synthetic rubber that has excellent abrasion resistance and good resistance to weak organic acids, alcohols, moderate chemicals, and ketones. It is contraindicated for ozone, strong acids, fats, oils, greases, and most hydrocarbons. Its temperature range is about –65°F to 250°F (–55°C to 120°C).

Chlorosulfonated polyethelene (HYPALON) resists acids, alkalis, and salts very well, but not aromatics or chlorinated hydrocarbons, chromic acid, and nitric acid. It offers excellent resistance against weathering, sunlight, ozone, oils, and commercial fuels like diesel and kerosene. It has a temperature range of 50°F to 275°F (10°C to 527°C).

FIBROUS MATERIALS

ASBESTOS Traditionally, compressed asbestos sheets have been indispensable material for soft gasket materials. These sheets are easy to cut into any shape, simple to use, easy to insert, and above all easily available. They are very tolerant over our mistakes in handling, use, and misuse. When bonded with different bonding chemicals, asbestos can be made into sheets, rolls, and ropes applicable to most temperatures and pressures used for almost all common applications. But this most popular gasket material is now driven to a backseat because of environmental and health problems. It is banned in many countries. A number of non-asbestos gasket materials have since been developed.

Compressed asbestos free (CAF) materials are developed without asbestos. With proper fillers and binders, they outperform conventional CAF gaskets, though costly. They are excellent for high temperature service and for a wide range of chemicals. Carbon fiber is slowly replacing asbestos in many applications. It can stand high temperatures in non-oxidizing atmospheres. It can be used in the pH range 0–14, again in non-oxidizing atmospheres. They demand a little better care.

Flexible graphite is essentially pure graphite without any filler, typically over 95% elemental carbon. The compressibility of flexible graphite makes it an excellent filler material for metallic gaskets. Flexible graphite may be used in services with temperatures up to 6000°F/3300°C, though it should not be used with strong oxidizers such as nitric or sulfuric acid. It is available in various widths and thicknesses, with adhesive back-up tape also, to easily stick it around a gasket surface. Its outstanding qualities make it indispensible in a wide range of gasket services and as a gasket repair material. It is also used as a filler material in spiral-wound gaskets.

Additional notes on flexible graphite tape can be found in *Other Types of Gaskets*.

Aramid offers high strength and excellent dimensional stability.

Cellulose is a natural fiber, used at low temperature and pressure.

Ceramic fiber makes excellent gasket material for hot air duct work with low pressures and light flanges up to approximately 2000°F/1093°C. It is also used as a filler material in spiral-wound gaskets.

Cork is a light duty material which can be used on damaged flanges because it easily flows into the flaws. Please mind the pressure, temperature, and chemical compatibility.

OTHERS

The following materials are more often used as fillers or add-ons to improve the qualities of the gaskets, rather than as straight gaskets. They are also good insulating materials. **Chart 12.1** summarizes various materials and their limitations.

Glass offers a good strength and moderate chemical resistance, and glass fibers are suitable at high temperatures.

Mineral wool or manmade mineral fiber (MMMF) is an inorganic metal silicate fiber for high temperature applications.

Mica (vermiculite) is a natural aluminum silicate which has very good thermal stability and chemical resistance.

Paper with suitable fillers and binders is frequently used in specific applications such as pump and compressor casings where thin sections of gasket material are required.

Felt is used as gaskets for atmospheric covers rather than as piping gaskets.

Chart 12.1
Some Common Gasket Materials and Their Limitations

Gasket Material	Used for Products	Maximum Temperature (°F)	Maximum Pressure (psi)	Maximum Temperature * Pressure (°F * psi)	Maximum Temperature (°C)	Maximum Pressure (bar)	Maximum Temperature * Pressure (°C * bar)
Synthetic rubbers	water, air	250	60	15,000	121	4	496
Vegetable fiber	oil	250	160	40,000	121	11	1,322
Synthetic rubbers with inserted cloth	water, air	250	500	1,25,000	121	34	4,130
Solid Teflon	chemicals	500	300	1,50,000	260	20	5,320
Compressed Asbestos 1)	most	750	333	2,50,000	399	23	9,068
Carbon Steel	high pressure fluids	750	2,133	16,00,000	399	145	58,036
Stainless Steel	high pressure or corrosive fluids	1,200	2,500	30,00,000	649	171	1,10,636
Spiral wound							
SS/Teflon	chemicals	500	500	2,50,000	260	34	8,866
CS/Asbestos 1)	most	750	333	2,50,000	399	23	9,068
SS/Asbestos 1)	corrosive	1,200	208	2,50,000	649	14	9,220
SS/Ceramic	hot gases	1,900	132	2,50,000	1,038	9	9,313

GASKETS BY SHAPE

PLAIN FLAT METAL GASKETS

Solid metal gaskets are used on valve bonnets, heat exchangers, hydraulic presses, and tongue-and-groove joints. They require a certain compressive force to make the metal flow into the imperfections; hence, it must be softer than the parent metal (see Figure 12.2).

Figure 12.2
Flat Metal Gasket

There are many variations and designs in the solid metal gaskets, including:

Round cross section solid metal gaskets seal by line contact on accurately machined and assembled flanges. They are generally made up of aluminum, copper, soft iron or steel, Monel, nickel, and 300 series stainless steels, and used on low pressures (see Figure 12.3).

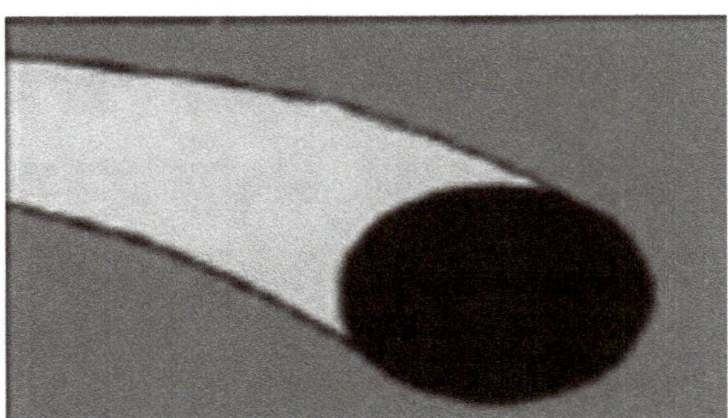

Figure 12.3
Round Metal Gasket

Ring type gaskets (API Ring type gaskets) are marvels of mechanical engineering. They may be carbon or alloy steel, depending on the service requirements. Frequently, the rings are of the same material as the flanges, except for lesser hardness. Either octagonal or oval rings may be used, ideally in conjunction with corresponding RTJ flanges (see Figures 12.4A and 12.4B, respectively). They are used in extremely high pressures in the range of 15,000 psi limited only by the metal temperatures. Basic cross sections are oval and octagonal. With standardized dimensions, they accept correspondingly grooved flanges. The octagonal rings are preferred as they are less likely to cause damage to the flange groove and for its higher sealing efficiency, but it cannot be used in the old round bottom groove. The newer flat bottom groove design will accept both cross sections. A pressure-energized effect increases the efficiency of the seal by a wedging action as the internal pressure of the system increases; in other words, the sealing effect increases as the pressure increases. The finish of the grooves should be about 63 micro-inches without any nicks, burrs, ridges, or tool marks. Dimensions for ring joint gaskets and grooves are standardized in ASME B1 6.20, API 6A, and ASME/ANSI B 16.5.

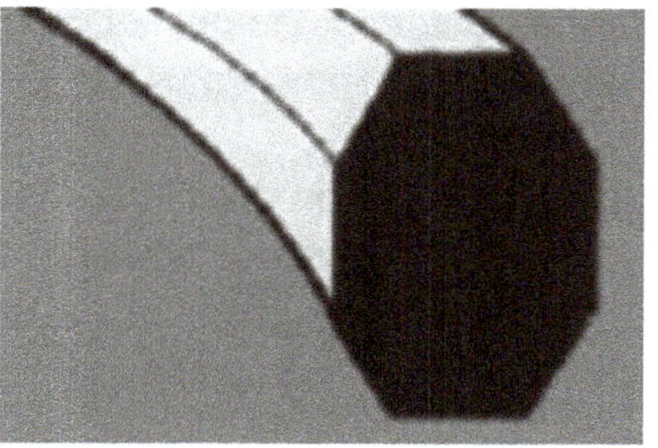

Figure 12.4A
Octagonal Gasket

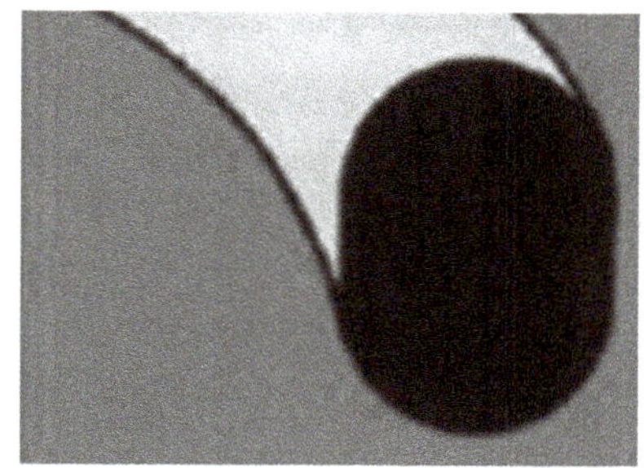

Figure 12.4B
Oval Gasket

RTJ Gaskets

RX and BX Rings are other variations of basic shapes. These gaskets must be used with the corresponding flanges. BX ring gaskets are square in the cross section and taper in each corner (see Figure 12.5A). They can only be used in API 6BX flanges. RX ring gaskets are similar in shape to the standard octagonal ring joint gasket but their cross section is designed to take advantage of the contained fluid pressure in effecting a seal (see Figure 12.5B). They are both made to API 6A.

Figure 12.5A

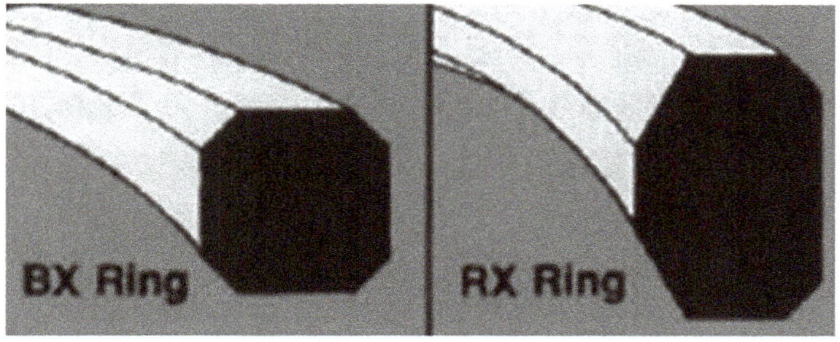

Figure 12.5B

Lens joint gaskets may be carbon or alloy steel, depending on the service requirement (see Figure 12.6). This high pressure gasket makes a line contact seal, but the sealing contact and effect increases as the pressure goes up. Hence they can take a bit of overload. They are reusable, but

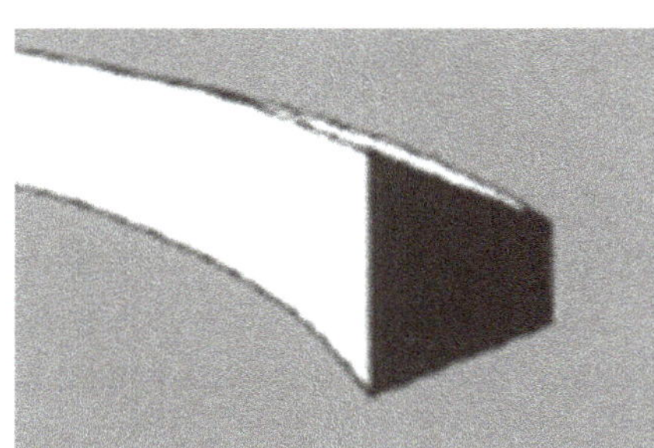

Figure 12.6
Lens Joint Gasket

require specially contoured flange faces, a spherical cross section at the gasket seating. (They can take small bolt contact load because the seating area is greater. They are used on extremely high pressure systems because of the elastic deformation of the gasket.

Delta gaskets are pressure actuated gaskets used primarily on pressure vessels and valve bonnets at very high pressures, even beyond 5000 psi. As the internal pressure increases, the metal will force itself against the equipment wall to affect a tighter seal (see Figure 12.7). Extremely smooth surface finish of 63 micro-inches or smoother is required when using this type of gasket.

Figure 12.7
Delta Gasket

Bridgeman gaskets are pressure energized gaskets specially used on pressure vessel heads and valve bonnets for pressures of 1500 psi and above. They are often silver or lead coated to make up for a softer gasket surface. As the internal pressure increases, the metal will force itself against the equipment wall to affect a tighter seal (see Figure 12.8).

Figure 12.8
Bridgeman gasket

DOUBLE-JACKETED GASKET

Double-jacketed gaskets (see Figure 12.9) are commonly used in heat exchangers and valve bonnets. They are available in virtually any material that is commercially available in 26-gauge sheet. They are also extensively used in standard flanges where the service is not critical. Because most double-jacketed gaskets are custom made, there is virtually no limit to the size, shape, or configuration in which these gaskets can be made.

Figure 12.9
Double-Jacketed Gasket

CORRUGATED METAL GASKET

A development over this is the corrugated gasket (see Figure 12.10A) where the corrugations provide an additional labyrinth seal and reduce the contact area, thereby improving its compressive characteristic. Some metal gaskets are corrugated with facing material often in the corrugations to improve sealing by offering certain resilience. These gaskets have better adoptability to flanges. A double-jacketed corrugated metal gasket with corrugated metal filler is frequently used at around 1000°F/540°C. Most often, corrugated gaskets of typically 22GA (.031") are utilized with a layer of flexible graphite or PTFE for better sealing.

At temperatures in excess of the range of 900°F to 10,000°F, where the standard soft filler is normally not recommended, a double-jacketed corrugated metal gasket with corrugated metal filler is frequently used (see Figure 12.10B). There are many more variations of this type of gasket.

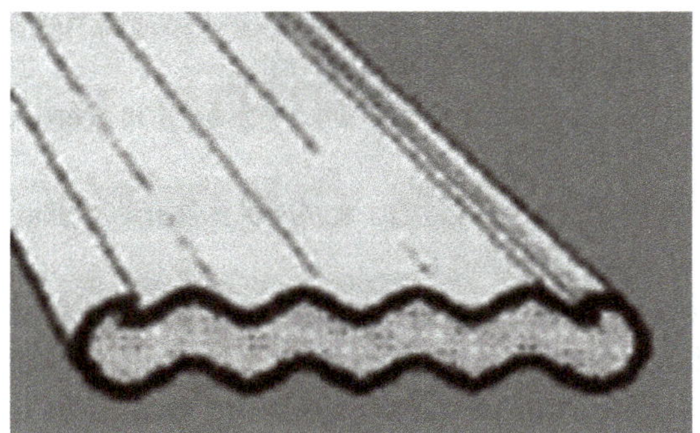

Figure 12.10A
Corrugated Metal Gasket

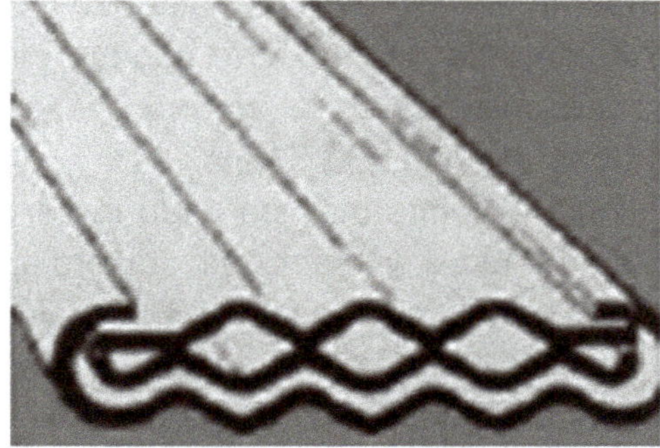

Figure 12.10B
Corrugated Metal Gasket with Metal Filler

METAL JACKETED GASKETS

Jacketed gaskets are made with aluminum, copper, brass, soft steel, nickel, Monel, Inconel, and various stainless steel types with non-asbestos high temperature filler. For still higher temperatures, flexible graphite or such similar materials are used as filler.

The following are not metal jacketed gaskets, but ...

OTHER TYPES OF GASKETS

ENVELOPE GASKETS Primarily used on glass-lined reactors and vessels, or such equipment and in certain services of corrosive acids, these gaskets are enclosed within two covers of Teflon with insert materials such as asbestos, non-asbestos materials, and rubber. They offer high compression, very good creep resistance, and resilience.

EYELET GASKET This soft gasket is enveloped or reinforced by a soft metal ring on the inner circumference which resists blow out and degradation, and offers better sealing.

EXPANDED PTFE Expanded PTFE is an excellent material which is very compressive and very flexible. It is a versatile trouble shooter for leaking, uneven, and corroded flanges. Available in spools or rolls, it has an adhesive on one side. Thus, it can be pressed and adhered into position (or formed) on the flange face. After a little overlap, the excess can be cut off and the DIY (do-it-yourself) gasket can be compressed between the flanges to make an effective seal. It is generally used on corrosive services or general services with low temperatures and pressures.

FLEXIBLE GRAPHITE TAPE This tape is essentially pure graphite, typically over 95% elemental carbon. It is available in reels and spools in various thicknesses and widths. It can be used in a similar way as expanded PTFE. It has excellent resistance to most acids, alkalis, and organic compounds. It makes a very good packing material even at high temperatures.

It has an adhesive on one side, so it can be pressed and formed into position on the flange face. After a little overlap, the excess can be cut off and the flanges tightened. This tape is overlapped onto the spiral wound gasket material as an additional protection for demanding applications. Do it — it will help.

Flexible graphite may be used in services with temperatures up to 950°F/485°C, though it should not be used with strong oxidizers such as nitric or sulfuric acid.

When I was introduced to this gasket material about 10–15 years ago, I was wonder struck. Before its launch, a successful fix up of floating head gaskets in the heat exchangers was a trick and at least three gaskets were wasted each time, apart from the waste of time in opening and reopening the covers many times over. With the introduction of flexible gasket, it was a pleasure.

SPIRAL WOUND GASKETS (ASME B16.20)

Spiral wound gaskets are by and large the industry favorite; they are made with alternating layers of a soft filler material and a formed metal wire (see Figure 12.11). Spiral wound gaskets are almost an industry standard for refineries, petrochemical plants, power plants, and most chemical plants. Extremely resilient, they can take care to a large extent of the flange movements due to pressure, temperature, line fluctuations, and vibrations.

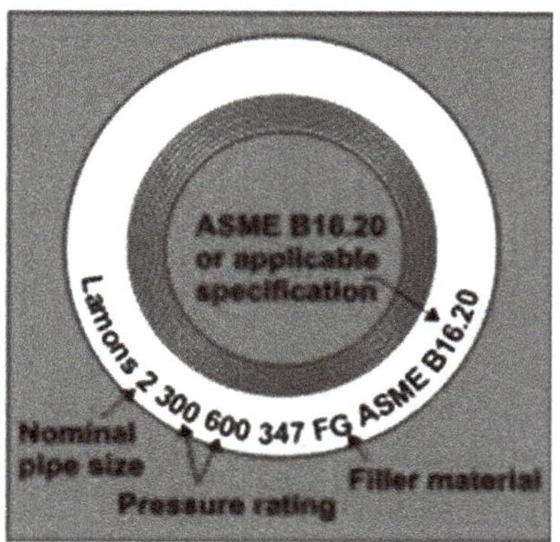

Figure 12.11

As the name suggests, they are spirally wound on alternate layers in a V pattern, and held by a metal ring. Low density and thick filler are used for lower pressure whereas thinner filler and high density are used for high pressures. V-shaped metal strip with filler in a phonographic pattern acts as a resilient, yet extremely compressible seal. Filler and metal strip can be tailored to suit individual specifications. Fluid characteristics and gaskets are made according to flange classes.

Metals can be steel, stainless steel, Monel, or Inconel, etc. As a matter of fact, they can be of any metal that can be fabricated as thin strips and welded, which makes them resistant to virtually all the fluids. Filler material is generally flexible graphite or PTFE. Asbestos is still used in some countries, though totally banned in others. The gaskets are manufactured to suit a complete range from cryogenic to 2000°F/1093°C, in all classes up to 2500. Spiral wound gaskets with PTFE filler are typically limited to a range of 500°F/260°C, whereas those with standard flexible graphite are rated to around 850°F/450°C. Integrating a mineral-based filler such as phyllosilicate into the design will increase the rating even well beyond.

An outer centering ring contains this metal strip and filler and also acts as a compression limiter. There are a good number of design variations with the spiral wound gaskets; these gaskets may also have an inner ring or both an inner and outer ring. An inner ring provides additional strength and buckling resistance; it also protects the actual gasket material from erosion and corrosion. When the gasket is made up of PTFE, an inner ring is a must (in accordance with ASME B16.20) to protect the spiral winding from this radial distortion (buckling). In some applications, such as valve bonnet gasket, limitation rings are not used and the spiral wound gasket straight away sits in the bonnet groove. The inner ring, filler metal, and filler gasket material should be compatible with the chemical used in the line, whereas the outer ring is typically corrosion protected carbon steel. As a general rule, the portion of the spiral wound gaskets exposed to the media stream (winding and inner ring) should be of the same material as the flanges. The gasket should be compressed to the limiting ring.

They can be used with the commercially available flanges finishes. They must be properly seated on the flange surfaces. Any protrusion into the flange groove must be avoided, which can create a serious problem. The thin strips of metal and filler will invariably unwind, travel and will take their safe lodging in a most inconvenient valve seat.

It is often a case with the pressure relief valves. The relief valve pops but does not seat back to the utter dismay of every one around. The culprit could be the metal strip of this winding protruding into the flange inside, which unwound onto the seating. The solution would be to shut down the plant, remove the relief valve, and recondition and fix back. Better be careful rather than sorrowful.

MAKING GASKETS

Ready-made gaskets are not available every time. Most of the industries keep stock of gasket sheets of various thicknesses and sizes. It is a practice to cut gaskets out of these non-metallic gasket sheets, wherever such gaskets are required.

1. It is a common practice to cut the gasket by keeping it on the flange edges and tapping or hammering it. This may seem ok because the gasket is soft and the flange hard, but it may eventually damage the flange.
2. A gasket cutter will do a better job.
3. The inside diameter of the gasket should not be more than the inside diameter of the flange because it will obstruct the flow. The gasket will get wet and spoiled.
4. Use correct thickness and grade. If the right grade is not available, higher-grade material is all right if it is compatible with the fluid, but not lower grade material.
5. Punch the exact number of holes for the bolts, not less or more.
6. Bolt hole in the gasket should be slightly larger than the bolt, generally equal to the bolt hole in the flange. It should never be larger than the bolt hole in the flange.
7. Store the gaskets away from sunlight, high humidity, temperature, and contaminants such as water, oil, and chemicals.
8. Do not hang the gaskets. They should be stored flat.
9. Large gaskets, particularly spiral wound gaskets, should be stored in their boxes until ready for use. Avoid mechanical damage.

Reuse of a gasket is not recommended as it deforms to set itself on to the flange and operating conditions to make a tight seal. Hence, it may not be able to adjust to the new conditions, even if the flanges are the same old ones. The cost of a gasket is much less than the cost of the leaking material and discomfort it can cause. Similarly, if bolts or studs are damaged due to corrosion, or otherwise, do not take chances. Do not take chances with lower grade bolts or studs.

To ensure good seal performance,

- Clean the studs and bolts with a wire brush to remove dirt on the threads. Check for the condition of threads, burrs, dents, and damaged worn out threads. Renew the defective studs and bolts.
- Clean gasket seating surfaces. Check for the condition of the faces. Check for scorings, pittings, cracks, and warpage. Flanges should be flat and parallel. 0.2mm flatness over the gasket seating width and less than 0.4mm total out of parallel across the whole flange are allowed.

- If there is left over material on the flange face, either from the gasket or the piped chemical, remove it completely. You can never achieve proper sealing on a dirty face! If material is to be removed from the grooves of tongue and groove joint or ring type joint, use a brass drift or similar tool.
- Use the right size, shape, and material of the gasket, which is free of mechanical defects. Keep the large gaskets in the covers until ready for fixing.
- Do not bend or buckle or play with the gaskets.

BOLTING UP PROCEDURE

The most common problems in cases of gasket leak are bad installation method and bolting. Though ideal, it is not practicable to tighten all the flange bolts simultaneously. Hence, a tightening sequence should be adopted while bolting up. Failure to follow this places an unequal load on the flange. As a result, flanges may cock and the joint may eventually fail. The gasket is actually compressed onto the flange faces by the bolting force. Gasket material flows into the imperfections of the sealing surface and makes a seal. Bolting and tightening should be just enough to deform the gasket into the surface where it is used to make a proper seal at the operating pressures and temperatures, but it should not be excessive so as to tear the gasket off.

Similarly all the bolts in a flange must be torqued to an equal value. Torque wrenches are not normally utilized on the piping flanges for routine use. Hand skill and calculation are used to give equal tightness to the flanges. In either case, the flange bolts must be tightened equally following a sequence, failing which the gasket will surely fail. This is particular the case for lines operating above 1800°F/1000°C. As elongation occurs in all metals, the gaskets tend to relax. The flanges must not separate under hydrostatic thrust of the media. A practice to retighten such flanges after the lines got heated up sufficiently is followed by some, even though the practice is widely debated.

Use the following procedures for a leak-free joint.

- The surface finish of the flange should be good enough to affect a seal. Apart from design surface finish, any nicks, burrs, and pitting that invariably occur due to operating and maintenance practices should be scrupulously removed to achieve a good gasket performance.
- Inspect the gasket for any mechanical damage. Make sure that the gasket is of the right grade and quality for the intended purpose.
- Place the gasket on the flange surface to be sealed; bring the opposing flange into contact with the gasket.
- Clean the bolts and lubricate them with a quality lubricant, such as an oil and graphite mixture. A number of specialty lubricants are presently available for this purpose and use of them will certainly help in maintaining the flanges. It should be understood that a non-lubricated bolt has an efficiency of about 50% of a well-lubricated bolt.
- Use of the right thread lubricant not only improves the performance, but also helps in removing the bolt. It will be very easy to open up a bolt after several months or years

in high temperature service after it was effectively lubricated with a proper thread lubricant. There are a number of thread lubricants in the market. Proper selection will help.

- Place the bolts into the bolt holes and finger-tighten the nuts. Use steel washers under the nuts.
- Start tightening the bolts following the tightening sequence in the diagrams here. See Figure 12.12 for 8-bolt flanges, Figure 12.13 for 12-bolt flanges, and Figure 12.14 for 16-bolt flanges. Failure to do so will result in cocked flanges. It will be difficult to bring the cocked flanges into normal position. The joint will surely leak. When it does, it will be extremely difficult to retighten to arrest the leak.
- During the initial tightening sequence, tighten bolts to 30% of the recommended bolt stress. Check for the gasket and flange condition for proper alignment.

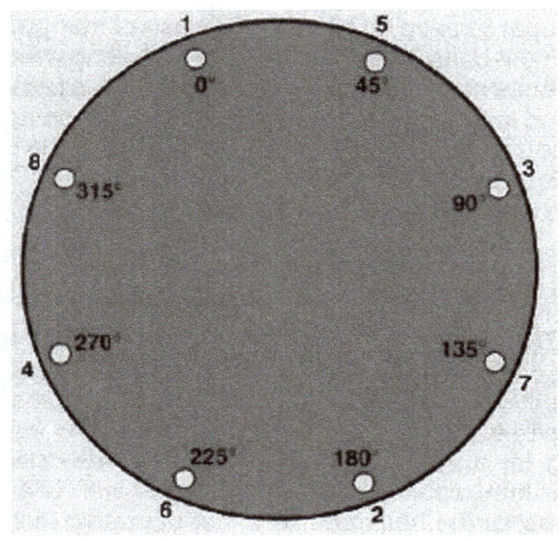

Sequencial Order	Rotational Order
1-2	1
3-4	5
5-6	3
7-8	7
	2
	6
	4
	8

Figure 12.12
8-Bolt Flanges

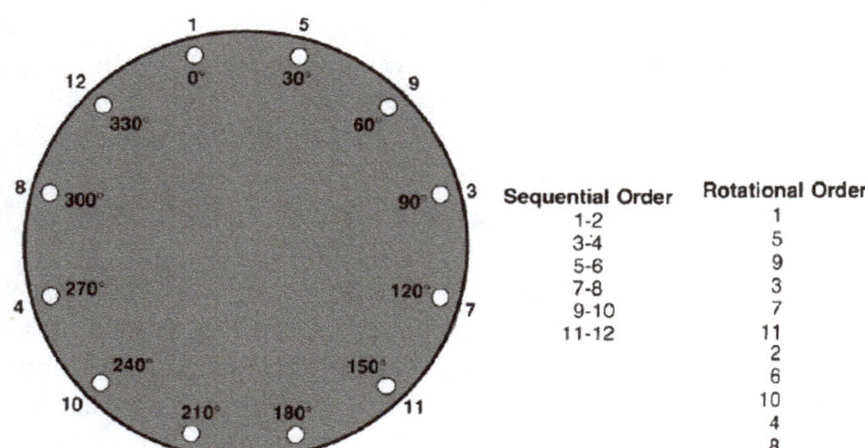

Sequential Order	Rotational Order
1-2	1
3-4	5
5-6	9
7-8	3
9-10	7
11-12	11
	2
	6
	10
	4
	8
	12

Figure 12.13
12-Bolt Flanges

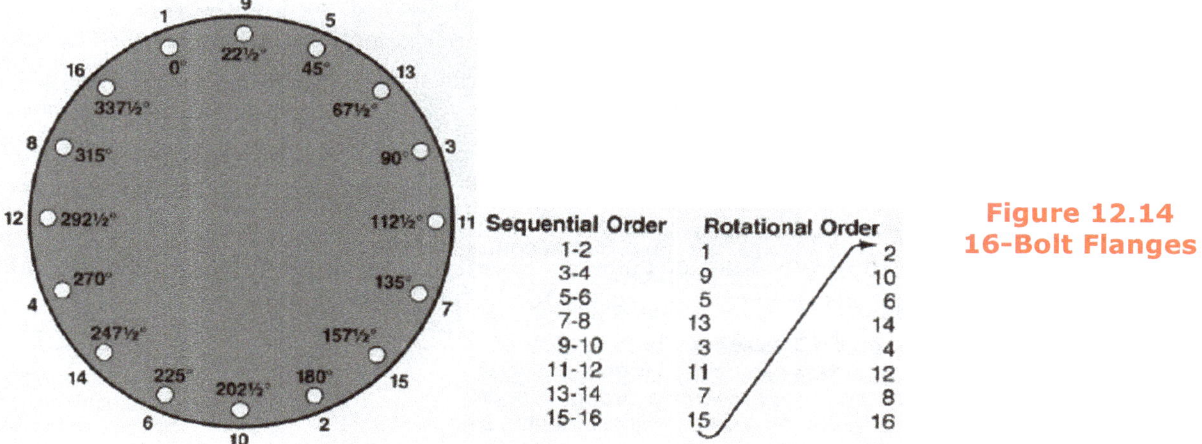

Figure 12.14
16-Bolt Flanges

- Increase the torque to 60% and then reach the torque values in three cycles.
- Then tighten the bolts by rotational method.
- Make a bolt-to-bolt torque check to make certain that the bolts are tightened and stressed evenly.
- Adjacent bolts do get relaxed. So make sure that the proper values are reached.
- Torque values are determined by diameter of the bolt, condition, and thread lubrication. The tightening sequence should still be used even if torque wrenches are not used. Proper and experienced judgment should be made for uniform tightening load.
- Plain gaskets or those without a centering ring should be installed and tightened such that initial tightening or later thermal expansion would not crush the gasket.

TROUBLESHOOTING LEAKING JOINTS

Joints fail, not just gaskets! Low bolting torques, over-tightened bolts, wrong bolt selection, use and its materials, insufficient bolt and nut lubrication, bad flange design or materials, poor gasket cutting or storage, and bad set up and fit up practices all contribute to seal failure.

One of the best methods for determining the cause of joint leakage is the careful examination of the gasket where the leakage occurred. It is good practice to inspect the gasket whenever a line is opened for routine maintenance or because of gasket failure.

1. If the gasket is found corroded, a better material compatible to the fluid in the lie should be used.
2. If the gasket is flown out or extruded badly, it means that a gasket with better load capacity and cold flow characteristics should be selected.
3. But if the gasket is crushed, it may be that excess bolt loading is used. If bolt loading is correct, select a material with better load capacity. Use a compression limitation ring on the gasket such as found on spiral wound gaskets.

4. If mechanical damage is observed on the gaskets, it may be due to improper positioning of the gaskets in the flanges, or bad centering of the gaskets or a wrong size.
5. If the gasket is not found compressed at all, the reason may be that a thinner gasket was used or the gasket selected is too hard or aged.
6. If the gasket is found unevenly compressed, the reason may be that a bolt tightening sequence was not followed, flanges are angularly misaligned, a thinner gasket was used, or the gasket selected is too hard or aged.
7. If the flanges are too wide apart or cocked, do not try to pull them by overstressing the bolts and the joint. Use straight spacers if they are far apart. Use taper spacers if they are cocked (see Figure 12.15).

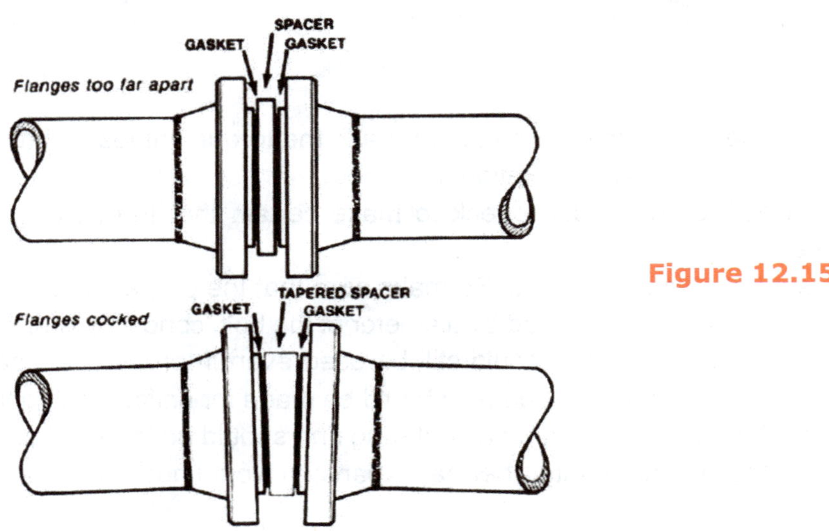

Figure 12.15

8. Many times a smaller gasket is used for want of a ready proper gasket. The mechanic centers this gasket initially on the flange faces with the hope that it will stay put in that condition. More often than not, this gasket falls down beyond the gasket seating and into the flow. It makes a bad joint, invariably leaks, and obstructs the flow.
9. Gaskets also do not seat properly if the flanges have parallel misalignment (see Figure 12.16).

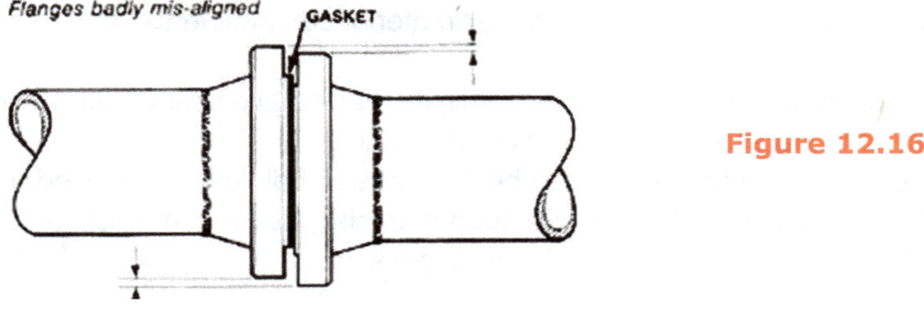

Figure 12.16

10. Ring gaskets must always be located and centered by the bolt circle. Use of smaller or bigger gaskets will spoil the requirement. Gaskets are made as per the flange classes and should be used as such. Pressure of the line fluid is only guidance.
11. Gaskets compress more on the outer periphery than at the insides. Here the flanges are actually bending down at the outer diameter to compress it more (see Figure 12.17). This happens when thin hand-made flanges are used, or when the seating surface is not enough. Reduce the gasket area and or use a softer material. The total out-of-parallel deviation is 0/015".

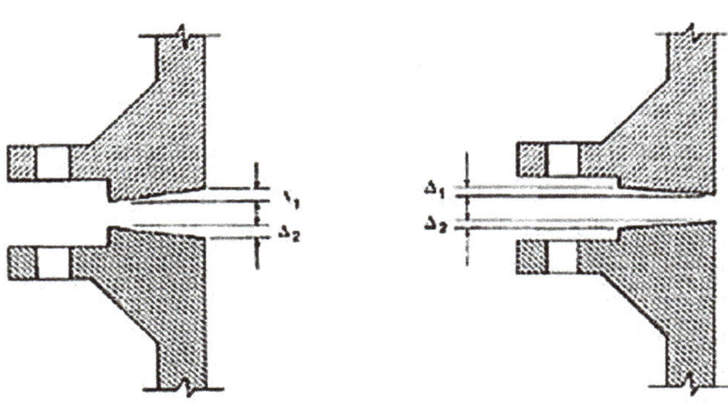

Figure 12.17

Total allowable out of parallel: $\Delta_1 + \Delta_2 = 0.015''$

Note — Deviation on right is less critical than deviation on left since bolt tightening will tend to bring flanges parallel due to flange bending.

MINIMIZE JOINT FAILURES

Leaks do occur in the plants. Many times they subside with a little more tightening of the bolts and some may require shutting down the line, and renewal of gaskets or bolts. There a number of reasons. Leaks may occur when bolts are under-tightened or over-tightened. Similarly they may occur when the gasket is undersized or oversized. Results may not be apparent in the beginning but disaster sets in slowly.

The following points should be kept in mind to reduce the joint failures.

Improperly or insufficiently tightened flange bolts are the main reason for most flange joint failures. Incorrect assembly may result in self loosening of studs or bolts, breakage due to fatigue or self relaxation, and eventual failure. On the other hand, over-tightening of fasteners can result in an over-compressed and crushed gasket, fatigue of the fasteners, and possibly stress corrosion cracking. Over-tightening may apply a higher load on the bolt than its ultimate strength.

Bad quality fasteners or under-designed ones will result in a failure. An example is the use of an ordinary run-of-the-mill mild steel stud in place of a B7 (ASTM B 193 Grade B7) or L7 stud. The result may not be immediate, but will eventually fail at elevated temperature or sub zero temperature.

Corrosion is the routine headache for any process plant manager. Atmospheric corrosion may create chaos in the process plants without giving any notice if the studs are not properly taken care of. It is better to use corrosion resistant fasteners like stainless steel ones in potentially corrosive atmospheres, for example, the vicinity of sea water or severe process effluents.

Incorrect selection of gasket material and its thickness is one of the primary causes of gasket failure. Aged gaskets, badly stored and handled gaskets, and reused gaskets may end up in the joint failure. Gaskets are reused occasionally to cut costs and save time, or due to lack of availability. Discretion is advised and should not result in a costly mistake. Initial over-tightening the flange bolts may force the gasket to creep. Retightening the flange bolts at elevated temperatures may overstress the gasket. This is often resorted to stop a leak. Bad flanges that are warped, misaligned, with surface damages, not cleaned, and corroded will no doubt result in a leak.

Surface finish of the flanges has profound effect on the gasket performance. Although the type of gasket dictates the surface finish of the flange, flange surfaces do get damaged, develop wavy finish, and let the flange joint leak (see Figure 12.18). If permitted, a softer material such as rubber could be used. Flexible graphite tape does a great job in filling out imperfections. Tape it on both sides of the gasket. If it is a corrosive service, expanded PTFE is available which is equally indispensible in such necessities.

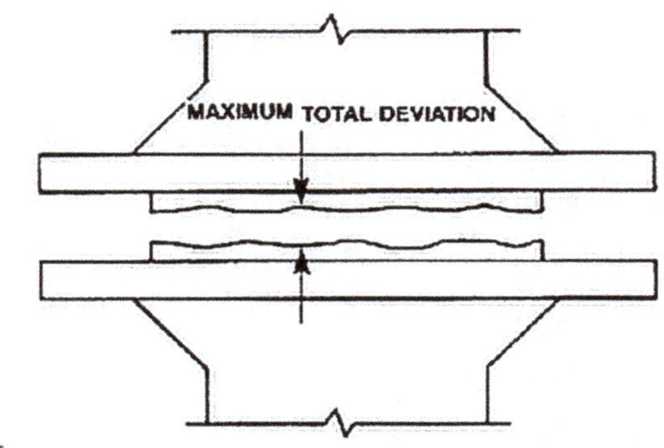

Figure 12.18

RETIGHTENING

Retightening is often a common practice in older systems. Professional managers routinely advise retightening of the flanges after 24 hours, 48 hours, or 72 hours in service. It is a highly debatable exercise. Fasteners and gaskets may set in to a relaxation mode soon after in service and the gaskets may creep. At elevated temperatures, the compressive forces on the flanges may decrease. This may result in a leak. To avoid this eventuality, flange bolts are retightened.

Extreme care must be exercised in repeated retightening of the fasteners as they may impose an unknown load and may ultimately damage the gasket. Proper material, proper sequence, and properly torqued fasteners will do well throughout their lifetime. It is not advised to retighten the flanges with elastomeric or asbestos free gaskets.

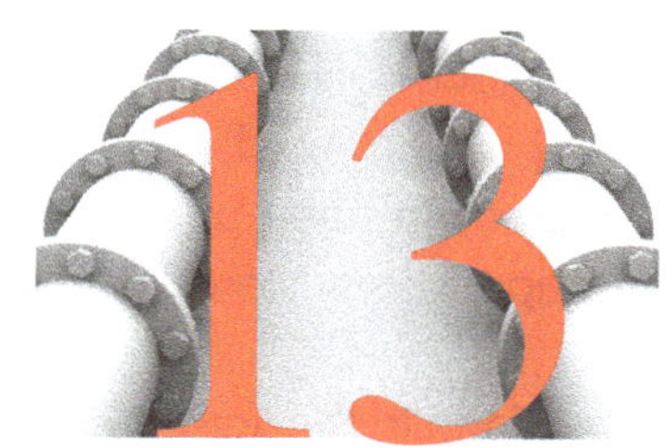

Opening Lines

Process requirements, problems, and maintenance activities necessitate opening and closing of pipelines. It is one of the major and most frequent jobs carried out by maintenance personnel. Most untoward incidents have happened while opening the pipe lines, the next being working on the opened, but long left lines. Even a gush of potable water falling on somebody can be quite nasty and, if caught unawares, can prove dangerous. Please find below some guidelines and rules governing opening of all lines, whether for blinding, alterations, or repairs. Industry norms may be diverse, but the normal safety measures stay good always. Many industries have elaborate procedures for the installation and assembly of flanged joints, but very few focus on the trials and tribulations of disassembly!

RESPONSIBILITIES

First and foremost, maintenance personnel should never start opening a line without express authorization by the operating or production personnel who are responsible that the lines are properly prepared, blocked off, and ready to be worked on. All lines or flanges to be opened must be clearly marked. The technician in charge of the job should personally make sure that the above practices are strictly adhered. Inspect the blinds, drains, and vents where the assigned jobs are carried out before actually opening any line. Responsible operating personnel should and would stand by at the job premises until a reasonably safe condition exists. Maintenance should make sure of safe condition by themselves rather than depending on the word of anybody else.

WHERE AND HOW TO OPEN LINES

Before opening lines or other equipment, the line must be depressurized, which means that the pressure in the line must be brought down to atmospheric or as near atmospheric as is practical. The liquid must be drained or the gas must be vented out to a safe level.

The personnel working on the lines must be informed of the hazards and consequences of the fluids in the pipe line or equipment; they must also be apprised of the safety precautions to be taken.

However, it is a cardinal principle while opening the lines that it should always be assumed that the line is full. *When anyone opens the lines with that guarded caution, chances of mishaps are less. Always double check. Do not depend on some other person's words. It's your life.*

Even with a little bit of passing of the block valves, the line may slowly pressurize and will come out from where it is open — and you will be exposed. *Please take care of that eventuality. Make sure that the valves are absolutely blocked.* In case of the slightest doubt that valves may pass, fix blinds in the line to carry out a safe operation. More information is detailed in the Chapter 15, *Blinding and Normalizing*.

In case of hazardous chemicals, fix blinds in the line. Disconnect the line. But that is easier said than done. Flanges have to be opened while the line is possibly under pressure. But blinding the line is a positive control mechanism from further ingress of fluid into the line where you are working.

Where it is not possible to blind the lines, double block and bleed valves may be incorporated in the line. But then this must be done during the design, installation, and commissioning of the line rather than now. However, if the job is carried out by depending on the double block and bleed valve mechanism, then the bleeder valve must be continuously monitored for leakage.

The point of safest and easiest breaking of the line must be determined before the start of the job. Similarly, safety worthiness of a line must be assessed before opening the line or the worthiness must be decided at the first opening before going ahead to disconnect at any other point. *Check for proper supporting.* Make sure that the line will have proper supports even after disconnection. If not, provide for temporary supports.

It is good practice to tap the union or nuts on the flange to be opened sharply with the hammer several times on all sides. Hammering releases the gripping effect of the threads and so helps to loosen a fitting. Then open the line at the flange wherever decided, before opening at other points. If it is not a flanged line, break it at a union or fitting.

Always open flanges at the **bottom**, or on the side away from you. Leave at least half of the opened bolts and nuts in the holes, until the flanges are parted.

Occasionally it will be observed that the flanges are actually springing back. A badly misaligned flange can even swing towards the persons standing on either of opposite sides. They open up along with the bolts. As the bolts are opened up, the flanges tend to load up the rest of the bolts. There may be a stage when finally one or more of the bolts get damaged. There is also a lot of unknown energy stored in these flanges. Then it is better to loosen all the bolts gradually, leaving the nuts in place. Some flanges sit tight even after opening all the bolts. They need to be pried open away from you, which is where all the skill lies. If there is any fluid left in the line, it will surely jump on to you if you are not careful.

Do not stand or let anyone else stand in a prone position that would be exposed to the flow from between the flanges.

It is always better to make a small opening before attempting to gas cut the line, particularly if the line is old and unattended for a long time, and contents are unknown. If trouble is encountered, plug or clamp the line immediately.

OBSTINATE BOLTS AND NUTS

Sometimes it will be extremely difficult to open some bolts with all the normal means, because of corrosion, galling, and wrong threading or over-tightening. More often than not, the nut gets rounded off. Iodine solutions, carbon tetra chloride, mild acids, etc., are traditionally used. But a number of penetrants have been developed and are now available; these can be of very good use. The present practice is to spray a little of these rust lickers on all the nuts before attempting to open them. But it is always a good practice to lubricate the bolt before tightening it. A lubricated bolt has more strength than an unlubricated bolt. Some very good thread lubricants are around, and experience has shown that bolts do come off easily when tried after years in service.

A variety of nut splitters are available. Use them with discretion.

A hacksaw comes handy. Cut the nuts in half and then wrench the bolt break loose. If there is gap between the flanges (there is enough often with two raised faces and gasket width), the bolt can be cut across in the center. Cut only half and break the bolt by wrenching it.

Then there is the chisel. Cut the nut by using the chisel. Sometimes after a few strokes, the nut tends to become loose releasing the galling, corrosion, etc.

In a worst case, nuts can be drilled right through the flange holes.

These are the methods to be adopted when hazardous conditions prevail. Take all safety precautions when using hand tools, particularly in hostile environments.

If everything fails, the last resort may be to cut those obstinate bolts on the flanges with a gas cutter. It can be performed:

- If the fluid handled in the line is not flammable or explosive
- If this operation does not create untoward vapor pressures in the line
- If the sparks from the gas cutter can be directed towards a safe area
- If the material of construction (MOC) of the flanges permit gas cutting (not plastics)
- If gas cutting is permitted by the production personnel in the area
- If spare bolts or studs are available

If all these points can be fulfilled, adopt the following procedure and remove the bolts or studs.

- Take all the precautions necessary for opening the flanges.
- Before trying to cut the bolts, try opening by heating it by gas torch. If the nuts get opened up, fine.
- Or, weld a good nut over the stud or the bolt head and use this nut to open the stud or bolt.
- Gas cut the nuts away from you, preferably from the bottom of the flange.
- Do not cut the flange.
- Wait and watch for leaks.
- Punch out the cut bolts on the flange and replace at least two bolts with new as they are punched out.
- Do not catch the cut bolts or nuts with bare hands.
- When everything is clear, go ahead as per normal practice.

PRESSURE ON LINE

If pressure on the line is observed when opened, or an appreciable amount of oil, gas, or any hazardous chemicals released, close the line at once. Report the condition to the production personnel and do not start work until the line is cleared.

SPILLS

Do not work in the area where the spillages of oil or hazardous chemicals that might have spilled over during opening of lines or from leaks. This also happens during the clamping of pipe leaks as a measure of temporary repair. Flush away all those contaminants and make a safe work place. Similarly, do not leave the spilled contaminants after the work is completed. If the spill is serious, the matter should be brought to the notice of all concerned. If the spillage is inflammable, take all the necessary precautions for fire protection.

Clothing that has been saturated with hazardous chemicals or oils should be removed at once. It is a health hazard and fire hazard.

Do not just open a line and leave it unless there is a specific reason and approval for it.

Most organizations have safety regulations laid out which require clear written permission to be taken before opening any pipeline. These are specified for the place and position of the work, time and safety precautions to be taken. Get permission in the first place. If the work continues for more than a shift or the stipulated time, get permission again and at each shift. Watch the area for unsafe conditions that have cropped up.

When removing valves for replacement or repair, the gate should be raised from the seat and left in that position to make sure that there is no stock locked up in the bonnet area of the valve. Alternately the bonnet bolts must be loosened if the valve stem cannot be operated. Wash the valve with water both inside and outside. There have been instances of serious accidents where valves were left without such care, when opened after a few days.

SPECIAL PRECAUTIONS

OPENING LINES WITHOUT CLEARING OF STOCK

There are occasions when the lines have to be opened, even when the line is not fully cleared, vented, or drained. They can be opened under the following circumstances:

1. The line is fully depressurized, which means that the line may have trapped fluid but not under pressure.
2. All the sources of pressurizing the line directly or indirectly are blocked.
3. Operating personnel are standing by the job until otherwise safe.
4. A plan was made as to which flange, drain, or vent should be opened for a safe discharge of locked-up fluid, and action was taken accordingly.
5. Use of personal protective equipment is available suitable to the hazardous chemical involved.

When preliminary loosening of several bolts indicates that the line has been depressured and valves in the system are holding, blinds must be installed at the block valves to prevent an ingress or egress of any fluid into the working line.

The same precautions shall be observed when removing blinds and normalizing the lines.

For taking up any major repairs or hot work, all lines must be cleaned of oil, gas, or vapors before opening for cold repairs or for any hot work. If the line handles inflammable or explosive substances, it must be purged with nitrogen before taking up further work.

Dead lines or long forgotten lines are hazardous as they can accumulate toxic gases and similar substances. Care must be exercised while working on these lines.

OPENING LINES WITH FLAMMABLE FLUIDS LIKE LPG AND PETROL

It is imperative that the pipelines and equipment containing flammable liquids and gases be cleared of all the stock — drained, or vented and purged out — before being opened.

Before breaking a line, make sure that there are no operating welding machines in the vicinity, and that this line is definitely not a part of the welding circuit.

All sources of ignition must be removed before opening lines. If there is a possibility that the flammable vapors be released through the flanges now being opened, switch off naked lights or flames within 60 meters (about 200 ft) radius.

Do not allow LPG or these flammable fluids to make skin contact or breathe the gases.

Do not permit workers to congregate around the job either out of curiosity or because of other jobs around.

Open such a line either for blinding or any other purpose, unless the line has been cleared of liquid or vapor, water washed or purged, and operating personnel are standing by.

Such flammable vapor lines like Liquefied Petroleum Gas (LPG) lines must be completely disconnected (not simply blinded) from the source of supply to prevent gas from being released into a line that is on maintenance.

PRECAUTIONS FOR SMALL LINES

Work on or around lines of one inch and smaller need special caution.

Do not stress or strain the line. Most likely, failure is near the joint connected to a larger line or vessel.

It is good practice to tap the fitting to be opened sharply with the hammer several times on all sides. Hammering a pipe fitting releases the gripping effect of the threads and so helps to loosen a fitting. While hammering, hold a pipe wrench or other heavy piece or metal on the opposite side of the fitting to give an additional support and remove the springing action of the line.

Use a back-up wrench in all cases of connecting or disconnecting on the opposite side. Without this holding wrench, the line may unscrew or screw at an unwanted place. Sometimes the line may break at the bends.

Never step on a small line.

BULL PLUGS AND CAPS

Loosen fittings slowly on a bull plug, pressure gage, lubricant screw on a plug cock, etc. Be careful while removing them. Loosen but make sure that it is holding until pressure is released.

Stand at a convenient and safe position while opening. There are instances when plugs just flew away, throwing toxic fluids on the face.

Bolting-Up

PREPARATION OF FLANGE FACES

Separate the flanges sufficiently for inspection. It is not always possible. Many times it becomes necessary to use flange splitters even to remove the gasket. Clean seating surfaces thoroughly with a scraper. Remove all left over gasket pieces and other foreign material from its mating surface. After some operation at elevated temperatures and pressures, gasket materials tend to get embedded in the flange. While trying to remove it, the gasket disintegrates and pieces stick to both flange mating surfaces. Even if a small unseen piece is left, the joint will surely leak. Scrap out all leftovers, but use a brush or drift softer than the flange.

To overcome this problem, various gasket release compounds have been developed. Use of such compounds help if they are not incompatible with the gasket itself. The good old and cheaper practice is to apply a graphite-based compound on both gasket surfaces and fix the gasket.

Extra effort must be taken in separating tongue and groove, ring type, and male and female flanges such that their grooves can be cleaned. The bad practice of leaving gasket pieces in the tongue and groove flanges (adding one more gasket into the groove) is often adopted; it should be discouraged. Ring-type flanges are generally in the higher pressure ratings and, hence, heavy. They do not leave much room for maintenance activity, but cleaning is paramount.

Inspect the seating surfaces for cracks, pitting, corrosion, and mechanical damage, or other defects will create leaks if left unattended. Tongue and groove flanges develop distortion, burring, and improper seating. Cracks and pittings on the flange faces can be repaired by filling, welding, and filing smooth. Minor pitting and cracks are repaired by depositing weld metal in the defect and then making smooth by filing or grinding. However, they can be ignored if they do not extend half way over and do not show signs of extension. But do not tolerate such defects in RTJ flanges (API ring-type joint flanges). Some maintenance compounds are now available which help in cold repairing the flange defects. They come in handy.

More of this discussion can be found in Chapter 32, *Emergency Repairs*.

Inspect the flanges for distortion, particularly if the outer edges contact each other at any place when the joint is made up. They must be replaced or repaired. Check also if the bolts are matching properly and are free in the holes.

INSTALLATION OF GASKETS

All flat metal, spiral-wound metal, corrugated metal, and asbestos gaskets must be renewed each time.

- Whenever there is a doubt as to the suitability of a gasket for reuse, it should be renewed.
- It is a good practice to renew gaskets in hot (450°F/230°C and up) oil service.
- Make sure that the gasket is the right material and size.
- Make sure that metal and metal clad gaskets are of the proper material for the service.
- Do not use multiple gaskets just to cover the gap. Gaskets cannot give adequate strength and are likely to blow out under unusual conditions such as fire.
- Orifice plates, blinds, etc. should be installed with a gasket on each side of the plate.
- Do not attempt to get a tight-flanged joint with flange faces which are not alike.
- Gaskets are not required between flanges of lined diaphragm valves and lined pipe because the plastic faces of each provide the sealing.

Gaskets do get damaged while in service, on removal, or while fixing. But some gaskets are still in good condition and can be reused under the following circumstances. Note, however: **Discretion is better than valor!!!**

- Asbestos composition gaskets (CAF) that have not been badly smashed or have not become brittle or hard.
- Steel clad asbestos ring gaskets for raised face flanges if they are not corroded or distorted.
- Small tongue and groove, large tongue and groove, and male and female gaskets if they are not corroded, dented, or otherwise damaged. Small tongue and groove gaskets that fill the groove and are not otherwise damaged need not be removed from the groove.
- API ring joint gaskets if the seating surfaces are not pitted or otherwise corroded or distorted.

BOLTING UP — GENERAL INSTRUCTIONS

1. Remove and clean all bolts so that the nuts turn freely. Check to see that there are no damaged threads and that the bolts are long enough to engage all the threads in the nuts when they are pulled tight. Coat the bolts with thread lubricant like Molykote or Neverseez or such thread lubricant.
2. Face the flanges together, install all but one or two bolts, and hand tighten the nuts. If the gasket does not stay in place while facing the flanges together, use a little waterproof grease or shellac on the seating surface to hold them. Do not use tape or string.
3. On plain and raised-face flanges, normally half the number of bolts plus one more can be inserted preferably on the bottom side. Then the gasket is inserted from top and the remainder of the bolts is installed after the gasket.

4. On tongue and groove, ring joint, and male and female flanges, the gasket must be inserted before the flanges are faced together (after all you cannot fix it afterwards).
5. If all of the bolts cannot be fixed due to flange misalignment, insert a spud bar in an open bolt bole and, using it as a lever, pull the flanges into position. While holding the bar, pull up two bolts which are opposite to each other snug enough to hold the flanges in position. Remove the bar and insert remaining bolts.
6. Check to see that the flanges are squarely faced. On tongue and grooves, see that the tongue is properly entered into the groove and on ring joints see that the ring is entered into the groove on both flanges. Check plain and raised-face flanges to see that the gasket is properly centered to insure the right size gasket has been used.
7. Tighten the bolts by working around the flange until all the bolts have the same tension. The use of any kind of a handle (pipe, etc.) to increase the length of the wrench is not permissible because of the danger of straining the wrench and the personnel hazards involved.
8. When flange faces are not quite parallel they must be tightened by pulling up on the bolt at the widest separation point first, and then proceeding to tighten all the bolts provided there is enough flexibility in the pipe to allow the flanges to pull together.
9. On short stiff manifolds such as at pumps, meter bypasses, etc., the flanges must fit squarely without cold pull or line strain. Otherwise they will leak or cracks may develop in the welds or in adjacent castings such as valve or pump bodies.
10. When stud bolts are used, check to see that all threads on both nuts are engaged. If they are not, loosen the nuts, center the bolt, and retighten.

EXTREME CASES

Flanges do get misaligned due to various process conditions, but more often it is an installation slip up. Flanges spring over to the misaligned position if ever opened and are difficult to bring back into the aligned position. However, many times flanges can be brought into proper alignment by only cold pull with the help of long bolts or with chain blocks. Use of tools like spud wrenches, drift pins, and pry bars would be helpful.

However there are cases where the flanges refuse to be aligned, or stay misaligned, putting strain on the associated piping and equipment. It is possible to bring the flanges back into position by heating and bending the pipes in situ, if heating is permitted.

BEND CARBON STEEL PIPE BACK INTO POSITION

1. Keep the bolt in place and, if necessary, use long bolts or cut long bolts from a stud rod.
2. Heat the pipe at about 2" to 3" behind the flange weld. Make it red hot in a 2" band all around the pipe.
3. Start tightening the bolts on the widest opening and continue until the flange faces are about parallel. Keep the heat on while bolts are being tightened.
4. Allow the pipe to cool and then loosen the bolts. Check if the flanges are in the alignment or have gone out as soon as the bolts are removed.

5. If they are parallel and close, change to regular bolts if longer bolts are used. Insert gasket and tighten in normal course.
6. If the flanges have not come into position, repeat the heating and bolting operation.
7. Cast iron pipes will surely break if such trials are made with heating or no heating.

Alloy steel pipes can also be brought into position like this. But check if they need special heat treatment after each job. If it is considered not possible to bring it back to position with the above method, it is better to cut or open a piece of pipe at the next weld joint or flange joint and remove it. Correct the misalignment externally, fix back and weld. This sounds easy but it will be difficult to re-weld very old and worked up pipes.

In all cases, take care of expansion during actual operation. It is good to make a decision before hand, rather than try one method and then the other.

CAST IRON PIPES

Cast iron flanges which are out of alignment cannot be pulled together as per the above procedures because they will surely break. And the break could be anywhere in the line, not necessarily near these flanges. See Figures 12.15, 12.16, and 12.17 once again.

To correct misalignment, use the following procedure.

1. Open up all the bolts and check for misalignment or, if they are out of parallel or just far apart.
2. If they are misaligned, examine the nearest support for the cause and correct it.
3. If the flanges are not parallel and the gap between flanges is actually tapered, measure the gap on all four sides.
4. Machine a cast iron spool of the same grade to the same taper. Machine OD of the spool to fit inside the bolt circle.
5. Fix this spool between the flanges matching the taper; install suitable gaskets on both sides.
6. Get or cut long studs of suitable length. Fix and tighten without causing any strain on the pipe line.
7. Many of the CI pipes have broken just because of over-tightening.
8. If the flanges are far apart, it is not wise to bring them close by the use of long bolts and a bit of a hard tightening of the bolts. Instead use a straight spacer and follow the same procedure.
9. If the gap is large, about one foot or more, it is good practice to renew one or both the pipes to the next length available instead of fixing a large spacer.

HOW TO ASSEMBLE A FLANGED JOINT

Bolting two flanges together with a gasket between their machined faces to assure a tight seal makes a flanged joint. The proper steps in making a flanged joint are shown below:

CLEAN ALL PARTS

Use a solvent-soaked rag to remove the rust preventing grease put on flanges at the factory. Next, clean off all dirt and grit particles. Then wipe off the gasket just to be sure it is clean.

ALIGN AND SUPPORT PIPE

When the pipe is in place, be sure it is properly supported. For example, a valve cannot hold up an unsupported length of pipe without undergoing great strain. The flanges should be accurately aligned by checking with a spirit level, both horizontally along the pipe and vertically across the flange faces. It is then ready to be bolted tight.

FLANGE FACINGS SHOULD BE ALIKE

Do not attempt to get a tight flanged joint with flange faces which are not alike. Steel valves and fittings normally are manufactured with raised flange faces. Piping materials in iron and brass usually have plain or flat flange faces.

The proper procedure is to join a plain face to a plain face — or a raised face to a raised face. The proper gaskets will assure a tight joint. But never should a raised face be joined to a flat or plain face.

When joining 125 class cast iron or any brass flanges to steel flanges, be sure to remove the raised face on the steel flange. A full-faced gasket is preferred for such a joint.

INSERT GASKET

With the flanges securely in position, slip in half the bolts at the bottom (half plus one more). They will hold the gasket in place.

The gasket should be coated with a little graphite and oil or other recommended lubricant before they are inserted. Then it will be easier to remove if the joint is opened later. Apply a thin, uniform coating (avoid "lumps" of lubricant as this may reduce the efficiency).

Carefully insert the new gasket between the flanges without damaging the gasket surfaces.

Make sure that the gasket is central in the flange and not dropped. This can be a problem if a wrong class of gasket is inserted wittingly or unwittingly. Take care that the gasket is not pinched or otherwise damaged.

BOLT UP

Finally, slip in the rest of the bolts, apply a little thread lubricant to each, and then turn up all nuts by hand, as far as they will go.

Lubricate fastener threads and all bearing surfaces (underside of bolt heads, nuts, washers) with approved lubricants. Clean the bolts and lubricate them with a quality lubricant, such as an oil and graphite mixture. A non-lubricated bolt has an efficiency of about 50% of a well-lubricated bolt.

Do not contaminate either flange or gasket faces with thread lubricant.

Bring the flanges close together, frequently checking that they are closing in true.

TIGHTENING BOLTS

Start pulling up with a wrench — not in rotation, but by the crossover method to load the bolts evenly and eliminate any concentrated stresses. Keep repeating by going over and across in four or five passes until the joint is uniformly tight. Torque wrenches are used, not necessarily for the smaller and lower class flanges. Uniform tightening can be achieved by skill and sheer experience on these flanges.

1. Tighten bolts loosely by hand in the first instance and then hand-tighten evenly.
2. Following a cross bolting pattern, tighten the bolts to 25–30% of the final torque value. Use of a torque wrench is advised now.
3. Check the flange condition to see if they are getting aligned properly or not.
4. Next, torque to a maximum of 60% of the full torque, following the crossover pattern
5. Then torque to the full torque value, still following the cross tightening pattern
6. Final check at full torque, in a clockwise direction on adjacent fasteners.
7. Torquing should only be done on the system in the ambient conditions, never while piping is at an elevated temperature.
8. Do not over tighten.
9. Failure to follow the tightening sequence will result in cocked flanges. It will be difficult to bring the cocked flanges into normal position. The joint will surely leak. When it does, it will be extremely difficult to retighten to arrest the leak.
10. Adjacent bolts do get relaxed. So make sure that the proper torque values are reached.
11. Torque values are determined by the diameter of the bolt, condition, and thread lubrication. The tightening sequence should still be used even if torque wrenches are not used. Proper and experienced judgment should be made for a uniform tightening load.
12. Bolting up plastic pipes requires extreme care or they will surely break.

A detailed discussion is also made in Chapter 12, Gaskets.

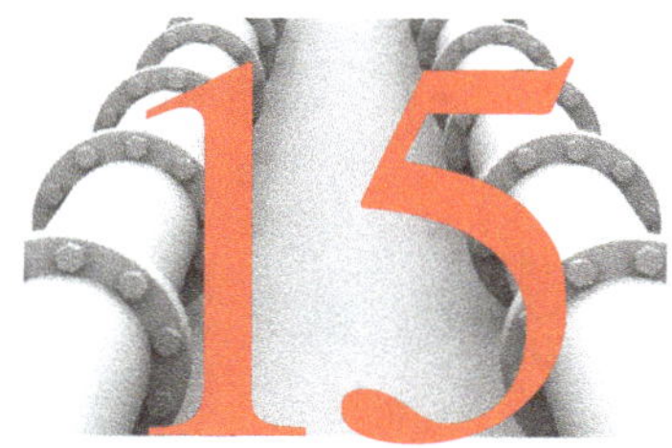

Blinding and Normalizing Lines

It is a normal practice in the industry to blank off or blind the lines for a positive shut off. Even though they are extremely engineered products which reached high level of reliability, valves still "pass." In the engineering jargon, "pass" means that they leak a certain amount of fluid though they are tightly shut off. This can be detrimental to the process or dangerous to the people working on the line elsewhere. Blinds are also inserted to hydro-test a particular line or vessel or heat exchanger, which needs to be shut off from other system. So a plate is inserted into the pipe line to prevent fluids from entering the other side of the lines, which effects a positive shut off. A blind makes it absolutely sure that the line is properly and securely blanked off with no possibility of

Figure 15.1
Blind Flange

any flow or leakage through a valve during maintenance (see Figure 15.1). Blind flanges are also used to close the lines in the end.

Two kinds of inline blinds are used: slip blinds and spectacle blinds (also known as figure 8 blinds). Spade blinds, or paddle blinds, are slip blinds that are made to suit the inside of the flange face within the bolt circle diameter. They are generally made 2–3 mm or 1/8" less than Bolt Circle Diameter(BCD) Or Pitch Circle Diameter (PCD) for easier fixing. Blinds should have a tail long enough to give an indication that they are in place (see Figure 15.2).

Figure 15.2
Slip Blind

Figure 8 blinds are just that — they look like the figure 8, which also look like spectacles (see Figure 15.3). Hence the name! They are also known as spectacle blinds. These are used when lines are to be frequently blinded and normalized. Either the blinded side or the open side will be in the piping.

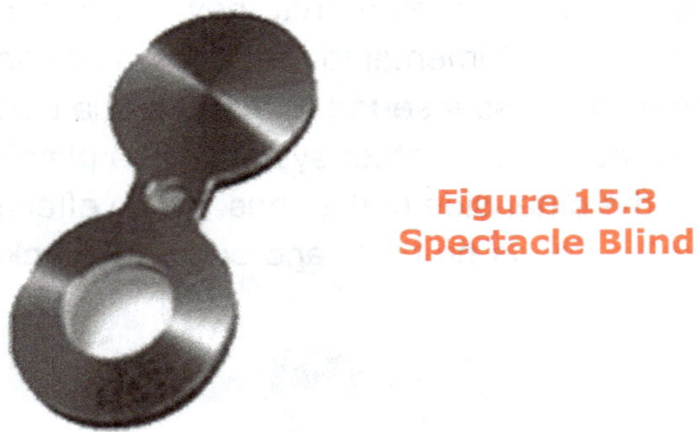

Figure 15.3
Spectacle Blind

Blinds are generally slip blinds in both cases, but full-face blinds are also made in both spectacle and spade types. Full face blinds are bolted the same way that end blinds are bolted.

Temporary blinds can be steel, but if they are used in permanent service they should be of suitable material, normally the same as the pipe line.

Follow the procedure suggested for opening the lines. Take all safety precautions. More often than not, blinding is carried out because a valve is passing.

Remove half of all the bolts from the flanges minus one. If the flange has eight bolts, completely remove three bolts in a row. Loosen all other bolts enough.

Separate the flanges sufficiently enough for insertion of the blind.

The blind should be of sufficient thickness to tackle the pressures involved. If the blinding job was carried out for a hydrotest, it is all the more important. Overpressure will surely distort the blind, may even puncture it. Then it will be very difficult to extract the blind from between the flanges.

Remove the old gasket completely and check its condition. Reuse of this gasket for the temporary purposes depends solely on the sight conditions.

Fix gaskets on both sides of the blind and insert it between the flanges.

Once it is certain that the gasket is sitting properly on the bolts, insert all the bolts and tighten.

Follow the bolting up procedure.

It is practice in many places in the industry to reuse the old gasket for the temporary phase of blinding and normalizing. The decision solely rests on the sight conditions. Also a practice is prevalent to use the gasket only on the pressure side. But this practice may spoil the flange face on the non-gasketed side.

For *normalizing lines*, the blind is removed and the flanges are boxed up. In case of spectacle blinds, the blind is reversed.

Follow the bolting up procedure.

While reopening the line, please take all the precautions that were originally taken for opening lines.

Do make sure that the line did not get pressurized by any chance or mistake.

Use new gaskets. Use gaskets on both sides; especially in the case of spectacle blinds, use gaskets on both sides.

Do fix gaskets on both sides of the flange, though there may not be apparent pressure on one side. Generally the removed gasket or a simpler or cheaper gasket is inserted on the non-pressure side. This is important to protect the flange faces from direct metal to metal contact.

16

PROCESS VALVES

Valves are indisputable leaders in the fluid power used all over to enable, stop, direct, or control flow or pressure. There are different types of valves like gate, globe, ball, check, butterfly, and needle valves that serve a variety of purposes. Gate, globe, and check valves are the most widely used. Each type of valve is classified by the pressure, MOC, and by the construction, such as flanged, screwed, socket welded, or butt welded. Valves generally have their pressure rating marked on the body, which should be at least equal or better than the design. Pressure and temperature ratings are interdependent. The ASME codes specify which types of valves and fittings are permitted to be used and their proper service applications in pressure piping. Valves are manufactured with a large number of different materials of construction suitable to the demanding applications. Material temperature limits for body and trim, pressure, and temperature ratings are given below.

Various types of valves are shown in Figures 16.1A to 16.1F. Figure 16.1A shows a class 800 gate valve, Figure 16.1B shows a class 800 globe valve and Figure 16.1C shows a class 800 check valve. These are generally smaller size valves, routinely available in Class 800, both in screwed and socket weld versions. Figure 16.1D shows a ball valve, Figure 16.1E shows a plug valve and Figure 16.1F shows a diaphragm valve. More detailed information about these valves follows.

Figure 16.1A
CL 800 Gate Valve

Figure 16.1B
CL 800 Globe Valve

Figure 16.1C
CL 800 Check Valve

Figure 16.1D
Ball Valve

Figure 16.1E
Plug Valve

Figure 16.1F
Diaphragm Valve

VALVE CLASSIFICATIONS BY FUNCTION

STOP VALVE

One of the critically important functions of a valve is to isolate the flow. It should offer tight shut off when closed; when open, it should offer minimum restriction to flow. Gate, globe, ball, plug, butterfly, diaphragm, and pinch valves are used, each with its own characteristic features and flow patterns.

FLOW REGULATION

When the flow of the fluid needs to be regulated between zero and maximum limits, globe, needle, ball, and butterfly valves are used. This is accomplished by introducing certain resistance to flow by way of the closing component, such as gate or globe. Globe and needle valves are better suited for this application.

PRESSURE REGULATION

Many processes require pressure to be regulated and maintained under varying flow conditions. The pressure regulator valve is basically a globe valve with balancing opposing forces of adjustable spring force and the feedback pressure. For accurate control, feedback is taken from the downstream fluid pressure. However, it is normally taken from within the valve for self acting valves. Pressure is hand controlled.

BACK FLOW PREVENTION

When it is necessary to prevent reverse flow of fluid, a non-return-valve (NRV) or check valve is used. The important requirements for this function are tight shut off against reverse flow, low resistance for forward flow, and fast response. The valve can be closed by gravity, fluid flow, or spring. Lift check valves or swing check valves are used for this purpose. There are stop check valves which normally act as check valves, but they can be closed and shut off by hand wheel. But the pressure drop across the valve is high, which must be considered.

PRESSURE RELIEF VALVES-SAFETY VALVES

Another important function of a valve for safety is the pressure relief valve. These safety valves are extensively used for relieving excess pressure that may cause damage or failure or pose a safety hazard. These are generally spring loaded but other designs are available.

RUPTURE DISC

The bursting/rupture disc is not a classical valve as such, but acts as a valve. Here a thin disc ruptures when a set pressure is exceeded and relieves the pressure. The fluid then escapes through the ruptured passage. Unlike other valves, this operation is irreversible and the system has to be shut down to renew the bursting disc.

VALVE CLASSIFICATIONS BY END CONNECTIONS

There are a number of methods of connecting valves into the piping systems.

- **Flanges:** The valve is provided with suitable rated flanges.
- **Wafer:** The valve is provided with suitable sealing faces and is trapped between line flanges.
- **Butt welded:** The valve is provided with butt weld end and welded into the piping system using high integrity joints.
- **Socket welded:** Socket welded valves are fillet welded, affording easy fitment into the piping system.
- **Screwed ends:** Ends can be provided with female or male screwed ends. The threads can be taper or parallel
- **Compression fittings:** Ends can be provided with compression fittings.

VALVE CLASSIFICATIONS BY PRESSURE CLASS

Steel and its alloy valves are made in seven pressure classes. Limits of operation of these valves are controlled by the pressure temperature parameters of the particular metallurgy and class. Cast iron valves are manufactured in 125 and 250 classes. Plastic and non-ferrous valves are only made for low pressure.

VALVE CLASSIFICATIONS BY METALLURGY

A good number of metallurgies are developed to cater to various pressure temperature constraints. Discretion, care, in-plant knowledge, and experience must also be exercised while selecting a proper valve for a particular service application apart from taking into consideration the standards, metallurgies, pressure, temperature, and flow conditions.

More detail is provided on this subject in Chapter 4, Materials for Valves and Fittings.

FLUID PROPERTIES AND OPERATING CONDITIONS

The properties of the fluid to be controlled have a major impact on the design and materials of construction of the valve. Over the years, the piping industry had developed a wide range of valve designs and materials to handle virtually any fluid being processed. Fluid viscosity, temperature, density, and flow rate should be taken into account for the selection of a valve. The valve must be suitable to withstand resulting corrosion and erosion of process flow and it should not harbor internal hold up of fluids. Important considerations include absolute internal and external leak tightness for any fluids, and definitely for toxic or explosive fluids. Some statutory regulations also include the need for a fire safe condition where the valve should maintain its internal and external integrity when the valve is surrounded by flames.

One of the most important considerations in valve design is the leakage of fluids into the surrounding environment. This is very vital when handling hazardous, corrosive, flammable, or nuclear materials. Typical leakage paths in a valve are the end connections like flanged or screwed ends, the valve spindles, and the body and bonnet joints.

Leakages from the end connections are best avoided by welded ends preferably butt welded (properly welded and inspected). At one point, industry was switching over to all welded construction. But the disadvantage is that it will be difficult to remove the welded valve from position for maintenance. Bellows-sealed valves and piston valves also effectively reduce the leakage across the spindles. Pinch valves and diaphragm valves are designed without glands, but they are limited in their use by the restrictions of elastomers used for diaphragms.

FLOW FACTORS

Valves have different flow characteristics and pressure drop conditions; these vary with each type and size. They are necessary to take into consideration while placing a valve in a new service or replacing the existing valve with a different type.

SHUT OFF CONDITION

Valves contain some of the most precisely machined surfaces at the seatings, which make them vulnerable to contamination and bad operation. These result in leakage across the valve (or passing) in closed positions or jamming in open, closed, or intermediate positions.

Valves are best placed in an upright position with the stem pointing straight up. Other positions are used but are compromised. It is not a good practice to install a valve with its stem down, as the bonnet now acts as a trap for sediment. This could be also dangerous in freezing temperatures as the liquid in bonnet may freeze and rupture it.

Valves should never be jammed in the open position. After fully opening the valve, it is a good practice to close the hand wheel approximately half a turn to prevent it from seizing. Many large valves are operated by gearboxes to reduce the hand torque required and improve the operating convenience and precision control.

The basic parts of most of the valves other than check valve are:

- Gate or globe or disc or piston, etc., (closing component)
- Body
- Bonnet
- Spindle
- Gland
- Packing
- Hand wheel or handle

GATE VALVES

Gate valves are the most used valves in the industry because of their ease of operation, full flow characteristic, and low pressure drop. They are basically on/off valves. They have a very large range of manufacture, in size from 6 mm to 2000 mm or more, in materials from plastics, cast iron, and bronze to a large choice in steel metallurgies to exotic metals. Gate valves are made in a variety of types, like knife edge-gate valves, slide valves, etc. Bellow-sealed gate valves avoid gland leaks. However, gate valves are not good enough for throttling. A throttling gate valve is detrimental to the life of a gate valve because it erodes the seat and gate. Angle or globe valves are preferred to control the flow.

The operation is simple: as the hand wheel is turned clockwise, it closes the disc or gate against the seat rings. These valves are made either with a rising stem, which indicates that the stem or screw rises as the valve is opened, or with a non-rising stem, where the stem does not rise externally. The sealing medium is the gate or a solid wedge, a flexible wedge that adjusts to suit the seal faces, or parallel discs which adjust between parallel sealing seats. A typical gate valve is shown in Figure 16.2A. Parts of the rising stem valve are shown in Figure 16.2B and parts of the non-rising stem valve are shown in Figure 16.2C.

Figure 16.2A
Typical Gate Valve

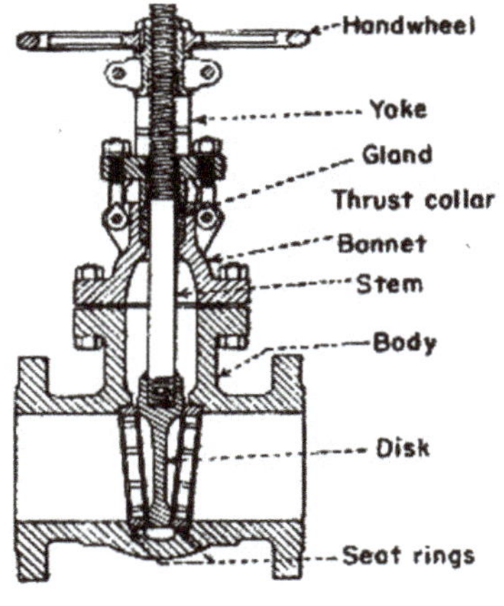

Figure 16.2B
Rising stem gate valve

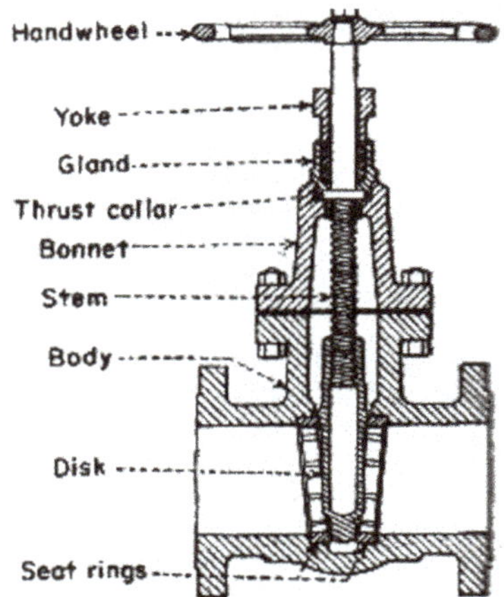

Figure 16.2C
Non rising stem Gate Valve

The valve should be inspected for disc/seat and seat ring corrosion and general sealing quality of the valve. Repeated and partial opening and closing of the valve can create a high velocity flow which can result in erosion and wiredrawing of the wedges and seats. A partially opened gate valve can induce a turbulent flow. Hence, this valve should be shut off when used for its sole purpose. If frequent operation is required, other types of valves may be used. Unlike a globe valve, a gate valve cannot increase or decrease the flow in proportion to the lift of its gate. So it cannot be used for flow control. Gate valves can be visually distinguished from globe valves.

GLOBE VALVES

Globe valves are essentially used for throttling flow. They are an industry standard as a bypass for control valves, also used primarily in high pressure service.

Apart from the body and bonnet, the basic parts of a globe valve are the disk, the valve stem, and the hand wheel. The body of the valve has a seating, a kind of an orifice through which the fluid flows. The seat is precision machined and has hardened surfaces. The seating may be soft in case a very tight seal is required. In some cases, the seats are replaceable. To throttle or stop flow, the valve is closed by turning the valve stem in until the disk is seated into the valve seat. This prevents fluid from flowing through the valve. Both the disc and the seat are very accurately machined to form a tight seal when the valve is closed. When the valve is open, the rate of fluid flow is controlled by the position of the disk in relation to the seat. Disks are available in various designs, and the flows in straight, angle, and cross-flows. Generally, the pressure is on the bottom of the plug. This keeps the fluid away from the gland packing when the valve is shut. Maintenance is easy with globe valves. Most often they can be repaired in situ without removing them from the end flanges. Figure 16.3 shows a typical globe valve; parts of the valve are shown in Figure 16.4. Picture of a higher class valve can be seen in Figure 16.5.

Figure 16.3
Typical Globe Valve

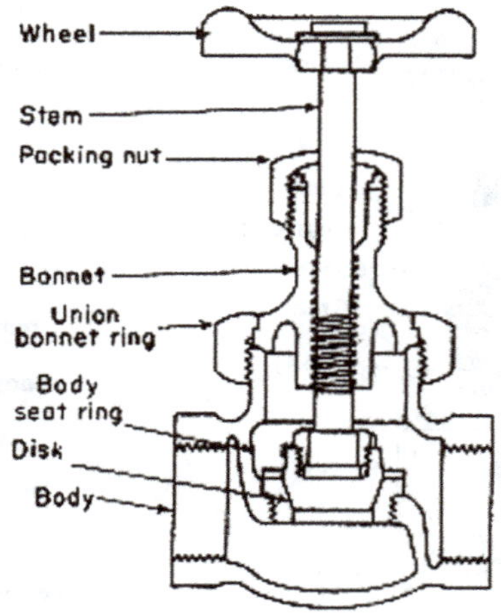

Figure 16.4
Parts of a Globe Valve

Figure 16.5
High Pressure Globe Valve

Unlike gate valves, the fluid takes a rather torturous path in globe valves, generally from under the disc seating to above and out. Hence, the pressure drop across globe valves is higher. Globe valves are available in a very large range of sizes and metallurgies. Globe valves are best suited for efficient throttling and accurate flow control, though pressure drop and cost are high. Globe valves operate in direct relation to the lift of the spindle. If ten turns are required to fully open the globe valve, then each turn increases the flow by 10%.

There are Y-pattern angle valves in which the fluids make a virtual straight path (see Figure 16.6). Pressure loss through angle valves is less than that through conventional globe valves. They provide tight shut off still with efficient throttling, especially in high-pressure and high-temperature applications. Seats are generally stellited. Bonnets are pressure sealed and discs are guided through out. Use of thrust bearings reduces operating torque.

FEATURES OF GATE AND GLOBE VALVES

A cut-away view of a typical globe valve is shown in Figure 16.7. Details of the various components are discussed here. A gate valve is similar except for the seats and the flow paths.

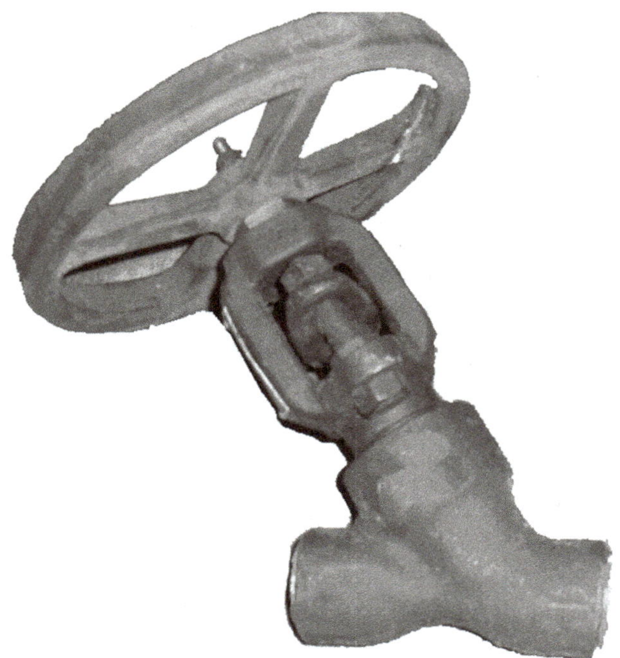

Figure 16.6
Y-Pattern Globe Valve

Figure 16.7
Cut-away View of Globe Valve

BODY-BONNET JOINT The body and bonnet are cast with standardized materials in single piece designs. The body and bonnet are normally bolted together with a gasket. In some smaller and inexpensive valves, these joints are screwed. Body and bonnets can be of full welded construction for specific services and higher pressures. On a pressure-seal valve, the higher the body cavity pressure, the greater the force on the gasket, which is generally a wedge shaped, soft iron, or graphite gasket wedged between the body and bonnet. Pressure-seal bonnets are used extensively for high-pressure, high-temperature applications, though normally not below ANSI class 600. Figure 16.8 shows a pressure sealed gate valve whereas Figure 16.9 shows a pressure sealed globe valve. A similar check valve is shown in Figure 16.10. Machining must be meticulous for smooth travel and accurate operation.

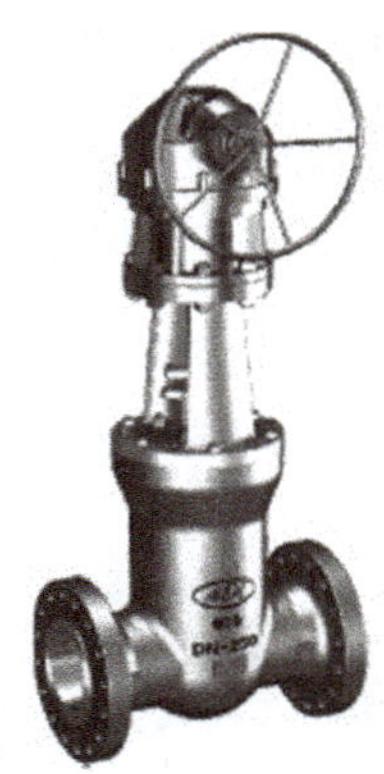

Figure 16.8
Gate Valve

Figure 16.9
Globe Valve

Figure 16.10
Check Valve

Pressure Sealed Valves

Yokes can be integral or bolted separately. The most common stem/bonnet design is the OS&Y (outside stem and yoke). In OS&Y design, the threads are outside the fluid area.

A body bonnet joint for lower class valves (Class 150 and occasionally for Class 300) is generally oval in shape. For smaller size valves, say 2" and less, it is generally square. Although this is not a standard, different manufacturers make their own standardized patterns. Classes beyond 150 have male and female joints between body and bonnet with clad spiral-wound ring gaskets. Above 600 class, valves may have ring-type joints.

SEATINGS There are a number of seating designs in gate valves, popular being solid wedge, flexible wedge, parallel wedge, or two-piece spring assisted wedges. A solid wedge is a single-piece wedge driven by the stem into accurately matched and machined seating rings. A flexible wedge is a two-piece wedge. Flexibility offers better sealing against wide temperatures and pressures. In the parallel wedge design, wedges are held tight against the seat by the help of a spring enclosed in between them.

They are designed to minimize lateral stem loading, adjust for minor misalignments, and offer a tight seating even under extreme pressure differentials. SS 410 is the most common seat and seat ring material. However, some are stellited for better wear resistance and some are PTFE coated for better sealing. Flow paths are machined to minimize flow resistance and reduce wear of the seat and seat rings or disc as the case may be.

SEAT RING Seat rings in gate, globe, and check valves are generally screwed and welded, or simply screwed. Screwed seat rings can be removed, repaired, and reused. To prevent screwed seat rings from being unscrewed while in service due to temperature fluctuations, corrosion, or vibration, they are sometimes tack welded. In many designs, tack welded rings can be removed. Seat rings have a wide seat area, whereas removable ones should have shoulder slots facilitate opening and tightening.

TRIM AND TRIM MATERIALS The word "trim" is often heard in the gate valve parlance which refers to the combination of seat, seat rings, and the associated stem. Bronze, brass, and plastic valves have the same trim materials as the body and bonnet, but steel valves can have a wide range of trims.

The most common trim materials are most 13% Cr. Seats are sometimes hard-faced with Stellite to improve their wear resistance. SS304, 316, Monel, and other materials are also used as trims. Where applications demand a positive seal, the valves are soft seated. PTFE inserts or combinations of metal and soft seats are employed.

STEM Stems or spindles in gate and globe valves are of precisely machined and polished single-piece construction. Threads are normally ACME and multi-start threads are used for faster operation. In gate valves, head formed as T at the fluid end engages with the corresponding T-slot in the wedge. The stem also features a back-seat shoulder which seats against the back-seat bush in the bonnet, when the valve is fully open. In globe valves, the disc is normally free-floating to facilitate ease of operation and alignment at all conditions. It is held on to the stem by a stem nut that allows the disc to swivel.

There are two varieties of stems: rising and non-rising stems. If the stem is rising beyond the valve handle while the valve is opened, it is a rising stem. On the contrary, if the stem does not rise, it is a non rising stem. Rising stem valves require more head room, but they give a direct indication of the valve opening. However, the non-rising valves reduce wear on the packing.

STUFFING BOX AND STEM PACKING The stuffing box is the part holding the packing that prevents leakage of fluids into the atmosphere. The chamber is accurately machined and flat bottomed for proper seating of gland packing. It is packed with a number of packing rings, normally five or six rings. Sealing performance depends on dimensional accuracy, finish of the stuffing box and the stem, and packing materials, their resilience, and number. A gland follower keeps the packing in tight condition on to the stem.

Low packing pressure cannot hold the valve against leakage. Excessive pressure makes the operation of the valve difficult, wearing out the spindle and stuffing box. To compensate for the loss of packing pressure, and to prevent possible leakage, the gland follower or the packing nut should be tightened occasionally. This could be a nuisance if the glands have to be frequently tightened. Hence, Belleville washers or disc springs are installed additionally with the gland nuts, which provide kind of a live load or self adjustment. This application is used where flow temperature and pressure fluctuations are frequent and recurrent gland adjustments are warranted.

Packing materials and their combinations are simply enormous. Presently the favored packing is braided graphite suitable up to 1200°F/649°C, though asbestos and its combinations were extensively used until recently. There are a number of other materials used singly or in combination, mostly in braided form. Braided PTFE rings are also used up to 400°F/200°C.

Classes above 300 may adopt an additional lantern ring to improve packing performance. A metal ring is provided after two or three packing rings on the stem. Leakage from the lower packing rings is taken off from this point through a take-off plug to a safer place. Additionally grease or sealing fluid can also be introduced through this port. On the other hand, a lantern ring may also feature a leak off plug which permits connection to liquid or grease seal. However, the arrangement sometimes results in scoring of the spindle and so it is used necessarily on essential services.

BACK SEAT Another aspect of good valve design is a precision machined, back-seat bush threaded in the bonnet. When the valve is full open condition, the corresponding shoulder on the spindle makes a tight closure and prevents gland leak.

FLANGE DRILLING AND FACING FINISH The flanges are drilled as per the required standards with the required gasket facing (raised face, RTJ, etc). The standard finish on the raised faces is 125–250 microns in Ra (AARH) with phonographic finish.

YOKE SLEEVE AND YOKE BUSH Yoke sleeve or yoke bush with its threads facilitates the spindle travel. They are accurately machined and assembled. These bushes may need to be removed for renewal because the internal threads may wear out. Yoke bushes can be removed without dismantling the bonnet joint if the yoke is bolted to the bonnet. On integral yokes, the bonnet should be dismantled to remove the yoke bush.

BONNET GASKETS Asbestos has lost its preeminence as the premium gasket material for body bonnet joints. Currently most of these gaskets are spirally wound with SS304 or 316 with graphite as filler, or SS304 or 316 clad gaskets with graphite inside. PTFE fillers are used for low temperature service. Ring type joints are with soft Iron or SS304 or 316 rings. Plain metal gaskets are occasionally used.

BYPASS OR EQUALIZING ARRANGEMENT Sometimes a bypass or a connection between upstream and downstream valves is arranged, particularly in case of high pressure steam valves. This balancing device is a much smaller valve of 1/2’ or 3/4’ size fixed across the main valve. This valve reduces the high pressure differential across both sides of the valve, thus bringing down the operating torque of the valve. This valve should be of the same, if not better pressure, temperature rating and metallurgy. It is attached to the main valve with both stems pointing parallel and up.

POSITION INDICATORS Position indicators are incorporated onto the valve, usually with a graduated scale to indicate the seat or disc position and, thereby. the valve opening.

LOCKING DEVICES Valves can be provided with locking devices which secure the valve in the fixed set position to prevent inadvertent, accidental, or unauthorized operation. They allow for the valves to be locked with a padlock.

CHAIN DRIVES Sometimes overhead and out-of-reach valves are fitted with chain wheels and a chain to operate the valve from the ground. They are fitted in addition to the exiting handle with a guard to guide the chain.

GEAR DRIVES Most of the large size valves are equipped with a gearbox to facilitate ease of operation. They are generally bevel gear operated, and fully enclosed. But some large or older valves may have open gears. Worm drives are possible.

ACTUATORS With the increase in demand for automation, actuators or operators for the valves are developed. They can be electrical, pneumatic, or hydraulic. They can operate in any position and are weather proof. Explosion proof and fire safe designs are also available. They normally have an overriding manual control. A typical pneumatic actuator is shown in Figure 16.11.

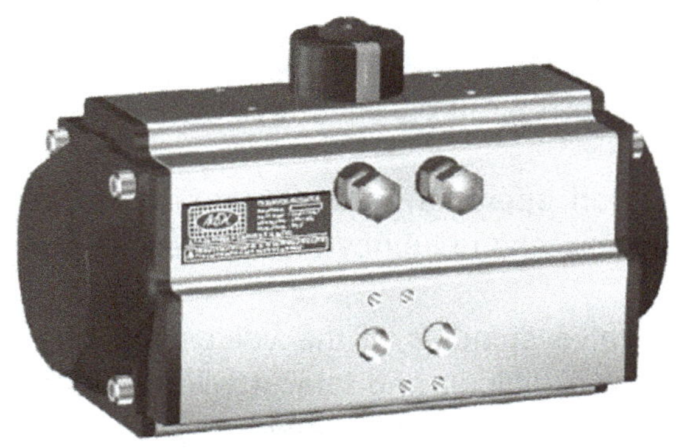

Figure 16.11
Pneumatic Actuator

Diaphragm or spring operators are pneumatic controlled. Diaphragm operators use low pressure air (generally 3–15 psi) to push the diaphragm up or down depending on the design requirement. So diaphragms can be made to close with air or open with air, which again gives certain flexibility in plant automation. It is also useful in the safety system design. Depending on the system, safety valves can be made to fail-safe operation by selecting "Air to open or air to close." This arrangement can control the valves excellently and precisely. They are standard piece of equipment for control valves.

On the other hand, piston operators use high pressure air and do away with pressure regulators. They can be double acting. Older type systems use pneumatic cylinders, directly controlling the valve stem.

Electric operators are more expensive and may require stringent enclosures against weather and explosion. However, they are useful where pneumatic supply is not available or where low ambient temperatures may affect air supplies. Hydraulic units are used where additional motive power required is required. Hydraulic controllers are not self contained as such as they need additional equipment like pumps, valves, etc.

Limit switches are appended to these operators to warn, shut down, or activate the systems.

PROBLEMS AND SOLUTIONS

As a principle, valves should be backed off by half a turn, after they are fully opened, to prevent them from jamming. Valves get jammed in the closed position. Though they should not be rammed tight into their seats, passing valves are forced shut by using leverages on the hand wheels. Sooner or later, this practice becomes a habit for the valve and the operator as well.

Operators use what are known as "valve persuaders" to close the valves tight and hard. Then it will be difficult to open them and you would need to persuade the valve to open again. These F-shaped devices are stashed all over the plant in the bottom corners of towers, etc., which are no doubt helpful in persuading the valve to open or close. But they are the culprits in jamming the valves.

Changes in temperature and pressure, and dirt in the process stream add to the problem. Over-pressurization by the fluid entering the valve cavity on one side can also create the same problem. A small bypass valve across the inlet and outlet of the valve can solve this problem, which equalizes the pressure when opened.

Galling of the mating parts, seat and seat wedge, or disc, and in the guides can be a reason for jamming in closed position. This problem crops up more frequently in stainless steels and high alloy steels.

One of the tell-tale reasons for difficulty in operation is the over-pressed gland packing. Added to this are the dirty spindles, bent spindles, and unlubricated spindles. This results in jerky operation.

Sand or dust generally settles on the valve spindles. If the valves are operated without cleaning, it will make the operation of the valve difficult and later it will make its way into the glands making matters worse. Occasionally too small a hand wheel, (generally a replacement wheel) is the reason.

Glands should not be tightened as a routine. Operators with small persuaders in their pockets and mechanics with a small slide wrench or channel lock pliers in their overall are dangerous workers!! Gland nuts should be tightened by quarter turns evenly until the leakage just stops. Pressure on the packing is applied by the packing nut or nuts sometimes by the gland flange bolting, depending on valve design, which bears down on the packing rings. Make a few turns of the hand wheel and apply a few drops of oil applied on the stem just above the packing, which will make its way through the packing. If the glands are tightened unevenly, the spindle may jam in any position. There are occasions of breakage of gland followers when the nuts are not tightened evenly.

If leakage cannot be stopped, the valve should be repacked. In properly designed valves, the glands can be repacked by back seating the valve in open position.

It is a good practice, though not practicable always, to operate the valve periodically. This reduces the chances of internal and external build up.

Body bonnet flange leak is another problem often encountered. As a matter of fact, piping designers tried to dispense with the body bonnet flange joint, particularly in the high pressure systems, to remove latent leakage sources (just as any flange joint), saving cost and weight in turn. All-welded valves are developed and used in specific services. However, the bonnet body joint stays in most of the valves due to a need for internal maintenance.

The reason for this joint leak is a bad gasket; a gasket not properly inserted, wrong bolts, and bolts not properly tightened. In extreme cases, any one of the flanges may be cracked, either due to bad material or improper loading of the flanges. The remedy is to inspect and rectify. Thin metallic or spiral wound gaskets are used. Check if they are damaged or compressed beyond their limit.

Nowadays a good number of maintenance compounds are available which can be used to cover the small scorings, undercuts, and pittings. Normally these two part compounds can be applied on the damaged area. After curing, the surface is filed or ground appropriately. Flange is boxed up with new gasket. Alternately, if allowed, the groove can be filled by welding, and filed.

Flexible graphite tape comes very handy in covering these imperfections. As a matter of fact, all maintenance engineers should keep a stock of different sizes of this tape. This tape has adhesive on one side which helps in fixing it. It is corrugated so that it can be shaped into round, oval or odd-shaped flanges. It can even be pasted onto an old gasket. Keep it on handy. It does a wonderful job.

Foreign material may be lodged under the seat rings and wedges. If the valve is opened and closed slightly several times, high velocity flow may push the foreign material out. But to leave the gate valve in this partial condition is welcome condition for valve failure as the same high velocity fluid can surely damage the seatings. In some cases, the foreign material could embed more or less permanently and prevent the full closure of valve. In such cases, the valve needs dismantling and rectification.

REPAIR

Valves are often required to be repaired as the valves are exposed to some of the toughest environments any piping component may face. But they are not worth repairing if the repair cost — whether done in situ or on a shop floor — is more than 50% of the price of a new valve. But still we would be forced to repair them due to long delivery time of a new valve, or very short downtime available for replacement.

Generally speaking, all bronze, cast iron, and plastic valves, smaller size class 800 valves, and valves below 12" size in class 150 are replaced rather than repaired. Iron valves, except for the largest sizes, are also replaced rather than repaired. However, stainless steel and alloy steel valves are often repaired because of their high cost and long lead time. It is a good idea to really repair a valve and keep it as a spare rather than buy a new valve every time. Most of the present day valves from reputed manufacturers respond well to repairs. Parts such as seats, seat rings, stems, and bushes are available with some manufacturers as spares for replacement and they do work well.

Most of the large diameter, welded end, and pressure-seal types are often repaired in the field as it is sometimes not economical or feasible to remove the valve.

There is the perennial problem of passing valves. The industry joke is, “Our valves pass, they do not fail.” To make the valves “fail to pass,” check for more obvious reasons such as foreign material.

Inspect the seat and seat rings of the valve. If the lapped surfaces of the seatings are damaged, scoured, or pitted, the valve will invariably pass. Minor scratches, and pitting, galling, wire drawing, grooving, or indentations of about 0.010” deep, can be re-lapped out and tight closure can be assured. But if the damage is worse, deeper than 0.010”, the seatings, seat wedges, and discs need renewal. In many designs, the seat rings can also be removed by chipping out the weld tacks and unscrewing them. They can be taken out, machined, lapped, and fixed back. Replacement parts are also available. In extreme case, the seats can be field welded, ground, and lapped in position but this job calls for precision, skill, and availability of certain facilities at site.

As a general guideline, the following procedure may be adopted for lapping. However, it should be noted that the valve industry abounds with a lot design variations and a suitable method must be adopted. Special fixtures and guides may be needed for lapping check, globe, and angle valves. Gate valves may need lapping straight through the flanges after making a fixture.

Apply a thin coat of coarse lapping compound on the seating surface. Use a fine grade if the damage is considerably less. Insert the disc and lap with an oscillatory motion. Withdraw after a few minutes of lapping and wipe the surfaces clean. The extent of the damage will now be more clearly visible. If machining is required, take a so-called clean cut and start lapping. If the damage is only minimal, use a fine compound and lap until satisfactory results are shown. Invariably, finish with fine compound.

Whenever a gate valve is disassembled for inspection or cleaning, make sure to mark the wedge for proper orientation with respect to the valve body. When reinserting the wedge back into the valve, make doubly sure that it is inserted in the same way. Failure to do so will definitely make the valve pass, and make a jarring and binding operation.

Inspect the seating (threads in the valve body) and seat threads in the gate, globe, angle, and check valve for any damage. Specific seat tools may be required for working on them as every manufacturer has its own design. It is a little tricky to fix them back as only few threads are available in a far away and confined space. Screw them and unscrew them and check for a proper continuous contact and a tight seal.

After machining, grinding, and lapping (wherever they are carried out), check for seating contact by gently tapping the wedge into position. Remove it and check for the contact mark. There should be even, smooth, and round markings on the seat and seat ring.

The valve can be checked before a hydro-test by smearing Prussian blue on the contact surface and lowering the wedge. Too much blue would smear all over rather than show up the contacts.

If the test pressures are not reached easily or if the valve passes during testing, the most likely reason is improper venting of air. The valve should be filled completely with test medium like water and all the trapped air must be vented out.

However, if the damage is extensive due to corrosion, erosion, or abrasion, it is worthwhile to renew the valve unless there are production and purchase constraints.

The write-up on this seating repair holds good for most of the check valves also.

When the valves are used repeatedly over the years, there is a likelihood of wear-out of the bush. The bush has internal threads through which the spindle moves. In some larger models, this bush can be removed without dismantling the valve because it is fixed in a bolted yoke. Spare parts are available for replacement. On the other hand, these bushes can be machined out in the maintenance machine shops in most of the industries. Threads are normally acme or square and generally multi-start.

In the gate and globe valves, the slot in the gate or disc where the spindle sits may be worn out or elongated. Then the spindle just turns without moving the seat up or down. The body bonnet flange must be opened to repair this condition. The gate should be released and should be taken off the seats first; otherwise, it will be difficult to lift the gate out of position. Check for the worn out part — whether it is gate, disc, or spindle — and take corrective action accordingly. These parts are replaceable. If they are not available, the worn-out portion can be suitably welded and corrected. But before attempting this, check if the actuator is working properly or not.

Then there is gland packing, a subject which is addressed in Chapter 17, Control Valves.

CHECK VALVES

Check valves are non-return valves which are used for an automatic prevention of back-flow or reverse flow in pipe lines. A pump can often run in the reverse direction due to reversal of flow. As a consequence, some parts like the impeller can work out loose. By a sudden reversal of flow, pipe lines can experience what is known as water hammer, a violent vibrating condition in the lines. Reversal can create flooding, mix up the fluids, break the vacuum, or force system shutdown. It is necessary to protect downstream equipment from mechanical damage. In operating principle, all check valves conform to one of two basic patterns. The valve is kept open by the flow while it is closed by gravity, by a spring force, or reversal of flow. Check valves are made in almost all sizes and with almost all materials. Following are some of the types of check valves.

SWING CHECK VALVE

Figure 16.12 shows a typical swing check valve; its parts are shown in Figure 16.13. Like in gate valves, flow moves through these valves just about in a straight line. A disc is hinged on to seat; hence, it swings. The disc is fastened to the hinge and locked in place by a lock nut and pin to prevent disassembly during service. The disc is normally free to rotate and the pin is wear resistant. Forces of the flow push it up and open, making a low pressure drop path. In case of lack of flow, the disc drops onto the seat by gravity, whereas reversal closes it. However, sudden slamming of the hinged disc may create turbulence and water hammer into the system. These valves can be used in the horizontal and vertical (flow upwards) position and they do not restrict flow.

Closing the valve externally is also possible in some models.

Figure 16.12
Swing Check Valve

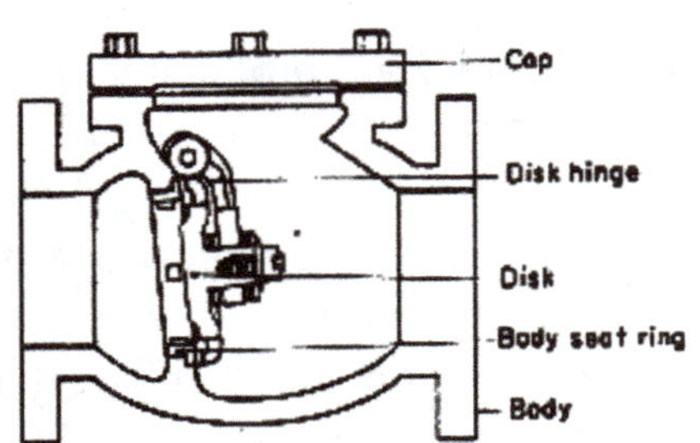

Figure 16.13
Parts of Swing Check Valve

WAFER DISC CHECK VALVE

Wafer disc check valves are a variation of swing check valves with an improved speed of operation and duty (see Figure 16.14 and Figure 16.15).

Figure 16.14
Wafer Disc
Check Valves

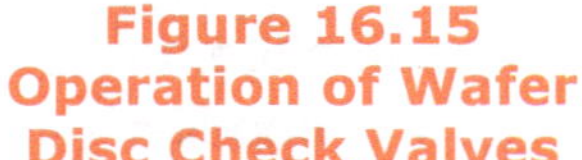

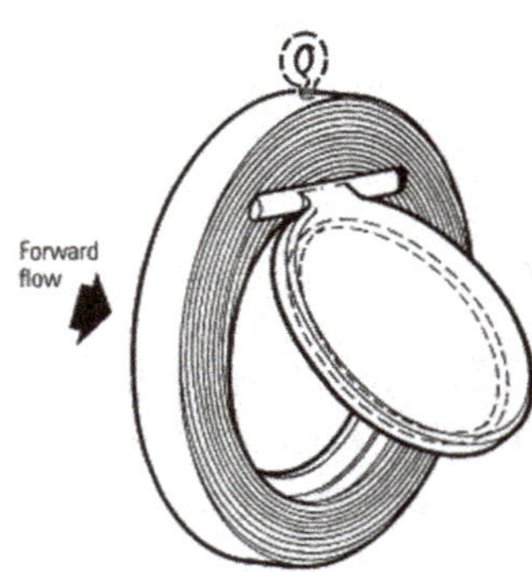

Figure 16.15
Operation of Wafer
Disc Check Valves

LIFT CHECK VALVES

There are variants in lift check valves. The variations are ball lift, disc lift, or piston lift check valves. They are lifted by the forward flow whereas reverse flow and gravity close them shut (see

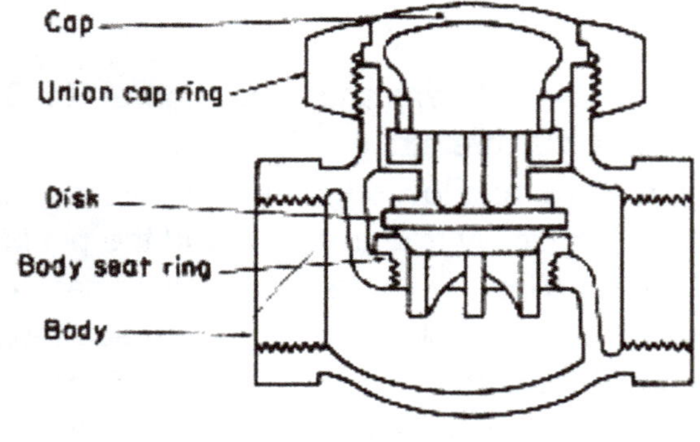

Figure 16.16
Lift Check Valve

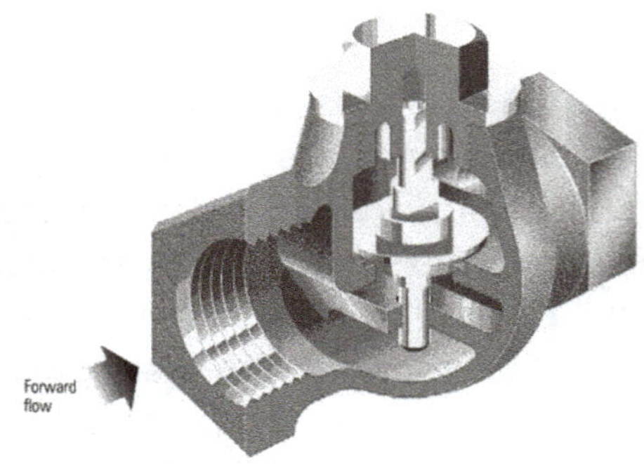

Figure 16.17
Operation of Lift Check Valve

Figures 16.16 and 16.17). They are used for more exacting services. But flow through lift check valves is in a changing course like that in globe and angle valves.

Lift check valves are like globe valves, with the exception that the operation is automatic in as much as the reversal or no flow condition closes the valve. A cone is retained in position on the seat rings by spring, where the flow in the right direction opens it up. It can only be used in horizontal position; large size valves are not available (typically limited to 3" NB).

BALL CHECK VALVE

A ball instead of a piston is positioned in the valve, which closes on the reversal of flow or by gravity. Often spring supports closure. Normal flow opens it up. It is a vertical valve, generally used in liquid applications. It is a routine feature on most of the injection pumps where they are positioned in the suction and discahrge nozzles. Figures 16.18A and 16.18B illustrate the closing and opening respectively of the valve.The lined ball valve is shown in Figure 16.19.

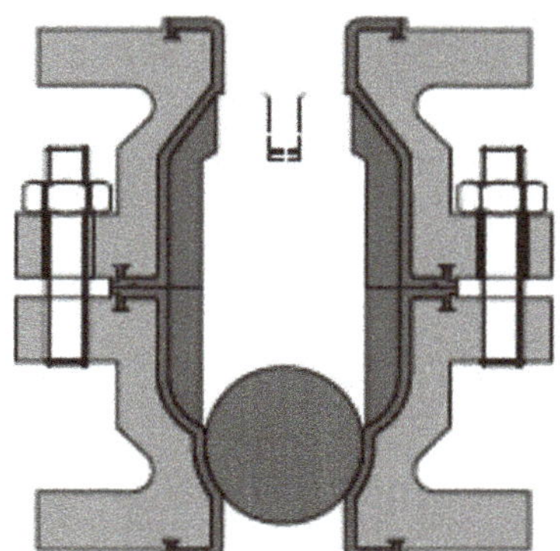

Figure 16.18A
Ball in closed position

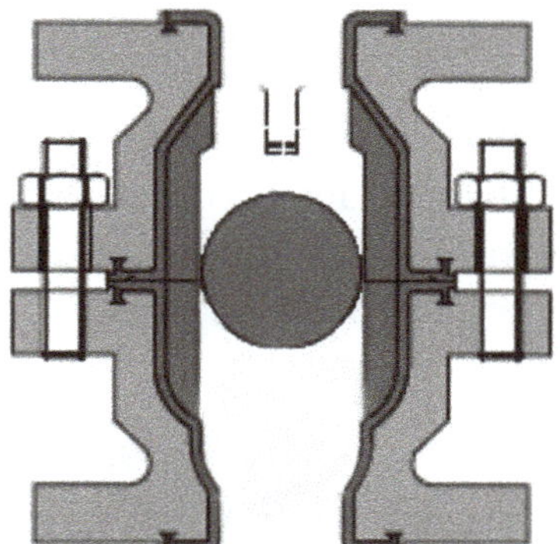

Figure 16.18B
Ball in open position

Figure 16.19
Lined Ball Check Valve

DISC LIFT TYPE

A disc is retained in position by a spring in this type of valve. When the fluid pressure is more than the spring force, the disc moves up and opens the path (see Figure 16.20). In case of no flow or reversal flow, the disc closes and restricts the flow. It can be installed in any direction.

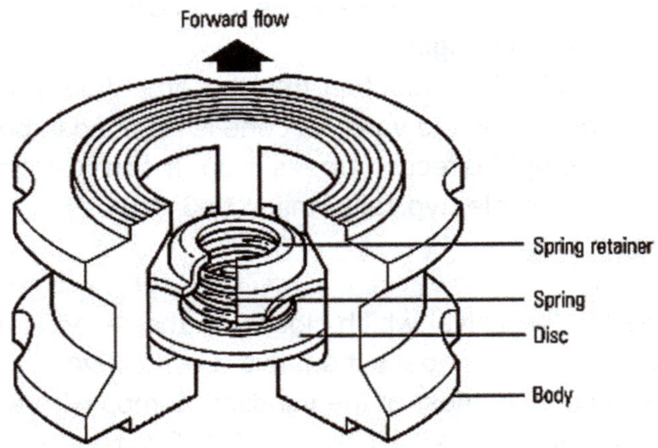

Figure 16.20
Disc Lift Check Valve

PISTON CHECK

The piston-type lift check valve is a variant of a lift check valve. A piston check valve makes use of a piston instead of a hinged disc (see Figure 16.21). Unlike swing check valves, these valves provide some cushioning effect while operating. However, they can only be used in a horizontal position. The flow path is similar to that of globe valves and, hence, pressure drop is high. Sometimes dash pots are installed in conjunction to dampen the surges due to frequent operation of these valves.

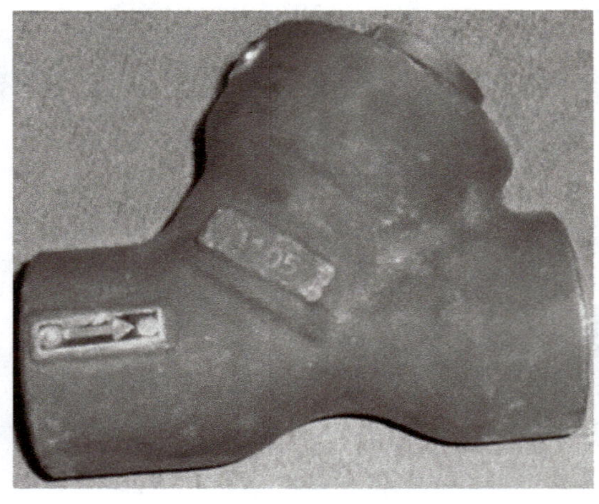

Figure 16.21
Piston Check Valve

DUAL FLAP CHECK VALVES

These check valves are of twin wafer design and fit between flanges. The valve discs are in two halves fixed side by side in the bore and hinged to a pin or shaft. A stainless steel spring holds the two halves in closed position. Flow opens up the wafers and reversal closes them swift and fast so much that water hammer is virtually eliminated. Figure 16.22 shows a general view of the valve and Figures 16.23A and 16.23B show its operation. An internal view is seen in Figure 16.24.

Figure 16.22
Dual Flap Check Valves

Figure 16.23A

Figure 16.23B

Operation of Dual Flap Check Valves

Figure 16.24
Internal View of Dual Flap Check Valves

AIR CHECK VALVES

Air check valves are specifically designed for compressed air services. The disc is a flat stainless steel and a spring holds it in position on the seat (see Figure 16.25). The spring closes the valve by holding the disc against the seat while the compressor is idling or unloading. This valve should be located at least 10 ft or 3 meters from the compressor.

Figure 16.25
Air Check Valves

STOP CHECK VALVES

These are basically lift check valves which work like any other check valve. Additionally they have a provision to close the valve shut by operating the handle. In effect, the disc is disconnected from the spindle so that it can act freely, open up upon the flow, and close at reversal.

NON-SLAM CHECK VALVE

In applications where the sudden closing of a check valve is not desirable, non-slam type check valves are used (see Figure.16.26).

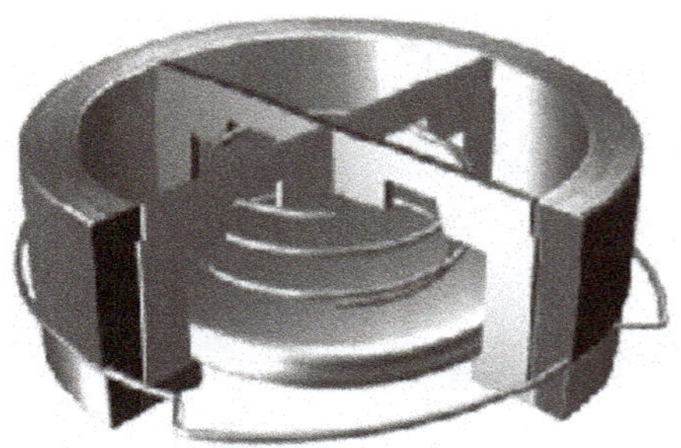

Figure 16.26
Non-slam Check Valve

FLAP CHECK VALVE

A flap-type check valve is simply a check valve with a rubber flap as the non-return medium (see Figure 16.27). It is generally used in low pressure and non critical services. A variation of this is the good old foot valve.

Figure 16.27
Flap Check Valve

PROBLEMS AND SOLUTIONS

Similar problems as in gate and globe valves occur in swing check valves. The flapper is hinged to a shaft inside the valve. Shafts that hold the flapper or the swing may break, wear out, or get jammed. Broken shafts do not behave, whereas worn-out shaft misbehave. The top cover may be opened to check the internals. When lifted and dropped, the swing should fall back freely and squarely. The shaft can generally be removed by removing the plugs on both sides.

Check for scorings on the seatings that will pass the valve. Comparatively, check valves operate under more severe conditions because the disc or flap slams back every time there is a closure or reverse flow. Seats take the beating.

Debris is a common problem. More of it is collected at the small crevices of pin and disc movement. The valve gets struck, and may not open when required. This takes a lot of system hammering. If it does not close on the reverse flow, it will not fulfill the purpose intended. Debris and scaling pose problems more severely in the case of lift check valves.

Leaks from the top cover joint can be attended in much the same was as gate valves.

For repair of seat rings, please read notes on repairs in the previous section about problems and solutions for gate and globe valves.

PISTON VALVES

Richard Klinger is credited with designing the first piston valve in 1922. It maintains the flow pattern of a conventional globe valve, but the disc is replaced with a cylindrical piston and two elastic, replaceable piston rings that actually affect the seal (see Figure 16.28). It is virtually a tight shut-off valve and does away with glands; hence, it is known as a glandless valve.

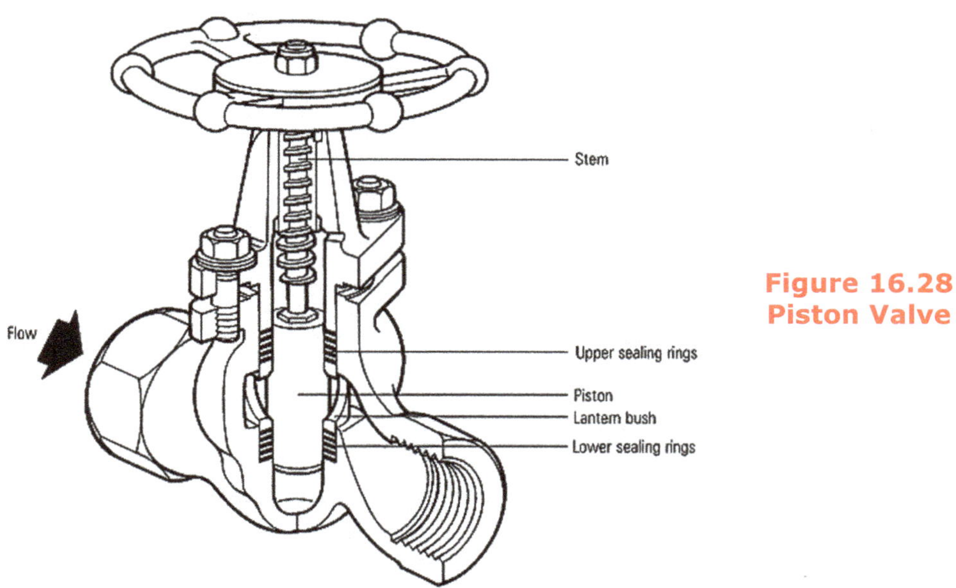

Figure 16.28
Piston Valve

These valves are also easy to operate and easy to maintain. The piston is moved through two sealing rings — the upper rings act like a gland, and the lower operates like a seating prevents the valve from passing. A lantern bush separates both rings and keeps them in position. Internals can be removed without dismantling the valve from position. These valves are available at temperatures up to 750°F/400°C at pressures of about 1000 psi or 75 bar. They are made in a number of metallurgies and materials of piston rings to suit the services. Replacement parts are available and both the seating rings can be renewed if valve passes.

BELLOWS-SEALED VALVES

Another design variation which does away with glands is the bellows-sealed valve, where flexible metallic bellows are used. Bellows are connected to the bonnet on one end and the other end to the stem. Bellows do what they always do, expand and contract with stem movement which in turn affects a positive seal, isolating the stem from environment. Bellows are prevented by a locking device to avoid rotation along with the stem. See Figure 16.29 for a cutaway view of the bellow sealed globe valve. Design of the bellows-sealed gland can be seen in Figure 16.30.

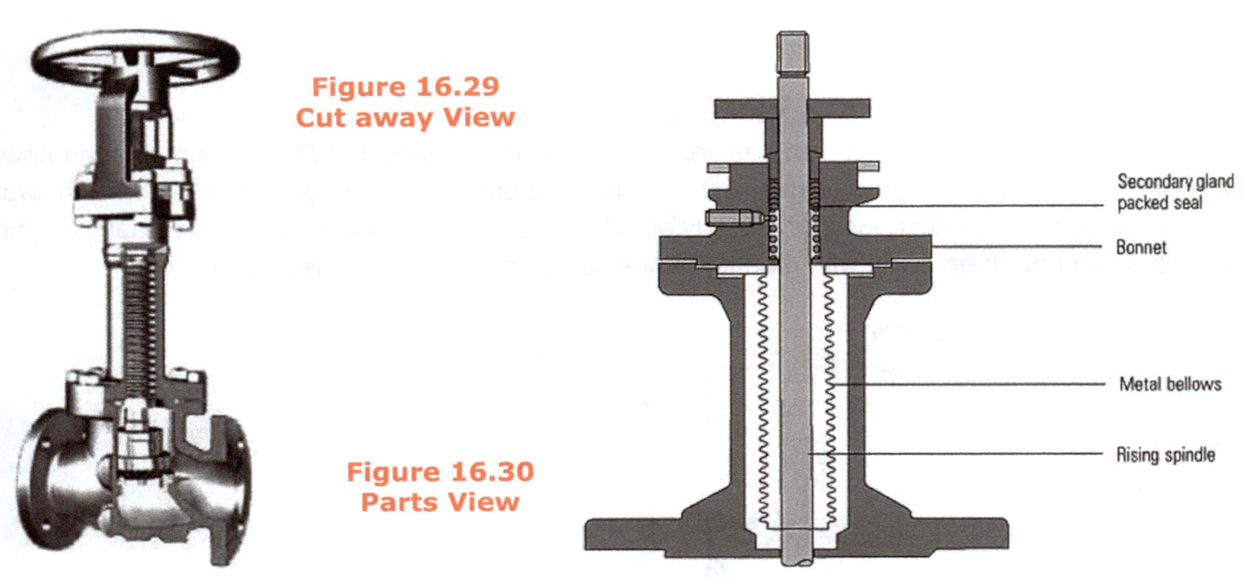

Figure 16.29
Cut away View

Figure 16.30
Parts View

Bellows-Sealed Valves

NEEDLE VALVES

Needle valves are some of the most widely used valves in the industry, aerospace, home, and laboratories. Similar in design and operation to the globe valve, a needle valve has a long tapered point at the end of the valve stem instead of a disk (see Figure 16.31A). Needle valves are a standard piece of equipment as a shut off valve for the pressure gauges, as they can deliver surgeless flow. Flow takes a 90° turn through the valve. The long taper of the valve seat and very fine threads on the stem permit delicate and precise control of the flow, which makes it more suitable as a throttle valve. However, it is better to avoid these valves for viscous fluids or slurries as the small flow orifice can be easily plugged with viscous fluids or solids. Valves are available in small sizes with metal-metal, plastic–plastic, or plastic–metal needles and seats. A cross-sectional view of a needle valve is illustrated in Figure 16.31B.

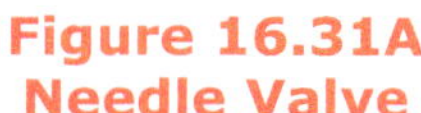

Figure 16.31A
Needle Valve

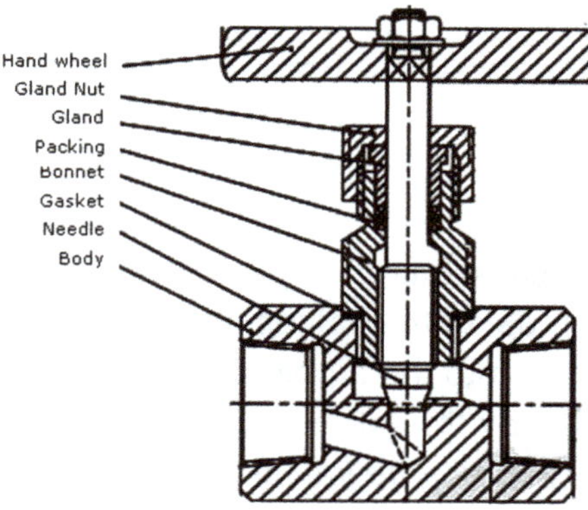

Figure 16.31B
Sectional View of Needle Valve

QUARTER TURN OR ROTARY VALVES

Plug valves, ball valves, and butterfly valves fall into this category. Quick opening, block in both directions, absence of gland packing as such, and tight shut-off are some of the characteristics of these valves. Because of straight through flow, the pressure drop is less. However, the quick opening feature may be disadvantageous as it may create certain water hammer in the system. These valves have a full port design which is the same as the line size or designated valve size. In the reduced port design, the valve inlet and outlets are reduced.

The plug valve is probably the oldest valve to be made. Brass valves with a conical plug with its hole in the center are a common site. Glass stop cocks in the laboratory are another example. Primarily an on-off valve, there are a number of variations in the basic plug valve. There is a cylindrical or tapered plug with a hole in the center. The plug rotates in the body. A quarter turn aligns the hole in the plug in line with the flanges, which permits the flow. Another quarter turn closes the valve. Valves are made with full bore and reduced bore. A handle is attached at the larger diameter end of the plug with a mechanism to turn it by a lever or gear reducer.

With this basic design, we have lubricated plug valves, lined plug valves, and multi-port plug valves. There is typically no bonnet, but a cover over the plug which holds the plug in position and allows free movement of the plug. There is a gasket for the cover and a sealing at the handle.

LUBRICATED PLUG VALVES

Lubricated plug valves are widely used in the petrochemical plants. A derivative of good old plug cock, in the lubricated plug valve, a special lubricant is pumped onto the walls of the plug, generally over grooves specially designed to seal. The famous Pascal principle is made use of and these valves effectively seal against a very high pressures in petrochemical services. The lubricant seals reduce friction by actually lifting the plug and prevent wear out across the sealing surfaces. The plug is designed with grooves that retain a lubricant to seal and lubricate the valve. It also provides a hydraulic jacking force to lift the plug within the body, thereby reducing force required for rotary operation. Sealing is effectively done by the presence of a sealing compound between the plug and the valve body. Figure 16.32 shows a flanged lubricated plug valve and Figure 16.33 a screwed valve. For parts of the valve, check Figure 16.34.

Good numbers of lubricants are developed specially to be compatible to the demands of various services. So if the lubricant is not meant for the service, the results may not be satisfactory. Older valves had a screw fitting on top of the stem to drive the sticky sealant lubricant into the valve, but presently most of the valves have a button head grease nipple. An internal check valve prevents process flows into the nipple area. A lubricant stick is fixed inside the grease gun and is pumped into the valve by the grease gun. Some of the grease guns are hydraulically assisted to pump sealant into high pressure valves. A sealant gun injects sealant down the center of the plug, out the radial holes. To prevent leakage through the stem or spindle, a special stem packing is screwed on to the stem. See Figure 16.35 to look at the injection of sealant into the valve.

Figure 16.32 Flanged Valve

Figure 16.33 Screwed Valve

Lubricated Plug Valves

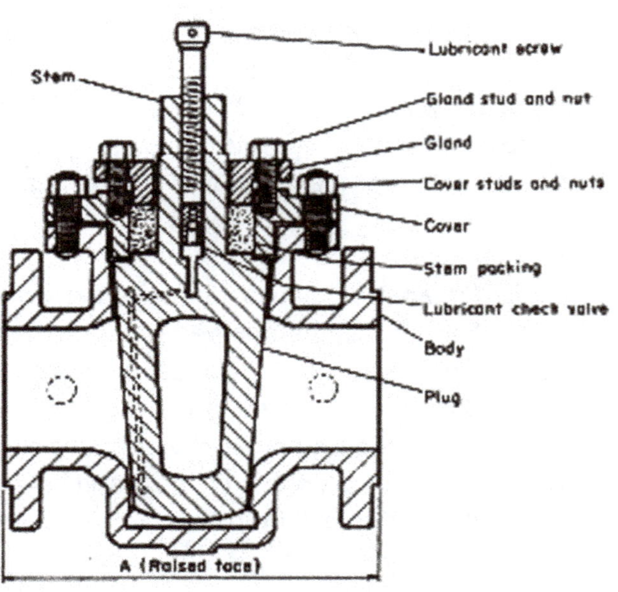

Figure 16.34 Parts View of Lubricate Plug Valve

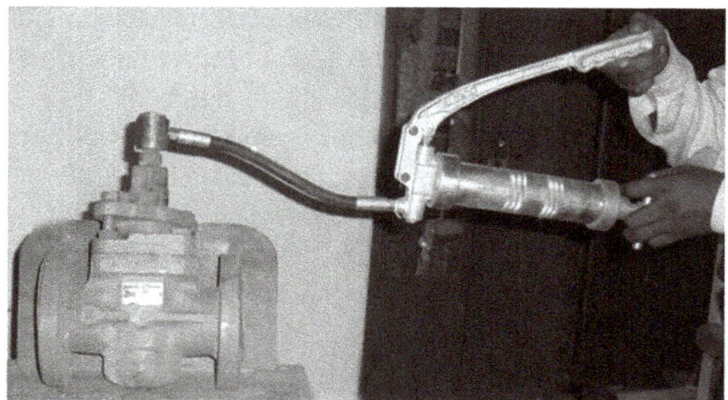

Figure 16.35
Sealant Injection

Lubricated plug valves are also available in multi-port design in three or four ways. In these valves, flow from any one port can be directed to any other port as per the requirement, or even connect all three ports together (see Figure 16.36).

Figure 16.36
Multi Port Lubricated Plug Valve

PROBLEMS AND SOLUTIONS (PLUG VALVES)

Generally these valves work reliably during their life time and sealant works wonders in these valves. Normally there are two bad conditions: passing of the valve or leakage through the stem.

If the valve passes, inject the sealant. Injection of the sealant can be done in the existing condition and does not require the valve to be closed or opened. In the worst case, a few shots may be required. Injection of the sealant also frees up the valve, requiring less torque for operation.

Occasionally the service medium may start leaking through the grease nipple. This means that the internal grease has solidified or washed away and the internal check valve is not holding. Carefully slide the sealant gun adopter onto the grease nipple and pump in grease until a satisfactory condition exists.

If the valve gets jammed, the same treatment will free it up. If not, in both cases, the valve may have to be dismantled from service. Open out the plug, thoroughly clean the flutes, and drill holes.

Please be careful as the plug valves can trap the service fluids effectively. If the valve is dismantled unwittingly, there may be splash of the fluids like ammonia or LPG. and may cause an accident.

If the drill holes are plugged, they have to be cleared. Clean the body. Clean the stem packing area and holes. Check and apply fresh sealant. Apply the same sealant that is used on the service. Assemble the plug back, carefully inserting each component back into its position. There are various designs with springs, diaphragms, and other components. Follow them properly and without fail. If there is a bottom jack screwing to position the plugs, it is wise not to disturb the setting. Fix the top cover with appropriate packing and bolt it. Check for freedom of operation. Apply the sealant again after fixing the button head. Apply stem packing. Stem packing is generally applied by screwing it.

If the plug or body is worn out, it would be better to renew the vale rather than repair it. Most of the reputed manufactures coat them with special materials.

There instances when the drive coupler cracks. When excessive pressure is applied on the valve stems by using extensions or cheaters, the drive coupler may break. The valve needs to be taken out of service, but not necessarily out of position. The top cover is opened and the coupler is renewed.

SLEEVED PLUG VALVES

Sleeved plug valves are extensively used in acid and slurry service where they are very effective, offer tight shut off, and full flow. Valves are generally made with anti corrosive materials like stainless steels to suit the service requirements. Unlike lined valves, here only a sleeve surrounding the plug is inserted tight into position and locked. The plug rotates by a quarter turn in this sleeve, which is more often Teflon. These are rarely available beyond Class 150. Smaller sizes are wrench operated whereas larger sizes are gearbox driven. These valves are also available in multiport design.

Figure 16.37 shows a sleeved plug valve and Figure 16.38 a cutaway view. A picture of a multi port sleeved plug valve is seen in Figure 16.39.

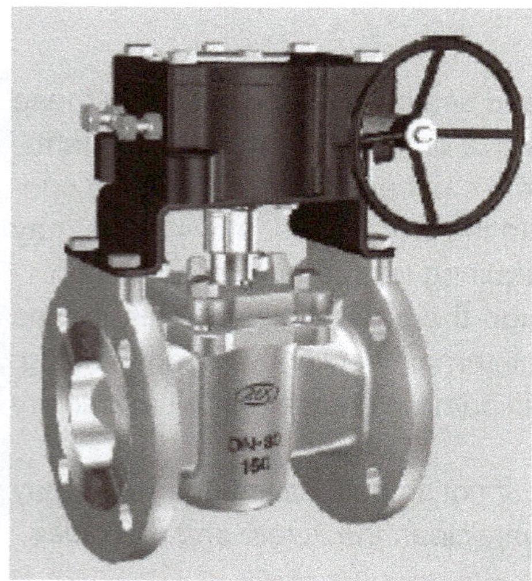

Figure 16.37
Sleeved Plug Valve

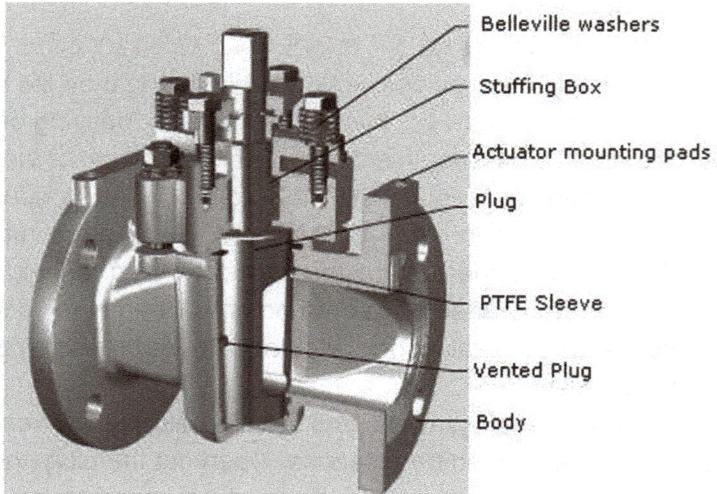

Figure 16.38
Cutaway View of Sleeved Plug Valve

Figure 16.39
Multi Port Sleeved Plug Valve

BALL VALVES

The ball valve is similar in concept to the plug valve but uses a rotating ball with a hole through it. It is becoming more and more popular with a greater number of valves sold particularly in the smaller sizes. It has virtually replaced gate valves in the instrument take-off lines. A ball inside the valve can be turned against soft seat rings or metal seats by turning a handle attached to it. This falls into a family of quarter turn valves which is closed or opened with a quarter turn of the handle. When the port or hole in the ball is in line with the valve ends, flow occurs. The valve is closed when it is brought perpendicular. The handle position helps in knowing whether the valve is open or closed. They offer perfect shut off, but do not allow for the flow control. On the other hand, operating the valve not in full shut off or full open condition is detrimental to valve seats. The ball may get locked in that position. As the ball moves in a wiping action across the seat rings, this valve handles suspended solids well, but abrasive solids can damage the seatings. They are made in one, two, three, or four piece construction, with top or bottom or even welded entry. Figure 16.40 shows a three-piece screwed ball valve. Figure 16.41 shows a three-piece flanged ball valve. Figure 16.42 shows a single-piece flanged ball valve.

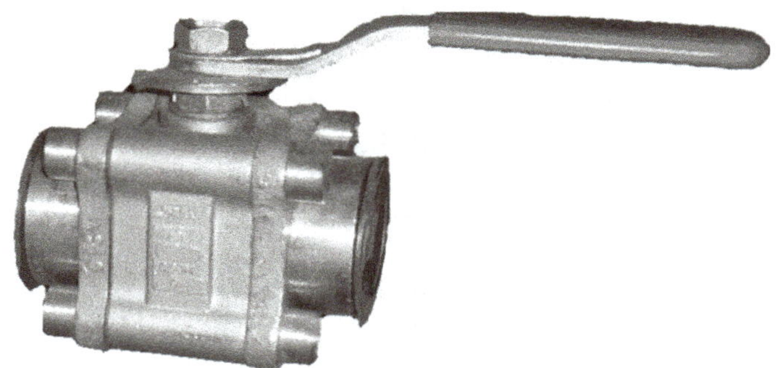

Figure 16.40
Three-Piece Screwed Ball Valve

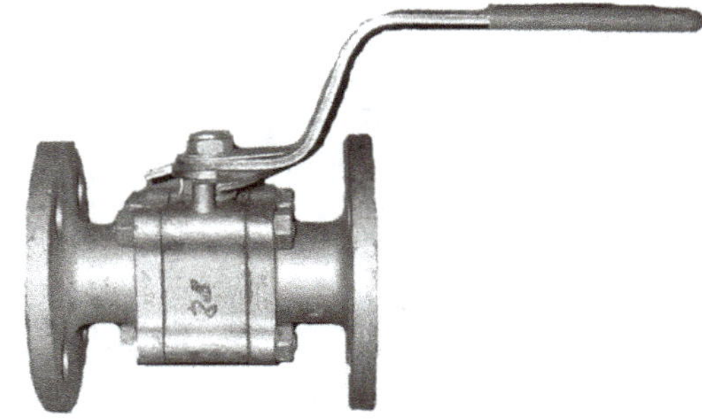

Figure 16.41
Three-Piece Flanged Ball Valve

Figure 16.42
Single Piece Flanged Ball Valve

Ball valves are available in various MOCs. Ball valves are easily made in multi-port designs. Three-way ball valves direct the flow to either one or the other or both sides, or are closed off completely.

Ports are available as full, standard, and reduced ports. In a full port ball valve, the hole in the ball is the same size as the pipeline. The valve is larger but the flow is not restricted. In a standard port ball, the valve ball is smaller and the port is also smaller. Though it is cheaper, it results in a slightly restricted flow with a higher pressure drop. In reduced port ball valves, flow is further restricted.

Ball valves are best suited for frequent and quick operation and tight shutoff applications. Low-pressure drop, ease in mounting and maintenance, and low cost are some of the features. Control of flow is not good. Ball valves are presently made up to very high pressures with metallic seals. They are compact, low cost, reliable, low weight, and easy to install and maintain.

Ball valves are made with a floating ball for normal duties and with a trunnion-mounted ball for heavier duties. See Figure 16.43A for floating ball design and Figure 16.43B for trunnion-mounted design. For corresponding pictures, refer to Figures 16.44A and 16.44B respectively.

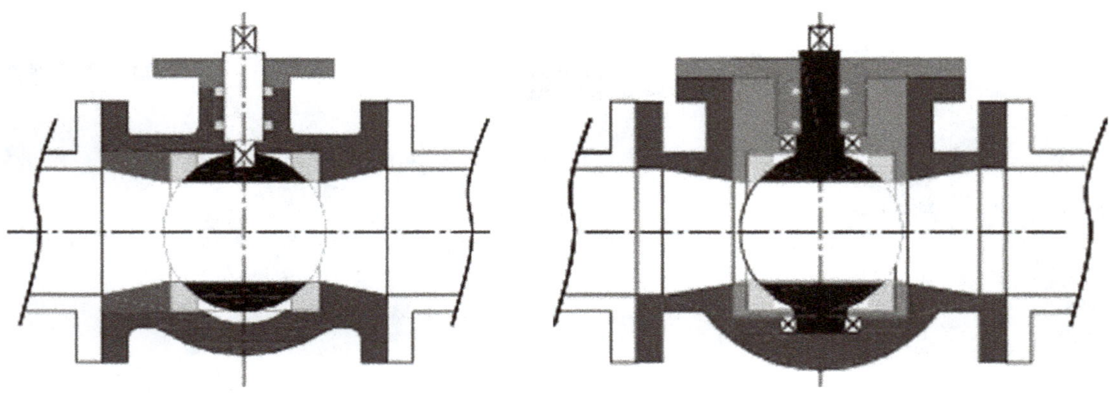

Figure 16.43A
Ball Valve–With Floating Ball

Figure 16.43B
Ball Valve Trunnion Mounted

Figure 16.44A
Ball Valve–With Floating Ball

Figure 16.44B
Ball Valve Trunnion Mounted

Ball valves are made in three designs basically specifying how the ball is inserted and can be maintained. Figure 16.45 shows expanded views of these designs.

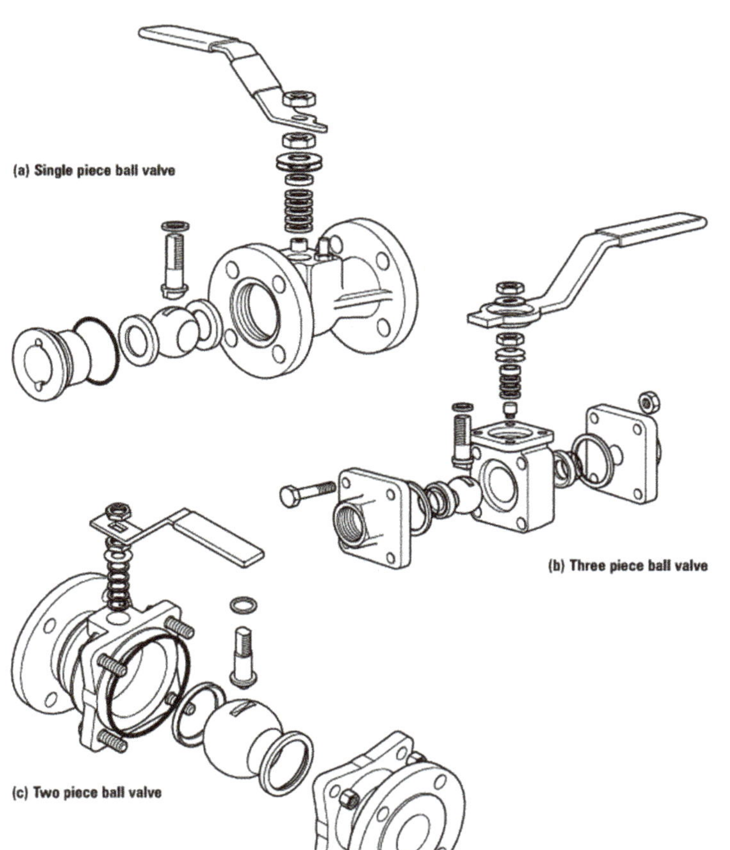

Figure 16.45
Expanded Views of
Ball Valves

In the single-piece design (a), the ball is inserted along the valve axis. One side of the flange is fitted with a screwed insert, which can be removed. Though this valve does away with additional seals, it needs to be removed from the line for servicing. In two-piece (c) and three-piece (b) design, the valve is split in two or three pieces along the plane of the flange. This design facilitates easy and in-line maintenance; though it requires additional seals on the split facings. In the valves with top entry design, the ball can be inserted from top.

Ball valves adapt very well to multi-port designs. Figure 16.46 shows a typical three-way ball valve. Ball valves offer excellent service in buried installations (see Figure 16.47). Please note the leak ports taken out, which indicate the flow condition.

Figure 16.46
Three-Way Ball Valve

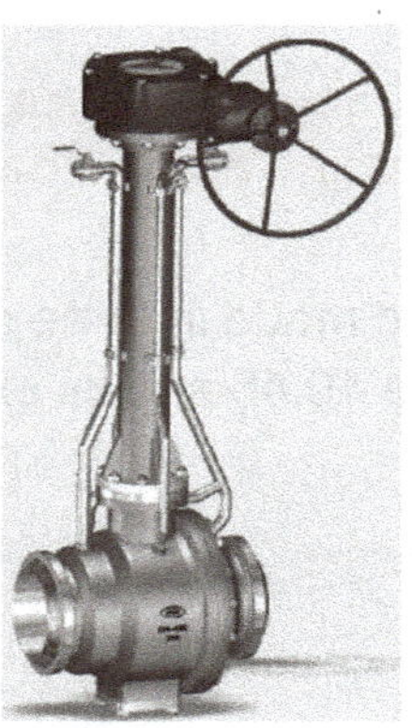

Figure 16.47
Ball Valve-Buried Version

PROBLEMS AND SOLUTIONS (BALL VALVES)

Ball valves rarely pass, but they do. Throttling the valve is not a good practice; it will surely damage the lips of the seat rings. The ball also may suffer from scratches and wear out due to the flow. In the partial operation, debris may also collect in the body, which would further spoil the valve.

Ball valves can be easily serviced. It is economical to service them because they are usually made with costly stainless steels; fortunately they respond well for repair. Generally the seat rings on both sides of the ball are the only parts that may need replacement when the valve passes, along with the gaskets. Please refer to the expanded view of the valve in Figure16.45.

In the single-piece design, the valve needs removal from service. Remove the end cover or insert with its gasket, which is generally screwed onto the end flange face. You may need a special tool to open it up. Remove the top bonnet cover with its gaskets, gland ring. Decouple and withdraw the drive coupler or shaft. The ball can now be removed from the side now with the seat rings. Renew the seat rings and assemble the valve back in with all its parts. Generally all the gaskets, seat rings, and packings also need to be renewed.

In two-, three-, or four-piece design, the valve needs to be removed from service. The numbers of joints are bolted with appropriate gaskets. It would often suffice to open any one-side joint.

Open them as appropriate and renew the seat rings, gaskets, packing, etc. Assemble back. In some designs, the valve can be left in place in the piping; the central ball holder with ball can be lifted out.

It is worthwhile to renew the soft packings because most often they are Teflon.

In some top entry designs, the valve does not necessarily need to be taken out of position, but must necessarily be taken out of service. By opening the top cover, the ball can be taken out and the seat rings renewed.

Most of the ball valves are now anti-static and blowout proof. These have small additional components which are seemingly innocuous. Do not discard them as their service is needed in prevention of static electricity build up and in blow out protection.

Balls can get corroded or eroded and worn out. They are worth renewal. However, they can be welded and rebuilt particularly in larger sizes if true round machining can be achieved.

BUTTERFLY VALVES

A butterfly valve is also a quarter-turn valve like a ball valve. A flat circular plate or disc is positioned in the center of the valve, which is connected to the handle through the shaft. Turning the handle makes the disc either parallel to the flow of fluid, allowing it to flow across, or perpendicular to it, thereby shutting off. Present day butterfly valves offer tight shutoff and very rugged design. With the advantage of simple design, construction, small size, and light weight, it is steadily replacing the gate valve. Once relegated to water services, butterfly valves are invading into other areas.

Butterfly valves are best suited for frequent operation for shutoff applications and for throttling because of their percentage control of flow. Low pressure drop, good control of flow, ease in mounting and maintenance, and low cost are some of their other features.

A handle-operated valve is shown in Figure 16.48A and a gear-driven valve in Figure16.48B.

Figure 16.48A Hand Operated

Figure 16.48B Gear Driven

Butterfly Valves

Butterfly valves have flexible rubber seats whereas some have a metal seat design. A flexible seat is either removable or permanent. These resilient butterfly valves are normally limited to a working pressure of 200 PSI or 15 bar. Metal-seated valves can handle pressures close to 1450 psi or 100 bar at high temperatures, whereas the valves with elastomers are rated at their temperature limits. Once relegated as passing valves, PTFE overcame most of the limitations of the elastomers. However, for operating at higher pressures and temperatures, metal seated valves were developed. There were extensive developments in the metal-seated valves; offset valves with bubble free leakage came into the scenery.

However the disc presents an obstruction to the flow of the fluid. At higher ratings, it is unfortunately thicker. If the butterfly valve is closed suddenly in liquid service, water hammer may result.

Butterfly valves come in two body styles: lugged or wafer. More commonly used wafer-style valves are installed between two flanges with bolts or studs extending across the flanges. Valves can be removed by removing half of the bolts and loosening the others. Lugged valves have threads at both sides of the valve body. These valves are fixed between two flanges using a separate set of bolts for each flange. This setup permits either side of the piping system to be disconnected without disturbing the other side. With this lugged valve, it is possible to leave the valve in place with its bolting to one flange, while the other end of flange and piping can be moved out.

A wafer valve is shown in Figure 16.49A and a lugged valve in Figure16.49B.

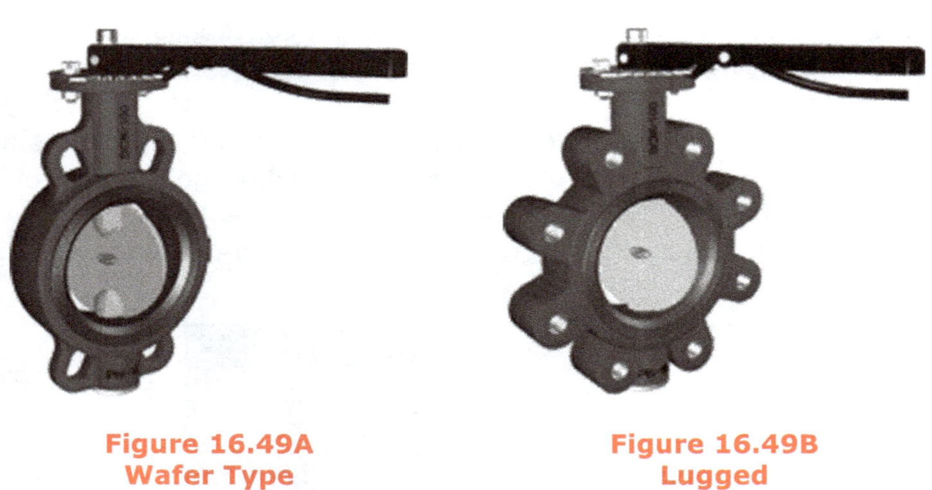

Figure 16.49A
Wafer Type

Figure 16.49B
Lugged

Butterfly Valves

High performance butterfly valves usually have a double eccentric design which can operate up to 725 PSI or about 50 bar. They do away with seat and disc contact in open or intermediate positions, and eliminate wear points. Tricentric butterfly valves can reach working pressures of 1450 PSI or about 100 bar. In the triple-offset butterfly valve, three separate offsets are designed. The centerline of the bore and the centerline of the disc/seat sealing surfaces are two of the offsets. The third offset — in the axis of the seat cone angle — minimizes rubbing of the seat/seal contact surfaces during operation and maintains sealing integrity during the life of the valve. Triple-offset valves have a high cycle life, low operating torque, and low maintenance costs.

High performance, double eccentric design valves are shown in Figure 16.50. Figure 16.50A shows the actuator-operated valve whereas Figure 16.50B is the gear-operated valve. Triple eccentric design valves are shown in Figure 16.51 where Figure 16.51A shows the gear-operated valve and Figure 16.51B shows the actuator-operated valve.

Lined butterfly valves are also available. See the section below on lined valves.

Figure 16.50A
Actuator Operated

Figure 16.50B
Gear Operated

Double Eccentric Butterfly Valves

Figure 16.51A
Gear Operated

Figure 16.51B
Actuator Operated

Triple Eccentric Butterfly Valves

PROBLEMS AND SOLUTIONS (BUTTERFLY VALVES)

By and large, butterfly and ball valves live without many problems. Occasionally the valve may be jerky or may require too much torque; in the worst case, the disc or ball may not turn at all with the movement of handle or the actuator.

The common reason for jerky operation is debris in the valve, dirty sealing rings. Opening and closing the valve several times will normally push out and remove the debris. But it is not a good practice to keep the ball valve in the intermediate positions. Packing that is too tight requires excess torque for operating. The shaft could be bent. Check!

If the valve is operated by a pneumatic actuator, the pressure may be low or the problem may be inside the actuator. The actuator mounting may be bad; and mounting bolts may be loose or missing. This misaligns the shafts, will lead to jerky operation or high torque operation. Operating under these conditions will eventually lead to broken shaft, keys, or separation.

If the valve is not operating totally, the possible reason is the key is sheared and, in the worst case, the shaft is sheared. A very high level of debris within the valve port is also a cause.

In the ball valves, the slot in the ball where the spindle sits may be worn out or elongated. The actuator may not be working properly.

In butterfly valves, the shaft key is the problem and so are the pins on the disc. The disc is locked with pins to the shaft. There are two shafts located normally, one at the top and another short one at the bottom. Both the spindles are pinned to the rotating disc where the drive shaft is keyed. The bottom shaft or spindle acts as a guide. There are also designs with through shaft or single shaft with pins locking the disc. In the lined valves, the pins are concealed in the elastomers or the lining. That gives us all the problem points. In all these cases, the valve needs to be removed from service for an appropriate corrective action.

A butterfly valve should be positioned in such a way that the rubber lining makes a full contact with the flanges. Therefore, slip-on flanges may not do justice and weld neck flanges may be better suited. The valve must be in closed position while being installed, and opened before tightening the bolts so that the disc takes its free position.

The lining material of a butterfly valve may peel off partially or fully, which creates the problem. Rubber may bulge and deform when left in closed position for long periods. This prevents the valve from opening. Small little debris collects inside the lining, particularly in the case of those insert-type butterfly valves, bulges due to peculiar corrosive actions. Rubber is lifted off at those points that prevent satisfactory operation.

A U-shaped liner is slipped over the body without any bonding. Such seats can be easily replaced. However, if the liner is made up of relatively rigid material like PTFE, the valve body is split horizontally to facilitate renewal of the liner. Experience shows that removable liners — particularly of the elastomer type — can bulge and harbor foreign material. Bonded liners do not suffer from this defect, but on the down side they cannot be replaced.

Passing of these valves can occur, again due to worn out seat rings, debris, and peeled-off elastomers. Peelings can sometimes hold up the valve in any one position.

If the valves are closed only partially, left near the closing position, most of the valves develop a kind of memory effect. Seats develop markings or scorings, plastic rings may deform to this position, and seat rings may erode due to the high velocity fluids and debris. This has full and final effect on the valves.

By and large, these valves last a life time. When they fail, it is worth replacing them with new ones.

DIAPHRAGM VALVES

The greatest advantage of the diaphragm valve is that it does away with the gland seals. They are principally used in the services of corrosive fluids, slurries, and viscous liquids. They can throttle the flow, but are not that good at low flow throttling. The main parts of a diaphragm valve are body, bonnet, compressor, and a flexible diaphragm. The diaphragm is connected to the compressor. The compressor, along with its threaded stem, pushes it down to close the valve. The body is lined with the same material as the diaphragm. Gaskets are not generally required between the flanges.

Valves are available at pressure ratings up to 200 PSI or 15kg/cm^2. Temperatures are limited by the diaphragms and body lining materials, range from –58° to 392°F (–50 to 200°C). Not all valves are suitable for low temperature service because at temperatures below freezing some lining materials become brittle. Similarly, gaskets are not required on the flange faces while using diaphragm valves. Needless to say that the diaphragms should be compatible with the fluid; a number of elastomers have been developed to suit various fluids being handled. Because of the construction of diaphragm valves, they may develop build up behind them if not used frequently.

They are widely used in both corrosive and abrasive applications in almost every industry. All working parts are effectively isolated from the service fluid, which increases its reliability. This valve adapts well to automation with its linear flow characteristics and on/off and control.

There are two different types of diaphragm valve based on the geometry of the valve body: Weir type and straight-through type. Both types are shown respectively in Figure 16.52A and 16.52B. Figure 16.53 shows this valve in operation. Figure 16.53A shows the Weir-type diaphragm valve open and closed whereas Figure 16.53B shows the straight through valve in open and closed position.

Figure 16.52A
Weir Type

Figure 16.52B
Straight-Through Type

Diaphragm Valves

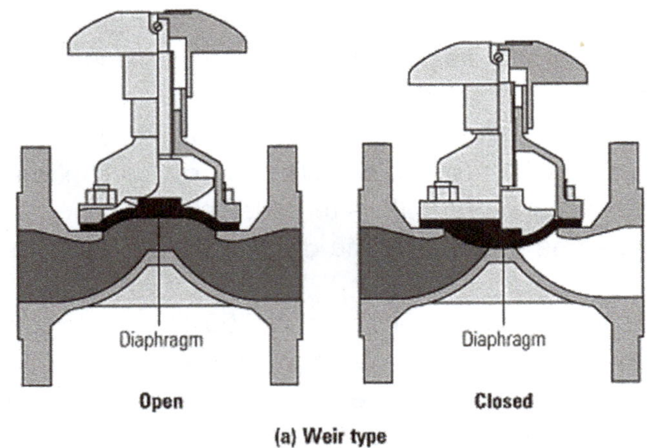

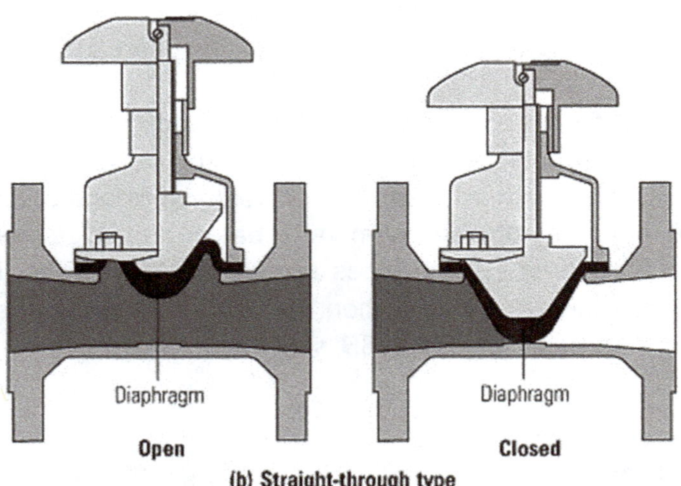

Figure 16.53
Operation of Diaphragm Valves

PROBLEMS AND SOLUTIONS (DIAPHRAGM VALVES)

Maintenance is easy because the valve is field serviceable. The bonnet can be removed out by opening its bolts and the body can remain in the pipe line. Follow the instructions available in the opening lines — special care should be taken while removing diaphragm valves as these are mostly used in corrosive services. Even after removing the valve from the service, thoroughly clean and neutralize it, and make it suitable for handling by human hands. Do not over-tighten bolts while fixing these valves because doing so will surely damage the lining. Do not test the valve beyond the working pressure. Similarly do not over-tighten the hand wheel just to be doubly sure that it is shut off. The maximum is half a turn after initial closure. Grease the hand wheel spindle.

It is easy to replace the diaphragms. However, follow these instructions in replacing the diaphragms.

1. Properly clean and start disassembly of the valve by removing bonnet nuts. Turn the handle a little more as if to close the valve. This will break the body bonnet joint, which may bind a little due to the action of rubbers. Lift the valve bonnet off body and turn hand wheel clockwise, partially exposing compressor.
2. Unscrew the diaphragm from the compressor by turning counter-clockwise.
3. Screw the new diaphragm into the compressor, hand tighten until bolt holes in the diaphragm and bonnet match, and then back off one-half turn.
4. Replace bonnet on body and finger tighten nuts.
5. Close the valve fully, back off one-quarter turn of the hand wheel, and tighten bonnet nuts with a calibrated torque wrench.
6. Open the valve fully for final tightening, but do not over-tighten nuts. This might compress the diaphragm.

LINED VALVES

Lined valves are used because of the corrosion resistance they offer apart from being compact in design, strong, shock resistant, and bi-directional. These valves are currently lined with a variety of plastics like Teflon, PFA, and PVDF. They are economical because they do away with exotic MOC. They are limited by temperature of the lining materials. Generally a sleeve of lining material like Teflon is pressed into the body and the plug is made to rotate in this sleeve. The plug and its seat are tapered to give a wedging action. Alternately, melt-processable PFA is lined and locked in the body and plug castings. Liners should not collapse in vacuum conditions and blow out in high pressure high temperature conditions. Valves should not harbor any cavities in open or closed positions to prevent accumulation of particulates because these valves are predominantly used in slurry and acid services. Blowout protection and anti-static arrangements are available in these valves.

Figure 16.54 shows a typical lined butterfly valve. Figure 16.55A shows a gear-operated lined plug valve and Figure 16.55B shows a hand-operated lined plug valve. You will see a typical lined ball valve in Figure 16.56 and a lined diaphragm valve in Figure 16.57.

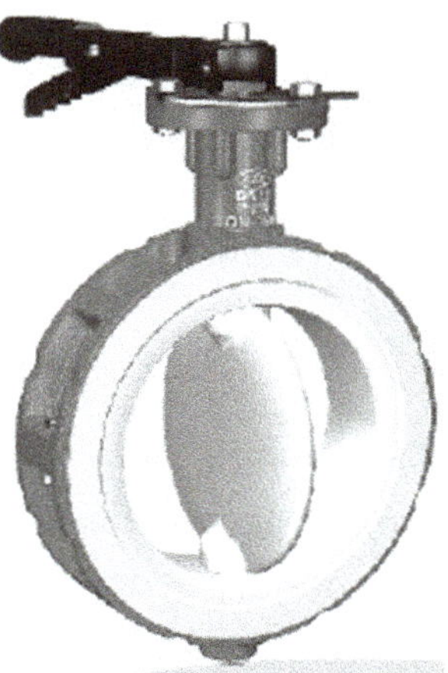

Figure 16.54
Lined Butterfly Valve

Figure 16.55A
Lined Plug Valve
(Gear Driven)

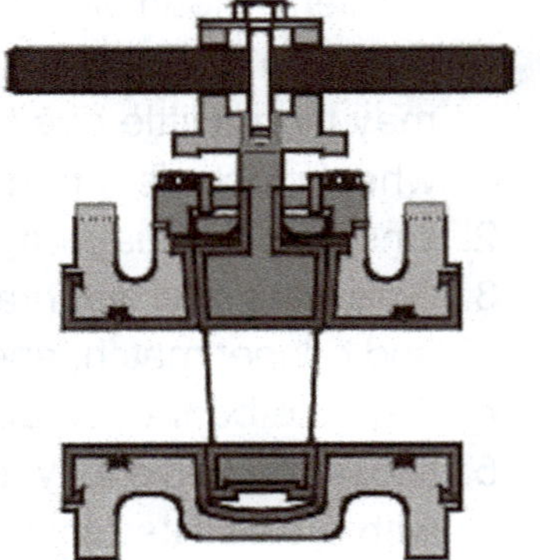

Figure 16.55B
Lined Plug Valve(Hand Operated)

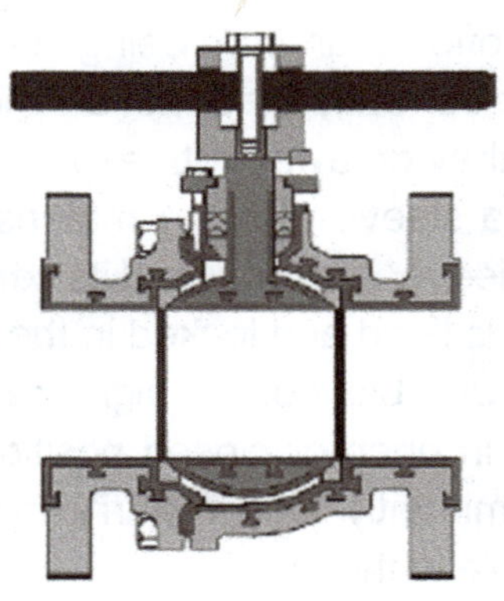

Figure 16.56
Lined Ball Valve

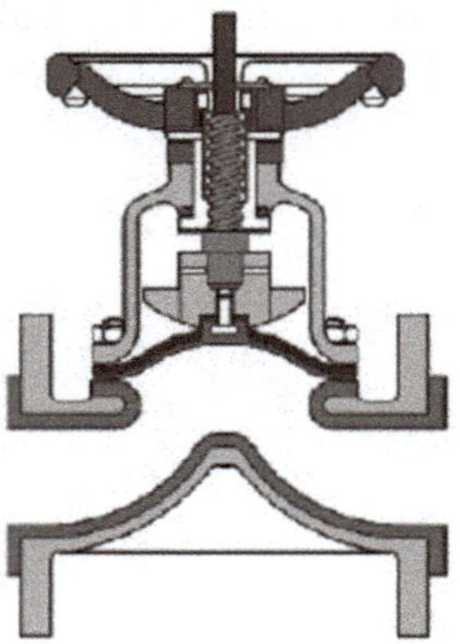

Figure 16.57
Lined Diaphragm Valve

Do not over tighten the flange bolts as this will surely damage the sealing faces. Because these valves are extensively used in slurry services, there is a tendency to steam clean valves. If steam is used, it should be below the maximum temperature limits of the liners. If the valves are stored for long periods of time, check. Elastomeric gaskets have a shelf life of five years. At low temperatures below freezing, some lining materials become brittle. Gaskets are not required on the flange faces while using most of the lined valves.

FLUSH BOTTOM VALVES

Flush bottom valves are used on reactors or process vessels for completely discharging the liquids without any entrapment in the valve body (see Figure 16.58). These valves will not allow for any dead space for unreacted mass to settle.

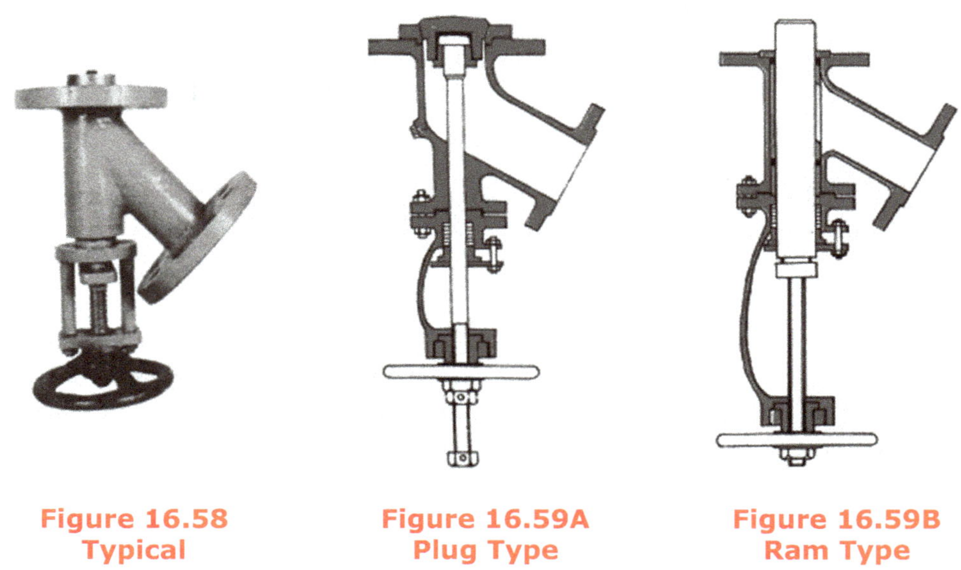

Figure 16.58 Typical

Figure 16.59A Plug Type

Figure 16.59B Ram Type

These valves have a modified Y-pattern globe valve. The stem is located in the straight branch and the flow path is through the angled branch. The disc either lifts up into the vessel or returns into the seat, making an opening. The valve seat stays flush with the bottom of the vessel, thus eliminating any entrapment. There are two types of valves: plug type (see Figure 16.59A) and ram type (see Figure 16.59B). They are also referred to as bottom drain valve, angle bottom valve, tank valve, and flush mount valve.

PINCH VALVES

Pinch valves do exactly what their name suggests — they pinch the flow. Inside the valve, a piece of elastomeric pipe piece is pinched or squeezed for a tight shutoff. No mechanical parts, no glands, and full bore flow. There is an outer pipe on this pinch pipe for a secondary containment. Valves are made up to 100 mm from 25mm at pressures up to 1450 psi or100 bar. Temperatures are limited by elastomers at –58° to 320°F (–50°C to 160°C). Figure 16.60A shows a pinch valve in open position and Figure 16.60B in closed position. These valves are more frequently used in mineral benefaction plants.

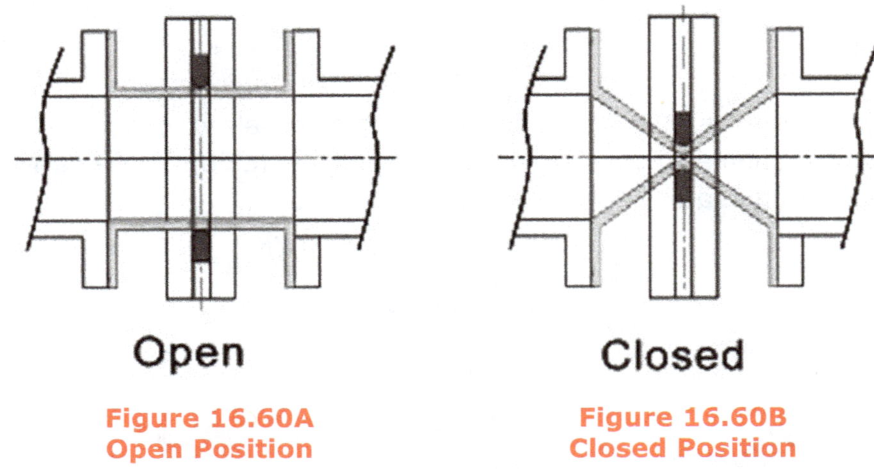

Figure 16.60A
Open Position

Figure 16.60B
Closed Position

Pinch Valves

KNIFE EDGE GATE VALVES

A knife edge gate valve is a variation of a gate valve designed for slurry, semi-solid applications. The wedge is shaped like a knife that can cut through semi-solid, pulpy and fibrous media. It offers a tight shutoff where any other valve would get jammed midway. A typical knife gate valve is shown in Figure 16.61.

Figure 16.61
Knife Edge Gate Valve

Knife edge gate valves are normally unidirectional valves and are available in a wide variety of metallurgies. They are indispensable in fertilizer, mining, steel, paper, and power industries in handling high temperature and abrasive slurries.

LOW TEMPERATURE VALVES

Low temperature and cryogenic services pose technical problems. Metal behavior is different; most of the fluids handled are flammable or explosive, which may also make fugitive emissions. LPG, LNG, ammonia, hydrogen, and helium are some of the gases. Low-temperature steels like LCB and LCC are suitable for services up to –50°F (–46°C) whereas stainless steels like CF8 and CF8M cover cryogenic temperatures down to –385°F (–196°C). These valves have extended bonnets to allow for the low temperature insulation. The length of vapor column in the bonnet keeps the cryogenic liquid away from the packing. A cryogenic gate valve can be seen in Figure 16.62, a cryogenic globe valve in Figure 16.63, and a cryogenic ball valve in Figure 16.64.

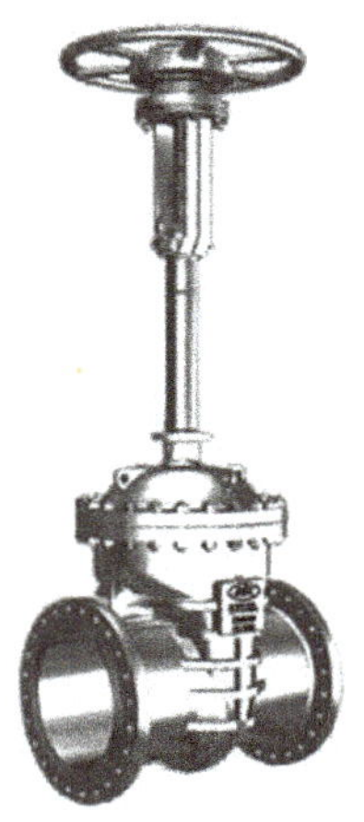

Figure 16.62
Gate Valve

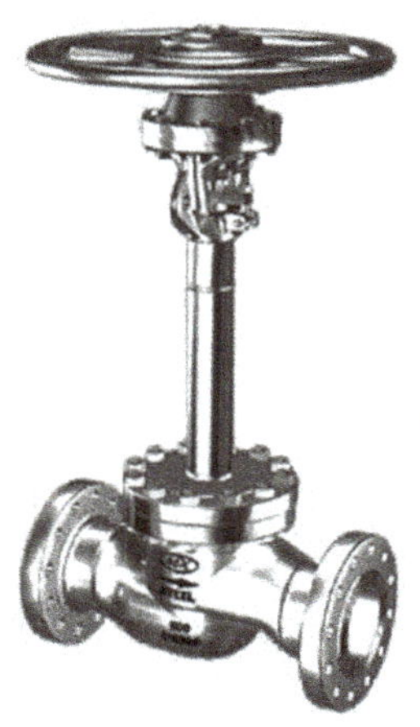

Figure 16.63
Globe Valve

Figure 16.64
Ball Valve

Cryogenic Valves

JACKETTED VALVES

For some services like bitumen and molten sulfur, the media should be kept hot at all times. Hence, jacketed piping is developed where the process liquid flows through an inner pipe which is jacketed or covered by a higher diameter pipe. Generally steam is passed through this outer diameter, which keeps the inner line hot and flowing. The jackets provide consistent heating or cooling of the flow media to prevent crystallization or seizing of flow media. Jacketed valves are also developed for this service. Although most of the valves can be jacketed, the plug and ball valves are the most common. They can be partially and fully jacketed valves. In most cases, partially jacketed valves offer uniform heat dissipation in spite of the insulating effect of the PTFE sleeve. However, fully jacketed valves provide maximum heating or cooling duties.

It is essential to size the inlet and outlet connections of the jacket, position them properly, and assure proper flow always. Jammed jacketed vale is difficult to work with and will be more difficult to free up again. Just putting back steam may not always succeed.

Figures 16.65 and 16.66 show the jacketed plug valve and ball valve respectively.

Figure 16.65
Jacketed Plug Valve

Figure 16.66
Jacketed Ball Valve

Chart 16.1 shows typical operating features such as pressure range, temperature range, and pressure drop of various types of valves.

Chart 16.1

Valve type	Size		Pressure range		Temperature range		Pressure drop[1]
	Min. (mm)	Max. (mm)	Min. (bar)	Max. (bar)	Min. (° C)	Max. (° C)	bar
Gate	3	2 250	>0	700	-196	675	0.007
Globe	3	760	>0	700	-196	650	0.590
Diaphragm	3	610	>0	21	-50	175	0.021
Ball (full bore)	6	1 220	>0	525	-55	300	0.007
Butterfly	50	1 830	>0	102	-30	538	0.120

[1] **Note:** Typical values for a DN150 bore valve passing saturated steam at 24 bar, flowing at 40 m/s.

GENERAL INSTRUCTIONS FOR VALVES

- Valves are generally protected during shipment. Flanges and bolt holes are covered with plastic or wooden shields. Do not remove them until the valve is about to be installed.
- Do not allow dirt, grit, or any other foreign material into the valve
- Check the valve seating surfaces, passages, and flanges face for any damage, even a little.
- Prior to installing the valve, flush the connected lines and clean them of all the dirt, rust, and scaling.

- Countercheck whether the valve being installed is suitable or not.
- Operate the valve a few times before fixing.
- Use only the designated gaskets.
- Confirm the direction of flow. Some valves are bidirectional, which means that they can be fixed in any direction. Many of them — like check, globe, and control valves — are not.
- Fix the valve in a convenient position so that the hand wheel is easily accessible.
- Inspecting valves once a year will yield fruitful results. If anything unusual is noticed, rectify immediately. But also investigate the causes and rectify them. Simple valves do give trouble at very wrong times.
- Gland leaks attended in time save precious energy and product, but also they save the valve from further damage.
- Valves are best placed in an upright position with the stem pointing straight up. Other positions are used, but are compromised. A valve with its stem down acts as a trap for sediment in the bonnet. At freezing temperatures, the liquid in bonnet may freeze and rupture it.
- After fully opening the valve, it is a good practice to close the hand wheel approximately half a turn to prevent it from seizing. The same is true while closing it.
- Do not use extensions or "valve persuaders" on the hand wheels. You may break them. Furthermore, the valve may develop a memory effect which forces you to make the habit permanent.
- It is a good practice, though not always practicable, to operate the valve periodically. This reduces the chances of internal and external build up.
- Valves are faithful servants which do not ask for much — only your care!

Nikola Tesla, the alternating current pioneer, invented a very simple one-way valve in 1920. Before that, however, Frank P.Cotter patented in 1907 a "simple self sealing check valve, adapted to be connected in the pipe connections without requiring special fittings and which may be readily opened for inspection or repair."

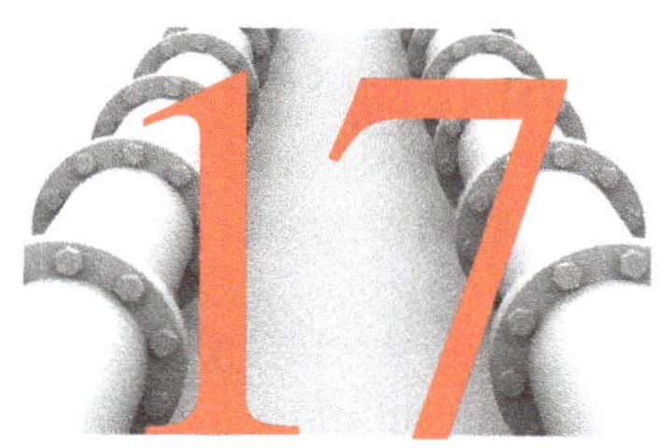

Control Valves

Present day demands of the process speeds and accuracy are huge. Changes in the amount of flow through a system must be made within few seconds, or less, for any modern plant to work satisfactorily. Hence, control valves are extensively used and are the heart of modern plant automation. A control valve is really an automated globe valve, a device which acts and makes corrections as signaled by the controller. Control valves are designed to control the flow accurately, which in effect control other process variables like pressure and temperature. Instead of a hand wheel, which moves the valve plug in or out, a control valve utilizes instrument air pressure or an electric motor to close or open the valve with the help of an actuator. Figure 17.1A shows a regular control valve and Figure 17.1B shows a diaphragm valve control valve.

Figure 17.1A
Regular Control Valve

Figure 17.1B
Diaphragm Valve Control Valve

Most generally, the control valves operate linearly, i.e., the stem moves up and down. Control valves also use rotary motion. Linear types include globe, gate, and slide valves. The flow path is tortuous and hence have higher pressure drop. They can control to very low flow rates and are better suited for high pressure applications.

Rotary types are used on quarter-turn valves such as ball valves, butterfly valves, and plug valves. These valves have a straight flow path and can handle higher capacity. Packing wear is less. They are more suitable for slurries and other abrasive stock. These valves can be designed specifically with specific balls, wafers, and seats such that they have a predictable flow pattern.

There are certain considerations for the selection of control valves.

- Action — air to close or air to open
- Need for tight shut off
- Mechanical strength
- Materials of construction — body and trim
- Importance of valve flow characteristics to the desired control effect
- Torque required for valve closure
- Nature and consistency of fluid
- Type of actuator

Positioners and other accessories improve the performance of the control valve.

ACTUATORS

Actuators use the motive power from instrument air, electricity, or hydraulics and operate the control valve. A manual actuator employs levers, gears, or wheels to facilitate movement whereas an automatic actuator has an external power source to provide the force and motion to operate a valve remotely or automatically. It may be impossible and impractical to operate some valves manually simply because of the sheer muscle power required. Power actuators are a necessity on valves in pipelines located in remote areas; they are also used on valves that are frequently operated or throttled. Some valves may be located in extremely hostile or toxic environments, which preclude manual operation. Additionally, as a safety feature, certain types of power actuators may be required to operate quickly, shutting down a valve in case of emergency. There are a few varieties of actuators; each has its own limits and merits.

PNEUMATIC ACTUATORS

Control valves with these ubiquitous pneumatic actuators are seen everywhere in the present day process industry.

DIAPHRAGM ACTUATORS

Pneumatic actuators with diaphragms using 3–15 psi air are the most common. Air pressure operates on a rubber diaphragm, which forces the valve stem and plug up or down. The air pressure acts against a spring; this balance of air pressure and spring forces holds the valve in a given position. A sensing instrument measures the pressure, temperature, or flow and sends a corresponding pneumatic signal to the controller which in turn sends a proportionate output signal to the control valve. The diaphragm accordingly adjusts the stem and the plug.

These actuators are single acting in as much as the air is applied only to one side of the diaphragm. They can be either direct acting with spring-to-extend or reverse acting with spring-to-retract. However in applications where its straight power is limited and wanting, a positioner is required. They are simple and cheaper, and can be maintained easily.

The diaphragm actuators are the industry's charm, even though their weight and size are large compared to others. Their benefits outweigh these disadvantages. Their greatest advantage is that they can safely be used in hazardous and fire prone environments. Diaphragm actuators are reliable and can be designed to fail-safe operation.

An Air-to-close control valve means that increased air pressure to the diaphragm closes the valve. Air-to-open means the opposite. Selection of this action depends entirely on the process requirements and fail-safe operation. The air-to-close control valve closes the valve on application of air, whereas loss of air supply opens it (fail to open). The opposite is the case of the air-to-open (fail to close) valve.

If the control valve regulates the flow of cooling water to the process, then it may be advisable to select an air-to-close valve so that no damage to the process equipment or material can result in high temperature should the air supply pressure fail. On the other hand, if the control valve is regulating steam or hot gas to the process, then an air-to-open valve may be necessary to safeguard the process or equipment from high temperatures or pressures if the air supply fails. Control valves should essentially be operated in fail-safe mode. In case of loss of air or the controlling signal, the valve should open or close, whichever is safer and better suited for the plant operations.

Control valves operate with 3–15 psi, which is an industry standard, but for specific applications, higher pressures are used. Control valve diaphragms are sized by the force required to open or close the control valve. Diaphragms are usually made of a plastic or rubber material and cannot take high temperatures.

PISTON ACTUATORS

When a larger stem stroke is required or when greater thrust is required, piston actuators are used. Both single-acting and double-acting actuators are available. They can operate at higher speeds, can withstand and need higher input pressures, and can exert higher torque, yet are compact and light weight. However, they are costly and they are essentially on-off actuators which need a positioner to really throttle the flow.

HYDRAULIC ACTUATORS

The hydraulic actuators are often simple devices with a minimum of mechanical parts, used on linear or quarter-turn valves. Sufficient fluid pressure acts on a piston to provide thrust in a linear motion for gate or globe valves. Alternatively, the thrust may be mechanically converted to operate a quarter-turn valve. Most types of fluid power actuators can be supplied with fail-safe features to close or open a valve under emergency circumstances. They are capable of much higher outputs and deliver higher torques at higher speeds. But they are costly, larger in size, and have complex front end machinery.

ELECTRIC ACTUATORS

The electric actuator has a motor drive that provides torque to operate a valve. Electric actuators are frequently used on multi-turn valves such as gate or globe valves. With the addition of a quarter-turn gearbox, they can be utilized on ball, plug, or other quarter-turn valves. They can deliver higher output, but their speed is slow which in turn limits their duty cycle.

POSITIONERS

Positioners and other accessories improve the performance of the control valve. High pressures, temperatures, high differentials, and certain media put additional strain on the control valve operation. Control valves are aided in their operation with the help of positioners and actuators to operate the sluggish control valves. If a control valve cannot be positioned precisely by controller air output, a booster or valve positioner may be required. Control valves handling viscous stocks, slurries, or hot oils that tend to coke up are usually equipped with valve positioners. Control valves whose action is sluggish (even though properly serviced and lubricated) should be reported for possible installation of a valve positioner. Positioners are used whenever accurate control of flow is required.

A valve positioner is installed on top or side of the valve and is mechanically connected to the valve stem to boost the operation of the valve. It will position the stem at the same place for a given input signal regardless of stem friction, differential pressure on the valve, diaphragm hysteresis, etc. It takes a low-pressure control signal and boosts it or amplifies it with the help of high pressure air. Thus the power may be increased 8-to-10 times. Precision and dependability of operation are gained from the positioner, especially with the larger valve sizes. Faster response and fine controls are added advantages. Electro-pneumatic convertors are incorporated into some positioners, which directly take the typical 4–20 mA input signal and control the valve.

OTHER FEATURES

In hot services, control valve bodies are provided with extensions or fins between the port area and the diaphragm. Generally, fins are provided when the valve temperature is above 230°C.

Valve packings and lubricants are selected in the same manner as conventional gate or globe valves. Usually, however, a lubricator isolating valve is provided for lubrication of the stem packing.

Many control valves are furnished with hand wheels so that the control valve can be manually opened or closed when there is some failure such as loss of instrument air or diaphragm rupture.

Materials of construction are selected just as for any other piping or plant equipment or material. Resistance to corrosion, erosion, temperature, pressure, and pressure drop are items to be considered in material selection. Hard facings such as stellite are usually used with high pressure drops or high temperature conditions.

Valves are either flanged or screwed. 1" and larger valves are usually flanged. Flange classes match with corresponding flange classes.

Control valves are available in single or double seat construction. Single seat bodies permit the thrust of fluid pressure to act over the entire bottom of the valve plug, thereby creating an upward force on the valve stem. The power unit must overcome this force to move or hold the stem downward. The larger the valve size and higher the fluid pressure, the more prohibitive this steam thrust becomes. See Figure 17.2 for single seat construction.

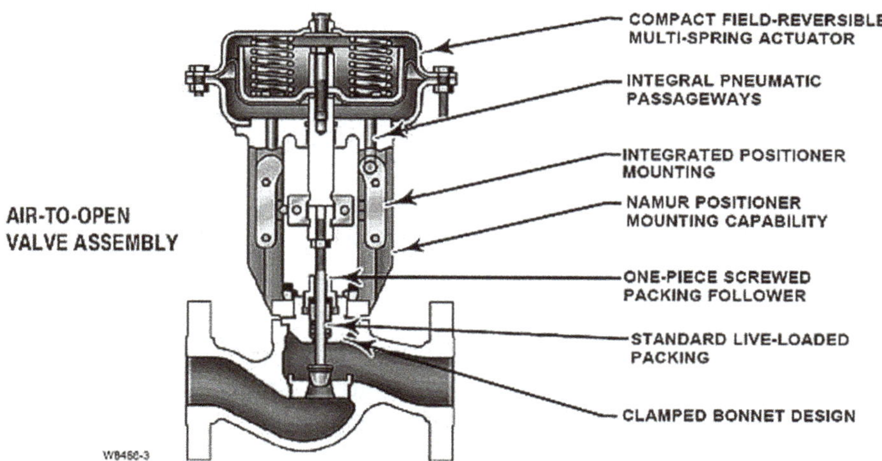

Figure 17.2

Double seat construction is designed to minimize the undesirable effects of stem thrust (see Figure 17.3). A double-seated valve has two valve plugs on a common spindle, with two valve seats. Inlet and outlet fluid pressures tend to balance themselves by acting both upward and downward on the two valve plugs. As the forces are partially balanced, valve seats can be made smaller. However the disadvantage of this construction is that with temperature variations, the stem portion between the two valve plugs contracts or expands linearly in a different amount than the valve body. In the closed position, therefore, one valve disc seats first and prevents the second disc from seating completely. A small leakage flow results. Where tight shut-off is required; therefore, a single point valve must be used.

Control valves cannot guarantee tight shut off and are not recommended thus. General arrangement of the control valve is such that there are two block valves upstream and down-

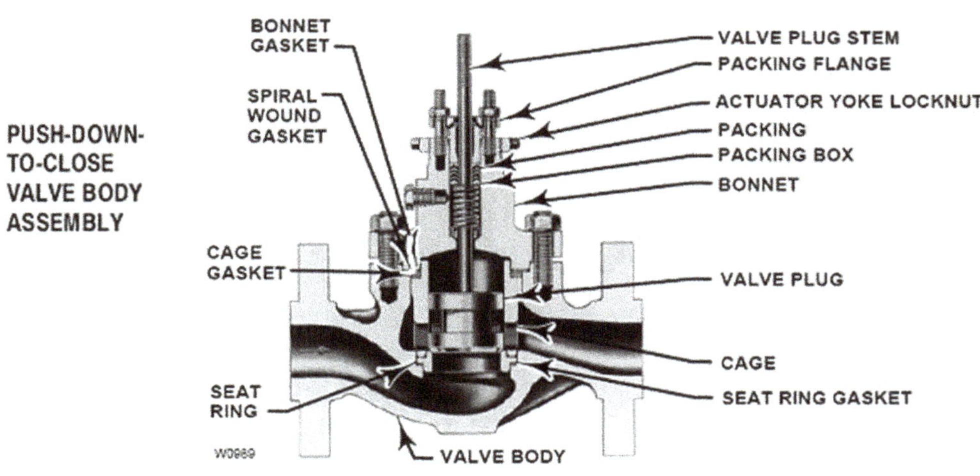

Figure 17.3

stream. A globe valve is arranged as a bypass valve to control the process flow in case the control valve fails to work.

At higher flow velocities, fluid pressure in the direction of flow through the valve tends to force the plug out of alignment with its seat and cause improper seating. Also, any sidewise movement of the valve stem resulting from this thrust would seriously increase friction on the stuffing box, due to binding of the stem against its bushing and the packing. This thrust is prevented from disturbing the stem alignment by the use of an extension stem on the bottom of the inner valve with bushings in the bottom flange of the valve body and in the bonnet. These are called top and bottom guides.

Frequently, butterfly valves are used for regulating the flow of gas or air, and particularly in large size lines. Butterfly valves are not as accurate or as good a regulator as conventional controls valves, but they are becoming increasingly reliable.

LEAKAGE CLASSES

Control valve leakage is classified with respect to how much the valve will leak when fully closed. Generally, it is higher cost for lower leakage. Control valves are not designed for perfect shutoff; they are designed to throttle. Various factors — like packing, guiding, seat material, actuator thrust, pressure drop, temperature, and the type of fluid — control the shutoff ability of the control valve. Single-seated valves shut off better than double-seated valves. ANSI is classified into six different leakage classes, of which two are important: Class IV and Class VI. Leakage classes of control valves are given in **Chart 17.1A**. Leakage rates for Class VI seats are given in **Chart 17.1B**.

Chart 17.1A

Valve Seat Leakage Classifications	
Leakage Class	**Maximum Allowable leakage**
I	
II	0.5% of rated capacity
III	0.1% of rated capacity
IV	0.01% of rated capacity
V	5 X 10^{-12} m^3/s of water per mm of seat diameter per bar differential (0.0005 ml/min per inch of seat diameter per psi differential)
VI	Not to exceed amounts shown in Chart 17.1B (based on seat diameter)

Chart 17.1B

Class VI Seat Allowable Leakage		
Nominal Port		**Allowable**
Diameter		**Leakage Rate**
mm	**in**	**ml/minute**
25	1	0.15
38	1 1/2	0.30
51	2	0.45
64	2 1/2	0.60
76	3	0.90
102	4	1.70
152	6	4.00
203	8	6.75

PLUG CHARACTERISTICS

All control valves have certain inherent flow characteristics depending on the plug and seat arrangement. Inherent flow characteristics of typical globe valves and rotary valves are shown in **Chart 17.2**. The physical shape of the plug and seat arrangement is often referred to as valve trim. Trim design decides the valve capacity and hence the flow as the stem travels in its complete stroke. This relation of the valve capacity and valve travel is known as the flow characteristic of the valve.

Chart 17.2
Flow Characteristics of Typical Globe Valves and Rotary Valves

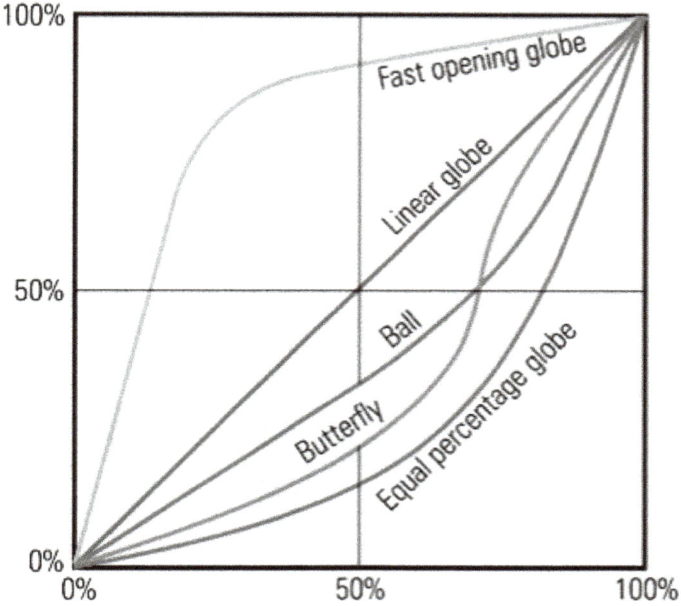

Inherent nonlinearities can be compensated by selecting a proper trim. Plugs are basically divided into three types, depending upon process control requirements. These are: fast opening or on-off, linear, and equal percent.

Each control valve manufacturer has its own terminology, shapes, and contours for control valve plugs in a large variety of control valve applications. However, all basically fall into one of these above categories.

Fast opening (or on-off) valve gives a large change in flow rate for a small valve lift from the closed position. Figure 17.4A shows a diagrammatic representation of a fast opening valve; Figure 17.5A shows the particular disc.

In the **linear type**, the valve plug is so shaped that the flow rate is in direct proportion to the valve lift, at a constant differential pressure. Linear type plugs are usually preferred over percent-type plugs because an operator, by viewing the operating position of the stem position indicator, can quickly sense the flow. This type of plug is less subject to jamming by foreign particles. Figure 17.4B shows a diagrammatic representation of the linear-type valve; Figure 17.5B shows the particular disc.

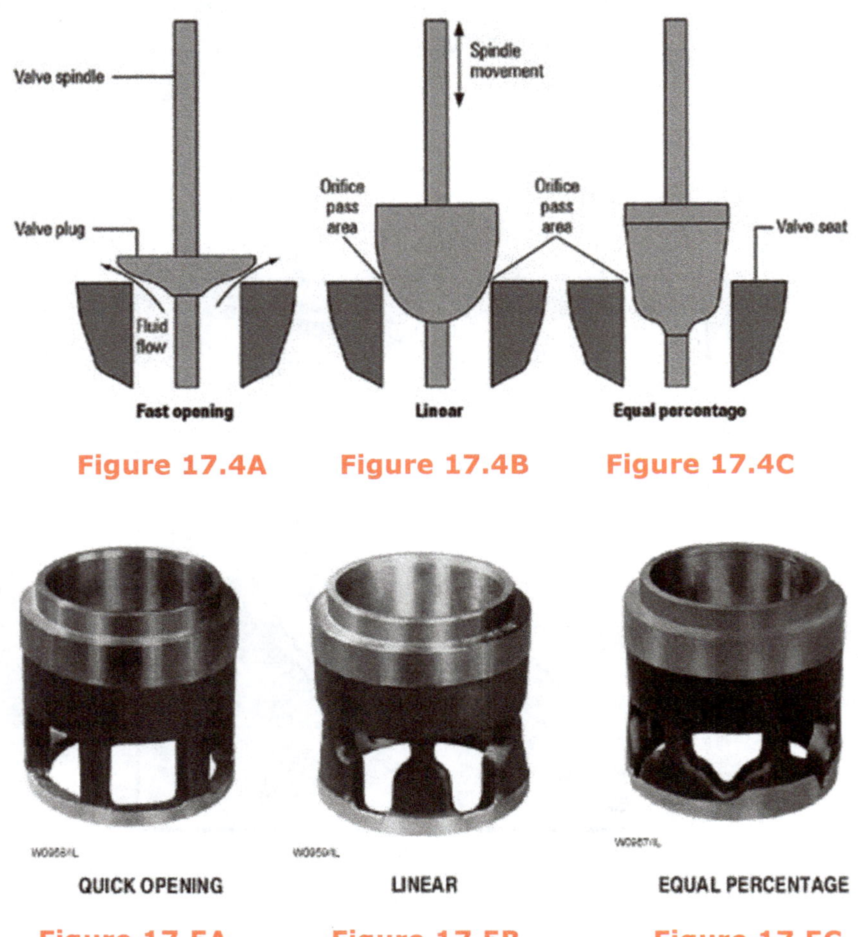

Figure 17.4A Figure 17.4B Figure 17.4C

Figure 17.5A Figure 17.5B Figure 17.5C

Equal percentage (or logarithmic) valves have valve plugs shaped so that each increment in valve lift increases the flow rate by a certain percentage. The relationship between valve lift and orifice size (and hence the flow rate) is not linear but logarithmic. Figure 17.4C shows a diagrammatic representation of an equal percentage valve; Figure 17.5C shows the particular disc.

CONTROL VALVE SIZING

A control valve is essentially a variable orifice. Sizing control valves are very important in achieving good system control. Control valves should always be smaller than the pipe leading to it. If not, the pipe itself tends to throttle the flow at high capacity loads and the controller is thus ineffective. If the valve is undersized, the system must operate at loads smaller than those for which it was designed. An oversized valve may result in poor control since, at low capacity, the valve would be controlling with its plunger close to the valve seat in the most non-linear portion of its performance. For linear valves, the size is selected for the normal flow at about 75% of maximum capacity. For percent characteristic valves, the size is selected for the normal flow between 50–70% of maximum valve capacity, which is approximately 80–90% of lift.

C_V or flow coefficient is a dimensionless value that indicates the valve's flow capacity. The C_V number is a critical factor determining the maximum capacity of the valve. As a definition, C_V is the number of U.S. gallons per minute of water which will pass through a given flow restriction with a pressure drop of 1 psi. A valve with a C_V of one has a capacity of one GPM of water at one PSI pressure drop. To calculate C_V inlet flow, inlet pressure, pressure drop, temperature, and specific gravity of the fluid under flow are required. Although C_V is the most commonly used term, there are other similar terms indicating the capacity such as K_V, K_{VS}, K_{VR}, A_V. See the description of K_V, and C_V in the **Chart 17.3**.

Chart 17.3
Flow Rate Definitions

K_V	Flow rate in m^3/h of water at a defined temperature, typically between 5°C and 40°C that will create a pressure drop of one bar across a valve orifice (More popular in Europe).
K_{VS}	The actual or stated K_V value of a particular valve when fully open constituting the valve flow coefficient, or capacity index.
K_{VR}	It is the flow coefficient required by the application.
C_V	The flow rate in gallons per minute of water at a defined temperature typically between 40°F and 100°F that will create a pressure drop of one pound per square inch. (More popular in the U.S. and some other parts of the world). Please be careful with this term as C_V Imperial and C_V U.S. exist. Though the basic definition is the same, actual values differ slightly because of difference between Imperial and U.S. gallons.
A_V	Flow rate in m^3 /s of water that will create a pressure drop of one Pascal.

OPERATION OF CONTROL VALVES

Because the final control element — the control valves — plays such a vital role in controlling the process, it is important that it operates properly. Proper design of the valve for the service specified is of vital importance. Excessive lag due to sticking valve stems is the main operating problem with control valves; it is usually the result of either high friction in the packing or low force of the drive diaphragm.

Valve sizing can be checked by its action on the system. If inadequate flow is obtained with the valve wide open, then the control valve is too small or the process pressure ahead of the control valve may be too low. If too much flow is obtained with small valve lift, the valve is too large or of the wrong design, or process pressure is too high on the upstream side of the control valve.

If a control valve is too small, some control can be obtained by opening the bypass and using the control valve to trim the flow.

A valve that does not assume the same position for a given air pressure on both increasing air pressure and decreasing air pressure indicates excessive friction in the stem guide or stuffing box. The effect on the process will be poor control. Free up the valve stem.

If the valve is sticky or jerky in action, it may need lubrication, the packing may be too tight, the stem may be roughened in the packing gland or, in the worst case, the stem is bent. Take appropriate action.

Excessive lubrication pressure and quantity do no good. Lubrication should be done moderately. If a small amount does not overcome leakage from the packing, adjust the gland nut by taking it up not more than one flat at a time (one sixth of a turn). Lubricate the packing at the same time to help it fit.

Excessive tightening of the packing tends to bind the valve stem. The valves will then stick and lag behind the controller. When it does move, it is by jumps, which upsets the process.

If a control valve does not respond to change in air pressure from the controller, the air line may be plugged or leaking. Control the process by means of the bypass valve, and loosen or disconnect the air line from the diaphragm housing. If air flow is free, stop the end of the air lead and check to see if supply air pressure is maintained. If the air system is functioning properly, the diaphragm is the culprit — it may be ruptured or too small.

If a valve does not seat tight with proper air signal, the stem may be too short, foreign material may be under the seat, or the plug or seat may be worn or corroded. Flush through the valve to remove foreign material. If not, get it serviced.

If the valve cannot be held closely by the normal instrument air pressure, it may be necessary to increase the force opposing the spring action. This can be done by increasing the diaphragm area, or by increasing the instrument air pressure.

Valve materials should be selected based on the pressure, temperature, and corrosion characteristics of the fluid handled. Keep the velocity low when handling abrasive fluids. Keep straight the length of pipe upstream and downstream of the valve to avoid choking.

REMOVING AND REINSTALLING CONTROL VALVES

Before removing a control valve from a system, follow the procedures outlined in Chapter 13, *Opening Lines*. In addition, see that any tubing removed from the control valve is stored in a clean, safe location. Tubing and tubing connectors are damaged easily. Handle them carefully. Small lengths of tubing can be tied together with a string or wire and hung close to the job. Carefully store and be sure to remember the location of any parts removed and stored.

As with globe valves, control valves must be installed correctly in relation to the direction of flow. Look for an arrow on the control valve body. Check the valve flanges. They are sometimes marked "IN" and "OUT".

Even if there are no flow direction markings on the valve body, it is still possible to determine the direction of flow by an internal inspection. A general rule for installation of control valves is that the pressure should not be on the packing gland when the valve is closed. The flow on a single-seated valve is into the face. In the double seated valve it is "IN" between the seats, and "OUT" around the casting. The "IN" side of a double seated valve is the side where you can see the seats.

If control valves are installed backwards, the valve will not regulate as evenly as it might if it is installed correctly, and the packing gland may develop a tendency to leak sooner.

GLAND PACKING

Gland packing is a sealing mechanism whereby the leak from the stem into the atmosphere is prevented in a valve. Although more sophisticated mechanisms and sealing solutions have taken over, valves still do with the humble gland packing. Certain deformable materials like Teflon, asbestos, graphite, and combinations of these are packed into a housing known as a stuffing box which holds tight against the stem and the box. They are essentially compressed in the housing, creating a pressure seal between the valve body and atmosphere. A number of rings are employed that can often be cut to size from a rope. They can be solid or split rings or be formed into special shapes such as a chevron. The stem finish should be in the order of 8-to-16 RMS and should be concentric in the stuffing box bore so that the packing is compressed uniformly. Anti-extrusion washers minimize extrusion of the packing. Gland packing creeps or hardens over a period of time, and has to be rectified by tightening or renewing the packing.

There are a number of packing materials to choose from; most of them are the same as the gasket materials re-formed to suit stuffing boxes. They should be able to exert minimum friction against the stem at the same time they have enough capacity to hold against the pressure.

Packing can wear out, extrude, consolidate or migrate, with the end result of leak across the stem. Packing can become consolidated, or get hardened due to aging or bad selection, resulting in leaks. Packing can extrude or cold flow, resulting in the loss of the sealing effect. Teflon is more susceptible to this problem. Packing or part of the packing rings can be caught by bad stem and migrate out of the stuffing box.

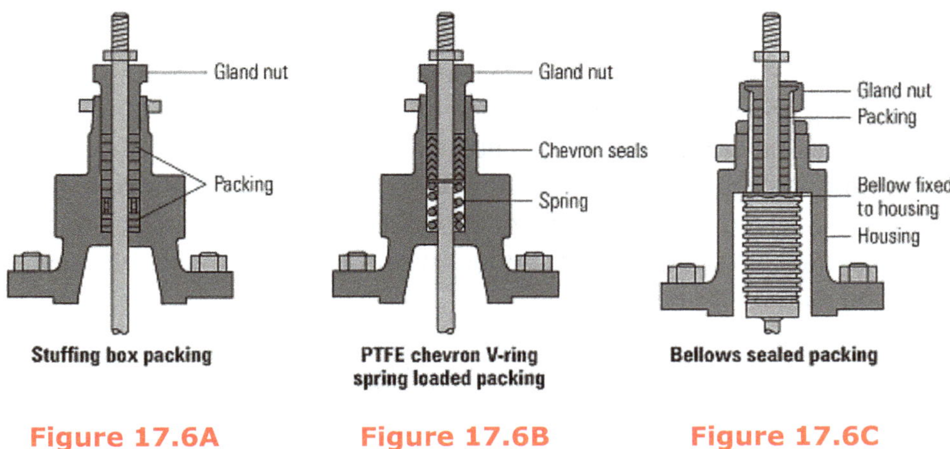

Figure 17.6A Figure 17.6B Figure 17.6C

To compensate for the loss of packing effectiveness, there are arrangements to live load internally or externally, or by spring loading. Spring loading adjusts the packing load as the packing wears out, and does away with manually tightening the gland. See Figures 17.6A, 17.6B, and 17.6C for various schemes of packing.

Synthetic, vegetable, animal, or mineral fibers are plaited, braided, twisted, or combined with all or any of these methods to form gland packing. Plaited and braided packings are sometimes reinforced with copper, steel, inconel, or aluminum to withstand high temperatures.

An effective gland packing must possess good anti-friction properties and should not spoil the stem. It should have chemical resistance to the contained fluid and also should not contaminate it, which is a very vital characteristic in the food, pharma, and beverage industries. It should withstand the pressure and temperature of the medium. It should be compressible, and yet maintain its resilience and not lose its lubricity. It has to work against many odds to maintain its leak tightness. A large part of tightening and repacking the glands is still played manually and the worker with a slide wrench plays an important albeit dangerous role in the plant operation.

PACKING THE GLAND

Figure 17.7 shows a typical packing gland of a valve with its parts.

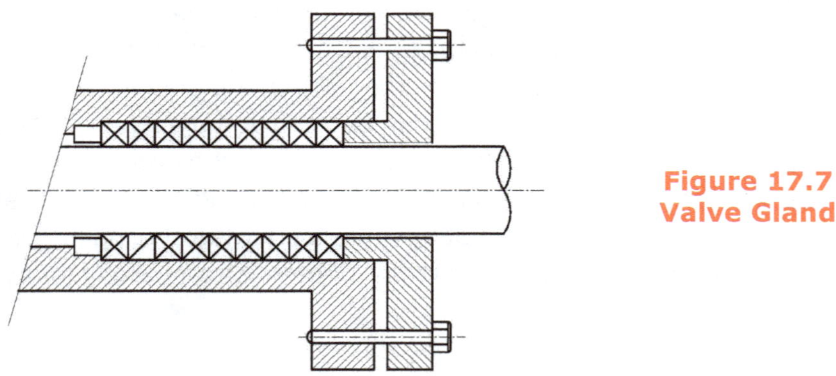

Figure 17.7
Valve Gland

Here is the usual practice followed for repacking the gland.

- Open the gland follower nuts and back off the gland follower.
- Remove all the old packing rings and clean the stuffing box.
- Take new packing recommended for the service and size suitable to the stuffing box.
- Wrap the packing rope on the stem or spindle and cut the packing to correct lengths.
- Some technicians cut the rings square whereas others make a cut diagonally, each claiming its own virtues.
- Packing rings are then pressed into the stuffing box individually by light tapping.
- Do not push all the rings at a time and stagger the joints. Fix the gland follower and check if it goes down easily.

- Tighten the gland nuts evenly and by rotation. Do not tighten the nut on one side and force the other nut later. The gland follower may break.
- Operate the hand wheel and check its free operation.
- Pressurize the valve and retighten the gland to obtain a leak-free seal. Now if excessive force is required to operate the valve, there must be something wrong. Check again.

Chart 17.4 summarizes some of the commonly-used packing materials and their temperature limits.

Chart 17.4
Packing Materials and Temperature Ranges

Material	Temperature Range	
	Deg F	Deg C
Jute, flax, hemp	32 to 140	0 to 60
Cotton	32 to 160	0 to 70
Rubberized cotton	32 to 180	0 to 80
PTFE	(-)420 to 430	(-)250 to 220
Aramid	(-)330 to 540	(-)200 to 280
Graphite fiber	(-)330 to 1110	(-)200 to 600
Asbestos reinforced st.steel or Inconel	32 to 1470	0 to 800
Copper mesh	32 to 1470	0 to 800
Aluminia silica filament/ Inconel reinforced	32 to 2190	0 to 1200

Relief Valves

The design of any pressure vessel is based on an assumed maximum operating pressure which, for economy, should be as near the normal maximum operating pressure as possible. However, practically every pressure vessel is prone to be subjected, sometime in its life, to an abnormal pressure. This may cause damage to the vessel, surroundings, workers, and materials unless means are provided to safeguard it. The cause of an excessive pressure may be due to any one or more of the following reasons:

Chart 18.1
Overpressure Scenarios
American Petroleum Institute (API) Recommended Practice 521

API RP 521 Item No.	Overpressure Cause
1	Closed outlets on vessels
2	Cooling water failure to condenser
3	Top-tower reflux failure
4	Side stream reflux failure
5	Lean oil failure to absorber
6	Accumulation of non-condensables
7	Entrance of highly volatile material
8	Overfilling storage or surge vessel
9	Failure of automatic control
10	Abnormal heat or vapor input
11	Split exchanger tube
12	Internal explosions
13	Chemical reaction
14	Hydraulic expansion
15	Exterior fire
16	Power failure (steam, electric, or other)
	Other

- Shutting off outlets to the vessel while fluid continues to enter, as in a receiver connected to a compressor
- Accidental incorrect manipulation of valves, permitting a higher pressure to enter a vessel from some source other than intended
- Increase of temperature of a vessel containing a volatile fluid, due to cooling failure resulting in increased vapor pressure (e.g., ammonia storage)
- Failure of automatic flow, pressure, or temperature control apparatus or normal relief apparatus
- Gradual accumulation of non-condensable gases in systems supposed to contain condensable fluids (e.g., refrigeration systems)
- Over-filling a vessel with liquid by a positive displacement pump
- Expansion of a liquid by a temperature rise in a container in which sufficient outage has not been left (e.g., water side of heat exchanger, when both inlet and outlet valves are closed, could develop enormous pressures due to liquid expansion)

The probability of some abnormal increase in pressure is so great that all pressure vessels should be protected by some emergency device somewhere in the system that can be relied upon

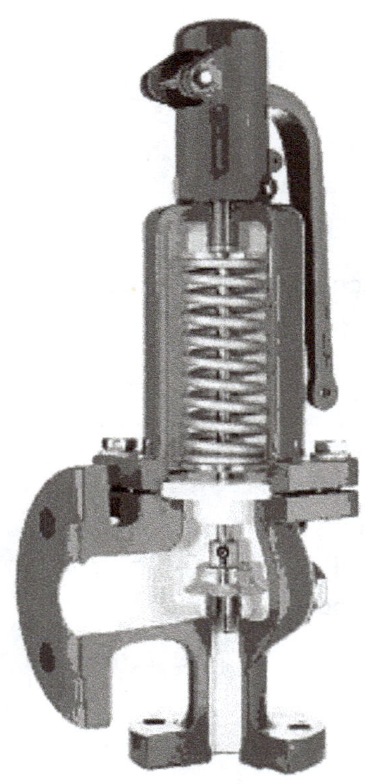

Figure 18.1A

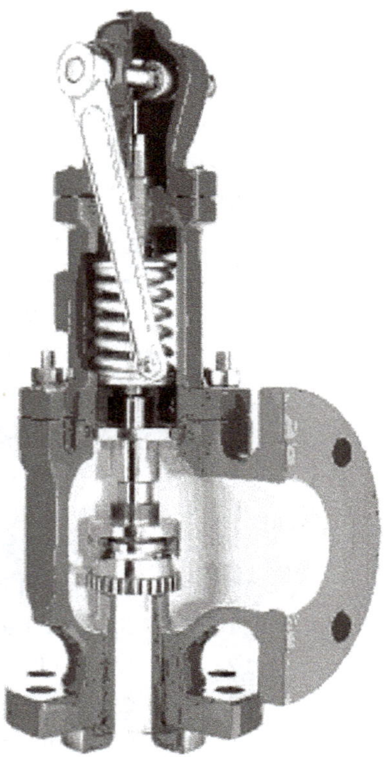

Figure 18.1B

Safety Relief Valves

to act in an emergency to prevent the pressure rising to the point where the vessel might rupture or damage may occur (see Figures 18.1A and 18.1B). Different devices are generally provided for this purpose. However, the safety relief valves are the most common instruments. These valves must open at the set pressure, discharge the rated flow, and seat back once the safe area of pressure is reached. They must be compatible with the fluids, withstand pressures and temperatures, and also maintain tight shutoff under normal operating conditions. They avert the system from becoming over-pressurized, rather than maintain the system pressure at a constant level. Relief valves are installed between the pressure source, such as compressor or the boiler and the first shut off valve. Most relief valves have an adjustment to change the set point, which is locked to prevent accidental change.

The American Petroleum Institute (API) published a checklist known as Guide for Pressure-Relieving and Depressuring Systems, better known as API 521 or API Recommended Practice 521. Every engineer must be aware of this list of overpressure scenarios, provided in **Chart 18.1**.

TERMINOLOGY

Dealing with safety relief valves, we come across some new terminology with which we should be familiar before we discuss their constructional features:

Relief Valve. An automatic pressure-relieving device actuated by the static pressure upstream of the valve, and which opens in proportion to the increase in pressure over the opening pressure. The relief valve opens slowly. It is used primarily for liquid service. There is no huddling chamber and no pop action.

Safety Valve. Characterized by the rapid full opening "pop" action. It is generally used for gas or vapor service. It is opened fully at 10% over pressure and closes tight within 4% blow down.

Pressure Relief Valve. A general term applied without distinction of safety valve, or relief valve. It is a spring-loaded pressure relief device that opens to relieve excess pressure and shuts back after normal conditions have been reestablished.

Set Pressure in psig. The inlet pressure at which the pressure relief valve is adjusted to open.

Balanced Safety Valve. A type of safety relief valve designed in such a manner that back pressure has little effect on the set pressure or operation of the valve.

Bellows–Type Valve. A balanced valve utilizing a bellows surrounding the valve guide and stem.

Accumulation. The pressure increase over the maximum allowable working pressure of the vessel during discharge through the pressure relief valve, expressed as a percent of that pressure or in psi.

Back Pressure. The pressure acting on the discharge side of the pressure relief valves.

Blow Down. The difference between the set pressure and the reseating pressure of a pressure relief valve, expressed in percent of the set pressure or in psi.

Built-Up Back Pressure. The pressure in the discharge header that develops as a result of flow after the safety relief valve opens.

Chattering. Rapid and continued opening and closing of a pressure relief valve.

Simmer or Warn. The slight leakage of safety or relief valves just before opening.

WORKING OF A SAFETY VALVE

The working of a safety valve can be better understood by a brief discussion. The basic pressure relief valve is of the spring-loaded design. It is a simple system with an inlet nozzle fixed with a suitable flange, a disc held against a seat with a spring under normal conditions, and all these parts enclosed in the valve housing of body and bonnet. The spring load is adjusted to make the set point at which the valve pops. When the system pressure goes up beyond the spring load, the beveled disc opens and releases the pressure out. As simple as that! See Figure 18.2A for a valve designed as per ASME, and Figure 18.2B for a DIN design.

That is where the simplicity ends.

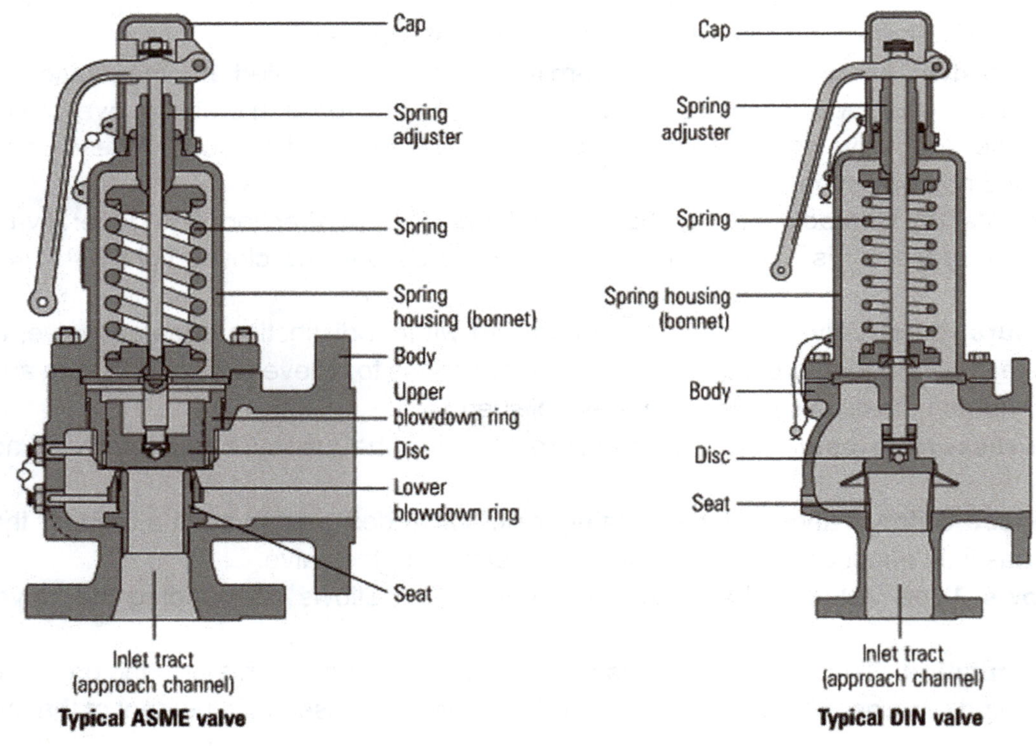

Figure 18.2A

Figure 18.2B

The discharging fluid velocities can reach very high to the point of 1500 feet per second, which can cause severe erosion of the valve seatings. So the seat should pop wide open at the high pressure, which is normally set at 10% higher than normal pressure. It should sit back at once when the pressure falls below and that pressure should be less than what it was sitting on all the time. The valve should afford a fast opening but should sit back as soon as the pressure falls to the set safe value. In normal operating conditions, the valve should sit tight and do nothing at all. It should not pass under all pressure, temperature, and flow conditions. It should then pop in spite of back pressure, if the system has any.

This sudden opening or "popping" is secured in all safety valves by an arrangement called the huddling chamber, which exposes a larger area of the disc to the action of the pressure as soon as leakage of the stream commences. This is done by making the disc somewhat larger in diameter than the sealing area and partially enclosing the seat in such a way that the vapor or gas is restrained from flowing out freely the seating surface. This is known as huddling chamber or control chamber (see Figure 18.3 A).

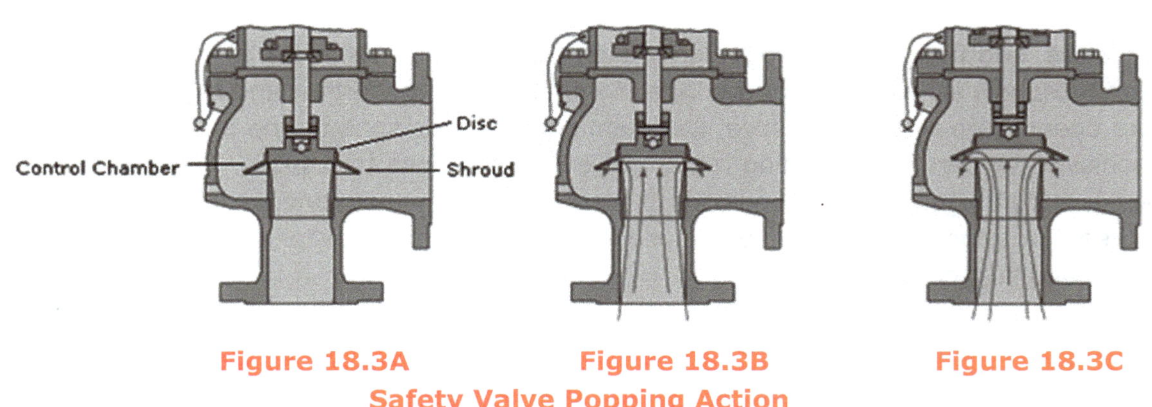

Figure 18.3A Figure 18.3B Figure 18.3C

Safety Valve Popping Action

In Figure 18.3B, as the valve starts to lift due to overpressure, fluid enters the control chamber. Now as a larger area is exposed to fluid pressure, the force increases. This overcomes the spring pressure faster, which results in a quick opening (see Figure 18.3C). The shroud also reverses the flow and this reaction force adds to the lift. As a result of these actions in the huddling chamber, the valve pops open as soon as an appreciable leakage takes place. Once the valve pops open and relieves the pressure in the vessel or system, it is desirable again to close the valve as soon as the pressure is reduced by the discharge to a point a little below the pop pressure. Because now a larger disc area is exposed to the system pressure, the valve will not seat back until the pressure falls below the working pressure. The design of the huddling chamber or the control chamber fixes the seat back point. To trim this point to suit individual system requirements, a nozzle ring is provided which can be adjusted (see Figure 18.4). Normally two adjusting rings are provided, upper and lower. The upper ring is factory set. The lower adjusting ring is adjusted to make fine settings in the overpressure and blowdown values.

The relief valve must, under all normal conditions, sit tight without any leakage. As a matter of fact, the valve sits in this condition for longer times under wide operating conditions of pressure,

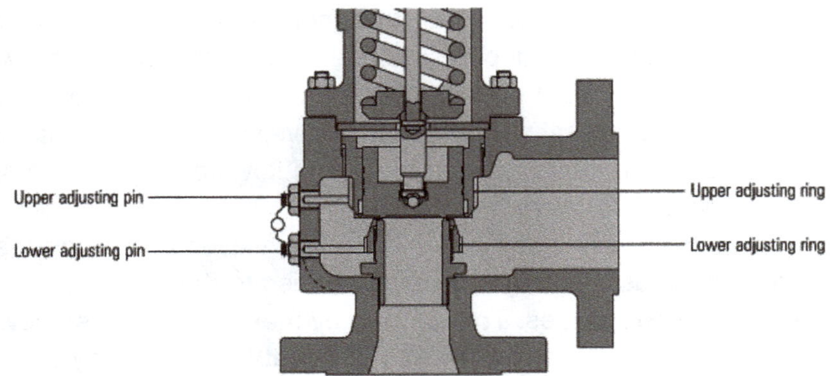

Figure 18.4
Safety Valve With Adjusting Rings

temperature, and flow exposed to severe fluids. Seat leakage will result in the loss of precious fluid, but also will slowly erode or damage the seatings where further and bigger loss and premature opening would be imperative. Excellent shut off conditions on the pressure relief valves are achieved by carefully selecting materials, extremely close tolerances, optically flat seating surfaces, and stringent control conditions in manufacture and maintenance. However, elastomer or soft seat arrangements can be employed in more exacting conditions if other pressure, temperature, and fluid conditions permit.

BACK PRESSURE AND BALANCED VALVES

Discharges of non-corrosive, non-toxic, clean, and expendable fluids like air and water from the relief valves are vented out directly into atmosphere, whereas all others are routed into closed system. These later systems exert a certain back pressure on the relief valves. Back pressure is an additional force that will add to the force of the spring; in turn, more pressure is required to open the valve. If the back pressure is 25 psi, it will require 25 psi more pressure to cause the valve to open. If the back pressure varies, it naturally has a significant effect on the discharge flow of the valve. As a simple solution, the spring force can be reduced, taking into account the back pressure from the system. However, in effect, this set point now will vary with the changes in the back pressure.

BALANCED BELLOWS VALVES AND BALANCED PISTON VALVES

To counter the variable superimposed back pressure, balanced bellows or balanced piston designs were developed. The bellows or piston has an effective pressure area equal to the seat area of the disc. The bonnet is vented to ensure that the pressure area of the bellows or piston will always be exposed to atmospheric pressure and to provide a telltale sign should the bellows or piston begin to leak. Variations in back pressure, therefore, will have no effect on set pressure. Back pressure may, however, affect flow. The bellows or pistons act to seal process fluid from escaping to the atmosphere and isolate the spring, bonnet, and guiding surfaces from contacting the process fluid. This is especially important for corrosive services. There is also built up back pressure — back pressure as a result of fluid flowing from the pressure relief valve into the downstream piping. It will not affect the valve opening pressure, but may show considerable effect on valve lift and flow.

CARE AND REPAIR

CARE

Safety and relief valves are the ultimate checks in protecting the equipment and piping from excessive pressure. So, it is absolutely essential to see that these are in good working order. Handle pressure relief valves as you would any precision pieces of equipment. They can be damaged easily. Also ensure that these safety valves operate at the desired set pressure for which they are intended. The following regular checks should be made on the safety valve.

1. When the relief valve is first put in commission, it should be operated manually by operating the lever provided for this purpose. This will blow out any dirt in the line, thus protecting the valve seat and disc.
2. Check these valves by gradually raising the system pressure and allowing the relief valve to blow, at the same time noting the pressure of the system. This is generally done on boilers. Realize that this sort of testing cannot be undertaken in processes where a pressure fluctuation upsets the process to a considerable extent. In such cases, the relief valves are periodically tested in the shop whenever a planned shutdown allows.
3. It is very important that no blinds or closed valves be left in between the system and the relief valves.
4. Relief valves are normally tagged with set pressure in psig, the position of the relief valve, and the date of last testing. It is the responsibility of the concerned department to keep record of these tests and see that these valves are tested periodically and honestly.
5. It is vitally important that safety relief valves are installed in their correct location. Safety relief valves are made of different materials for different services. The pressure setting is different for each valve. Work closely with the operating personnel when installing them. If you are instructed to remove a safety relief valve and there is no tag, place a paper tag on the valve and on the pipe flange with all pertinent information.
6. Wrong or badly fixed spiral-wound gaskets are found to be the major culprits in many instances. Due to non-availability of proper gaskets or just to save time, an oversized gasket is sometimes fixed with its windings protruding into the bore or the nozzle of the valve. When the valve pops, thin strips of metal unwind and make their way into the valve seat. The valve may seat back after the pressure goes down, but it will continue to pass. The only alternative would be to shutdown the system and remove the valve. You are warned!

HOW A SAFETY RELIEF VALVE IS REPAIRED

Sometimes, a safety relief valve will not close completely as it should after the pressure has been lowered. Safety relief valves may start to leak a small amount of liquid or gas or they may pop at a lower or higher pressure than their setting. Whenever a safety relief valve does not function properly, it must be removed from the system and sent to the concerned shop for repairs. Never make any adjustments or minor repairs to a safety relief valve while it is under pressure. Only the concerned and correct shop personnel should disassemble safety relief valves and make repairs.

If a safety relief valve does not close after it has popped, then either dirt or scale in the valve could be the cause. Usually, cleaning the valve will remedy this. If the safety relief valve begins to leak a little liquid or gas, the seats must usually be "lapped in" with grinding compound. If the valve does not pop at the correct pressure, correction must be made with the adjusting bolt nut.

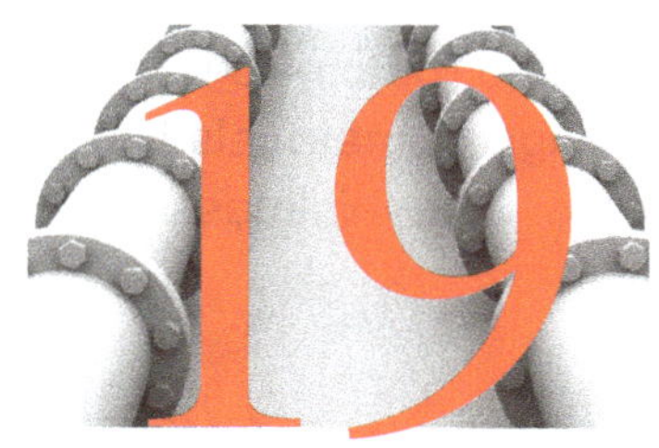

Rupture Discs (Safety Heads)

All pressure vessels must be equipped with a primary pressure relief device, which can be a relief valve, a rupture disc, or both. A rupture disc is a thin piece of stainless steel metal that is usually installed between a set of flanges in a piping system. It is a pressure-relieving device like a relief valve. However, unlike a relief valve, it is sacrificial, which means that it cannot be reused.

A rupture disc, as its name implies, is designed to rupture when the line pressure exceeds the safe operating pressure. It is basically pre-bulged metal disc made to fail at the rated pressure. All rupture discs have their rupture or pressure rating and their temperature rating stamped on the disc. It is important that the correct disc be installed in each piping system. The pressure rating for the support must correspond to the disc. If there is any possibility of the pipe line pressure dropping below zero psig, a vacuum support is also installed on the upstream side of the rupture disc. A vacuum support also has the pressure temperature stamped on it. Most rupture discs are designed to be installed in special flanges.

Rupture disks cost less than relief valves. They generally need a little or no maintenance. No calibration is needed in its life time. But it is a non-closing device which must be replaced after a failure.

When a disc ruptures, the full pipe line volume can be discharged, thus reducing the pressure quickly. This is an advantage rupture disks have over relief valves, which restrict the flow considerably. They have an exceptionally quick response. They should invariably be placed on the vessels that can run away (no pun intended), where relief valves may not really be quick enough to prevent a catastrophe.

When the rupture disc ruptures, the product continues to be lost until the flow is stopped by other means. If loss is not considered, it is the right choice for pressure release. However, an open valve is usually installed upstream of a rupture disc to restrict or stop the flow through the disc after the excess pressure has been reduced.

Certain highly viscous, corrosive fluids may actually damage the internals of relief valves. Some slurries, liquids with solids, and toxic and freezing fluids may plug them. Then a rupture disc downstream can protect the relief valve. The disc bursting pressure would then correspond to the relief valve setting, or slightly lower.

Rupture discs are pieces of precision manufacture and 20% of all discs are destructive tested. They can work under extremely high temperatures to cryogenic, very high pressures

(100,000 psi) to vacuum. They can handle routine-to-hostile fluids. Sizes generally available are from 1/8" to about 48" with a whole range of materials like aluminum, stainless steel, Hastalloy, inconel, Monel, and other exotic metals. Disc thickness may vary from 0.001" to 0.065".

INSTALLATION

Normal system pressure and temperature, media, pulsating gas flow or surging liquid, ambient environment, continuous pressure service or continuous vacuum service or alternating, discharge position, and location are some of the considerations for rupture disc installation.

All design codes require that the maximum rating of a relief device installed on a pressure vessel must not permit the operating pressure to rise more than 110% of the vessels maximum allowable working pressure (MAWP) in an unfired system.

There are three methods of installing a rupture disc: a) all alone, b) in series with a relief valve, and c) parallel to a relief valve. These methods are adopted best suiting the operating philosophy and system demand.

Figure 19.1A shows where a rupture disc is used alone as the pressure relieving device. It creates an unrestricted flow path but for its sizing. If discharge piping is used it must adequately be supported. In **Figure 19.1B**, the rupture disc assembly protects or isolates the relief valve from the operating fluids. The system is known as the primary relief. However, in a properly-designed system, the relief valve controls the flow but also provides shutoff after blowdown. A pressure sensing devise must be located between the valve and the disc assembly. **Figure 19.1C** is a free system where both act independently.

Rupture Disc Installations

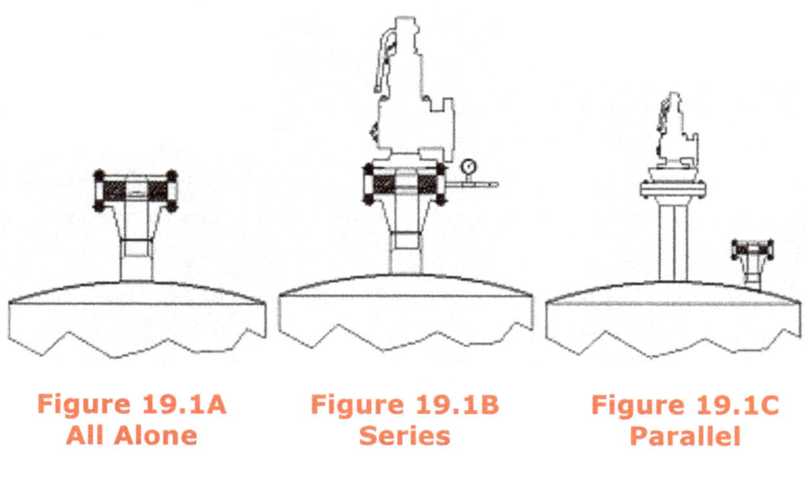

Figure 19.1A All Alone | **Figure 19.1B** Series | **Figure 19.1C** Parallel

TYPES

In all cases the relief device should be placed in such a way that it immediately and continuously senses the pressure and acts upon an upset.

Basically, rupture discs are metal or graphite. Metal discs are further classified as conventional or forward acting, compression or reverse acting.

Figure 19.2 shows the classical design of the rupture disc. Its prebulged design keeps its contour during normal operating pressure and temperature. It stays put in the condition until overpressure overcomes the mechanical strength of the thin disc metal that will eventually burst. As the pressure increases beyond the operating point, the bulge raises what will ultimately burst, dumping excess pressure. Rupture discs have a flat or concave surface. They are available with a scored pattern that will follow a designated pattern rather than just fragment. Figure 19.3A shows a typical rupture disc with scorings and Figure 19. 3B shows such a rupture disc after bursting. Compare this with the rupture disc in Figure 19.2.

Figure 19.2
Standard Rupture Disc

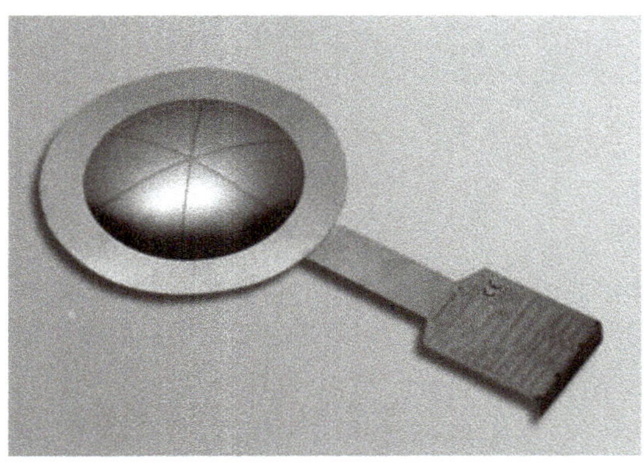

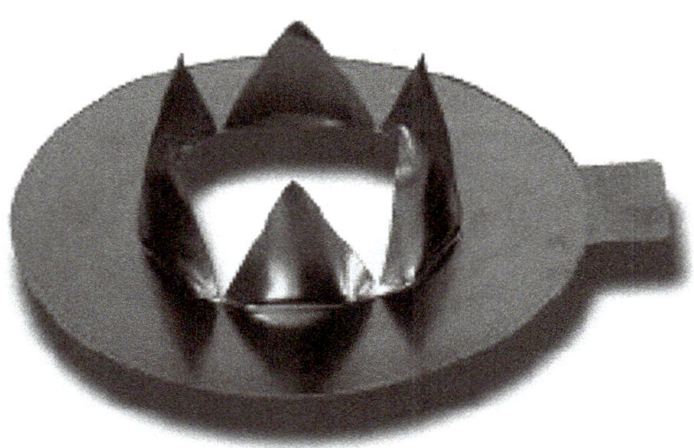

Patterned Rupture Disc

Figure 19.3A **Figure 19.3B**

Composite rupture discs generally have a seal membrane (usually in PTFE), a back up ring, and a main disc. Patterned slots and holes do not allow fragmentation, but make a disignated bursting path (see Figure 19.4).

Figure 19.4
Composite Rupture Disc

On the other,hand, in the reverse acting rupture disc the process fluid attacks the disc on the convex side. As the pressure increases beyond the limits, compressive forces turn the disc over or reverse it (see Figures 19.5A and 19.5B). In one design, knife blades are acurately positioned, which cut the disc open. In another design, a scoring pattern is made on the disc and the disc folloows the pattern while bursting. Reverse acting discs can operate up to 90% of their designed pressure. They are capable of full vacuum without additional backup disc. They can better handle pulsating and surging conditions. They can be made with thicker materials for better corrosion resistance.

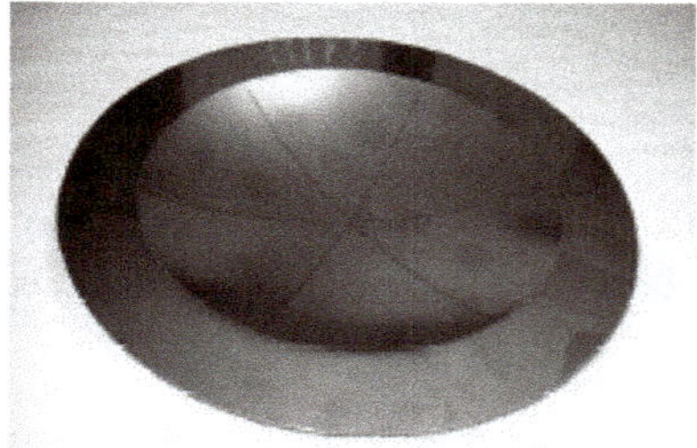

Reverse Acting Rupture Disc

Figure 19.5A **Figure 19.5B**

Graphite discs are used at low and medium pressure in contact with aggressive fluids, even at high temperatures. They are normally used at low pressures. Graphite discs are made of high

Figure 19.6A

Figure 19.6B

Graphite Discs

purity graphite impregnated with resins in order to make the product non-porous. The opening is total, but with fragmentation. Vacuum support may be needed in most cases (see Figures 19.6A and 19.6B).

Rupture discs should be placed in the designated holders and bolts should be accurately torqued for a reliable performance of the devices. The holder model must be properly selected according to the disc model. Some models of rupture discs can be placed across standard flanges, whereas most of them are not. Discs and holders are available in a number of engineering materials such as carbon steel, stainless steel, nickel, Monel, Inconel, and Hastelloy (see Figure 19.7).

Figure 19.7
Rupture Disc Holders

CARE

- Install the rupture discs properly, fully following the manufacturer's instructions. Discs often tell the story of failure, leaving telltale marks.
- Check the condition of the disc before installtion. Check for inappropriate bulging, dimpling, slipped back up discs, etc.
- Do not touch the rupture disc with your hands. Hold only by the tag handle.
- Damaged flanges, damaged holders, uneven bolting, under or overtightening will surely play spoil sport.
- Always use a torque wrench and torque the bolts to the specified torque. Over-tightening will crush the disc or separate the plastic seal whereas the disc may slip off if not tightened properly.
- Flange and holder surfaces should be absolutely clean free of foreign material, nicks, and burrs. You may use a fine emery cloth to smoothen the surfaces, but never attempt to machine them.
- Check for corrosion, pittings, and cracks. The outlet flange may be covered with a thin plastic sheet to prevent atmospheric corrosion damaging the discs.
- If there is any little chance for vacuum in the line, vacuum support should be installed. Check if they are properly attached.

- Store the rupture discs properly in their boxes along with the instructions supplied with each disc. Because the metals are so thin, they can be hurt easily. If they are hurt, you will fall ill at one point of time!
- Check for leaks if the rupture disc is used in series with a relief valve. Back pressure is likely to build up, impairing the operation of the disc. If a leak is observed, replace the disc as soon as possible.
- Fix the disc with proper orientation. A reversed disc ruptures at much higher pressure, defeating its very purpose.

Strainers

A strainer is a device which strains or mechanically separates foreign particles from a flowing fluid in the pipe line, generally with the help of a mesh, perforated sheet, to protect the downstream rotary equipment like pumps, compressors, and other process equipment like reactors, convertors, and heat exchangers. The strainers should invariably be fixed upstream of steam traps, control valves and flow meters. The most common types are cone strainers, Y strainers, basket strainers, and duplex strainers.

Temporary strainers or cone strainers are basically designed for use during line start up, where a considerable amount of line debris of weld metal is expected. The straining element can be a cone, a cylinder, or a plate, with a heavy wire mesh (see Figures 20.1A and 20.1B). In case of fine mesh, it is backed up by perforated screen. These are fixed in between two mating flanges with the mesh cone in the line, generally within a spool piece a little longer than the cone strainer is used. These are generally found in the suction lines of pumps, often left in place after initial trials.

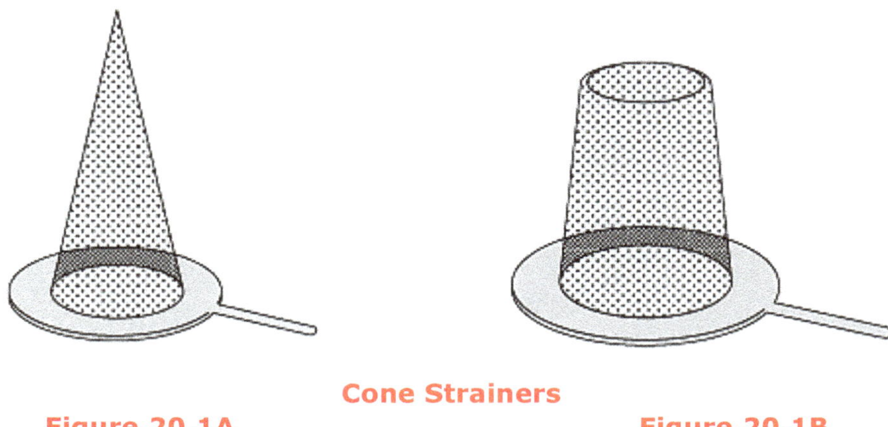

Cone Strainers

Figure 20.1A

Figure 20.1B

Y strainers are routinely used in the pipelines and standard piece of equipment in steam lines (see Figure 20.2). They are available up to 6000 psi working pressure and higher metallurgies like chrom-moly steels must be used for handling higher pressures and temperatures. The direction of flow is marked on the body (see Figure 20.3). The strainer element should point downwards. However, on steam and gas lines, it should point horizontal so that liquid or condensate does not collect, causing erosion and corrosion. Figure 20.4 shows the correct orientation of strainers.

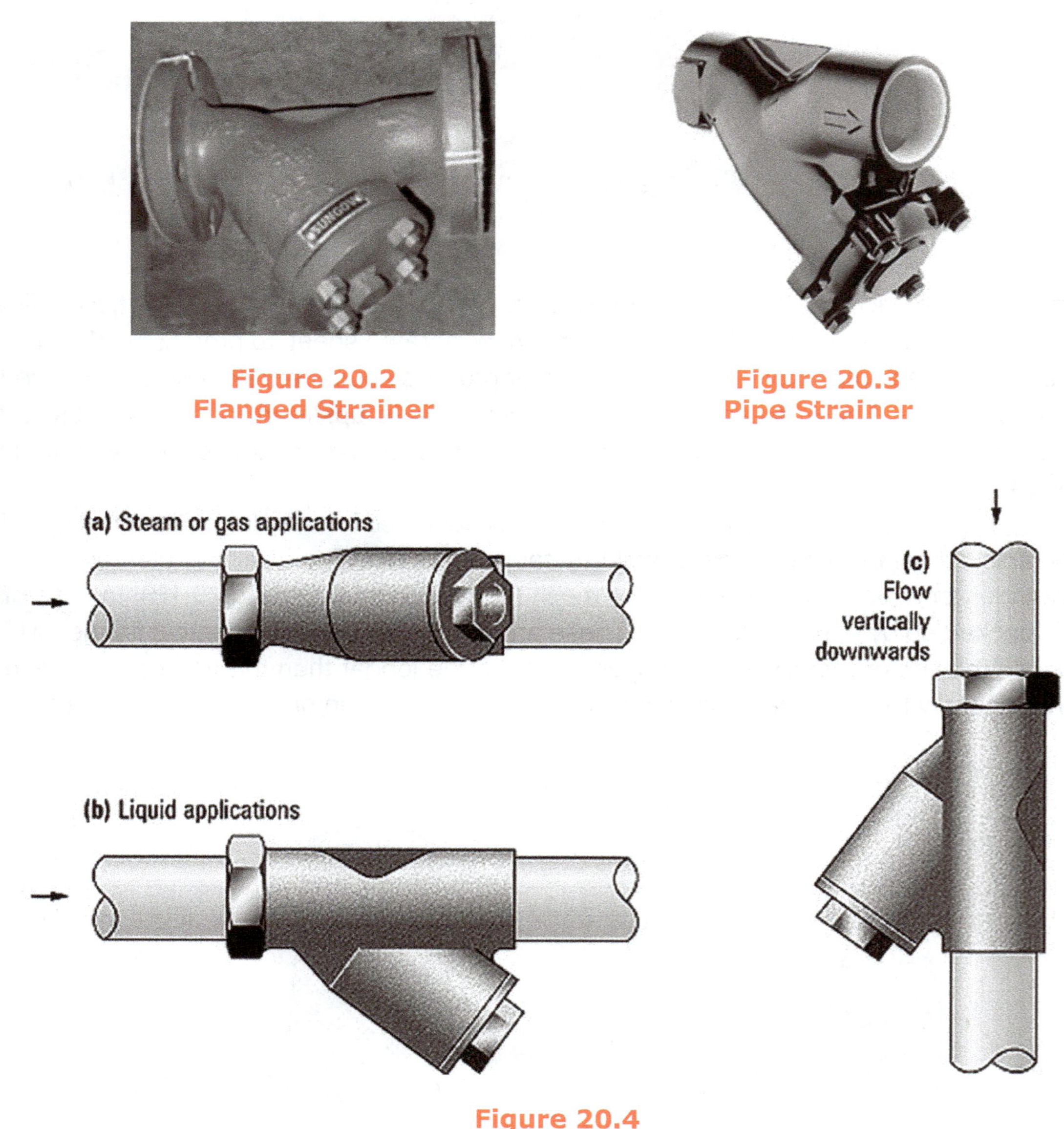

Figure 20.2
Flanged Strainer

Figure 20.3
Pipe Strainer

Figure 20.4
Correct Orientation of Strainers

The element can be removed, cleaned, and fixed back. Self-cleaning provisions can be made that may not require shutdown of the line. A blowdown valve can be fixed on the strainer cap; it can be used to remove the collected debris without shutting down the line. Line pressure then helps to push out the debris.

Normally Y strainers have less debris-handling capacity than basket type strainers and also have a higher pressure drop. Figure 20.5 shows a basket strainer.

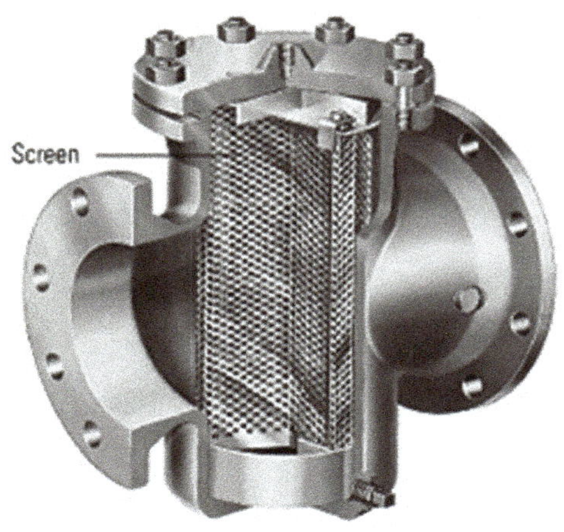

Figure 20.5
Basket Strainer

Duplex type basket strainers make the maintenance easy. Debris can be removed and filters cleaned without shutting down the line, but they can be fixed only on horizontal lines. In a duplex strainer, two strainers are fixed parallel with suitable valve arrangement, such that any one of the strainers can be taken into line, other into maintenance or cleaning (see Figure 20.6). The drain plug is provided for strainers on steam lines to remove any accumulated condensate from time to time. These strainers can be automated.

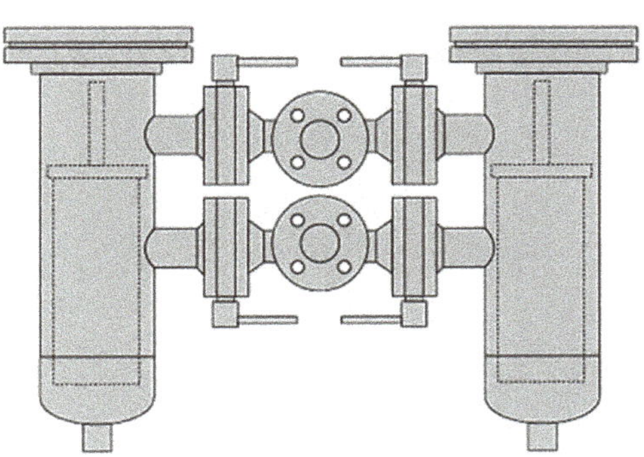

Figure 20.6
Duplex Strainer

With suitable valve arrangement, the strainer can be backwashed, where liquid flowing through the strainer is directed in the reverse and debris is collected down. This arrangement is often found near the heat exchangers where the cooling medium is like sea water containing a lot of foreign materials. Self cleaning strainers are also available.

The mesh sizes and metallurgies are dictated by the process parameters. In some cases, a magnet is fixed in a suitable corrosion-free enclosure to collect magnetic particles like rust, etc.

STEAM TRAPS

Contrary is true in the case of a steam trap. A steam trap actually traps condensate rather than steam, air, and gases, and removes them away from the steam system. They are vital components in the steam piping and indispensable in maintaining proper steam temperature and dry condition. We have large varieties of steam traps catering to individual needs of flow from vacuum services to very high pressures. Steam traps must be selected and tailored to suit individual services and maintained as such. Like other piping components, they are subjected to extremes of pressure, temperature, and vagaries of flow conditions. Unlike other piping components, these traps actually save energy, prevent corrosion by reducing water logging in the system, and return some of the costliest chemicals back to the boiler. Traps are used with heat exchangers, heating coils, turbines, absorption chillers, and wherever steam is spent. A steam trap has a whole set of jobs to do.

The trap must remove the condensate as it is getting developed in the steam system. Condensate may have to be discharged at the steam temperature. When the condensate load varies, as in the case of heat exchangers, the steam trap must discharge varying loads of condensate under changing differential pressures across the trap. Air is everywhere but it must be effectively evacuated from the steam systems; proper space must be created for the steam to act and do its job. The system must also be purged of any other gases. That brings us to the dirt, which may be carried all the way from the steam drum and pipe lines with all the chemicals adding to its load. Corrosion is a perennial problem. If the traps discharge to gravity or open to atmosphere, the back pressure is zero. But if they drain into a wet return line or if they should lift condensate to an overhead return line, there will be a back pressure against which they must work. When steam and condensate mixes, water hammer can occur; this can damage sensitive parts. The design of the equipment from which the condensate is trapped plays an important role in the selection of the trap. Steam should not be lost as it is a precious commodity that should be made as clean and dry as possible.

It is often difficult to select the right trap, with each type having its own benefits, disadvantages, and characteristics. Water hammer, steam condition, maximum and working steam pressure, system back pressure, temperature, and flow rate, reliability of type, and make dictate the selection of the right steam trap and its size (see Figures 21.1, 21.2, 21.3, and 21.4).

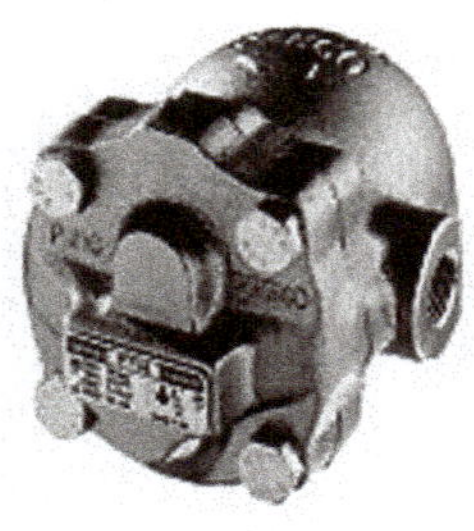

Ball float type

Figure 21.1

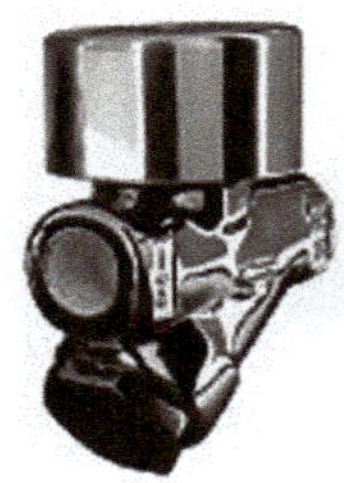

Thermodynamic type

Figure 21.2

Thermostatic type

Figure 21.3

Inverted bucket type

Figure 21.4

TYPES OF TRAPS

All steam traps fall into three main categories, also standardized by International Standard ISO 6704:1982. They are thermostatic, thermodynamic, and mechanical traps, a brief description of which follows.

THERMOSTATIC (OPERATED BY CHANGES IN FLUID TEMPERATURE)

A thermostatic trap operates on the principles of expanding liquids and metals that shut the valve or open it. The temperature of saturated steam is determined by its pressure. As steam gives up its heat, condensate is produced at steam temperature. As the process continues, the temperature of the condensate will fall. As long as air and condensate are cooler, the thermostatic element contracts and the trap discharges them. When the condensate gets hot with the ingress of steam, the element expands and closes the valve. It stays thus until cool condensate contracts it again.

Figure 21.5A shows a simple thermostatic trap and Figure 21.5B shows its cutaway view. Figure 21.6 shows its internal arrangement. An oil-filled element expands when heated to close the valve against the seat. The adjustment allows the temperature of the trap discharge to be altered between 140°F and 212°F (60°C and 100°C), which makes it ideally suited as a device to get rid of large quantities of air and cold condensate at start-up.

Figure 21.5A
Thermostatic Trap

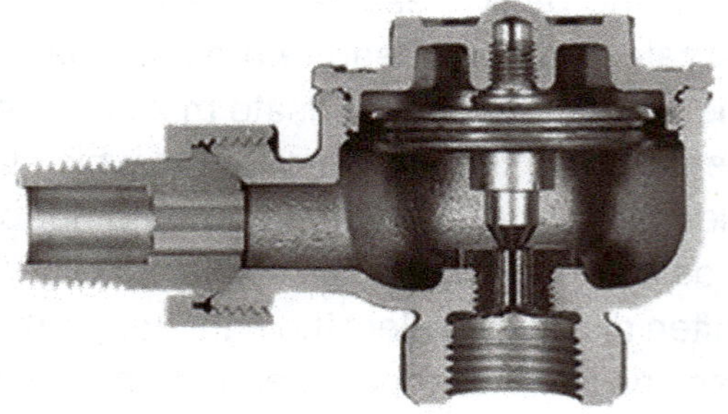

Figure 21.5B
Cutaway View Of Thermostatic Trap

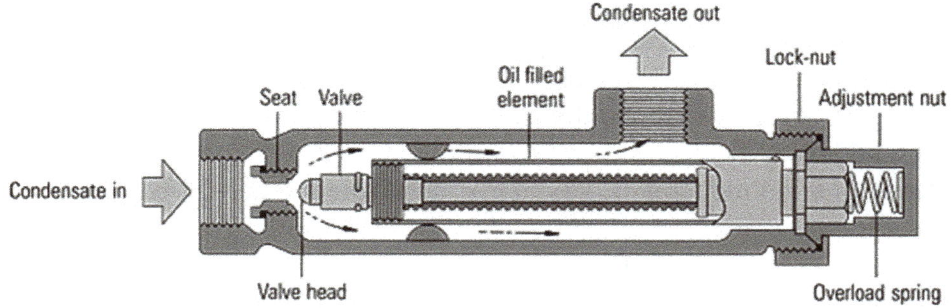

Figure 21.6
Internal Arrangement of Thermostatic Trap

Temperature adjustment within certain limits below 212°F/100°C is a blessed advantage for this trap. The trap is fully open when cold, such that it can vent air and discharge condensate at start up. Obviously it cannot be used for immediate withdrawal of condensate from steam space and cannot be used as such. It can withstand vibration and water hammer. However, the sensing element is delicate and corrosion can play its role and destroy it. In freezing conditions, the trap must be insulated.

THERMODYNAMIC (OPERATED BY CHANGES IN FLUID DYNAMICS)

Thermodynamic traps take advantage for their operation with the difference in velocity or kinetic energy between steam and condensate passing through a fixed or modulating orifice. Downstream flash steam also helps in its operation. These traps are simple, robust, and reliable; they can operate up to very high temperatures and pressures (see Figure 21.7). They are found extensively all over the industry on the steam lines every tens of meters. Thermodynamic, disc, impulse, and labyrinth steam traps fall into this group. They generally last a lifetime with a fairly large capacity for size. Stainless steel construction makes them impervious to corrosion. With only one moving part; their maintenance is easy and can be carried out without removing from the line. Vibration, water hammer, and freezing temperatures do not affect it, but they may not

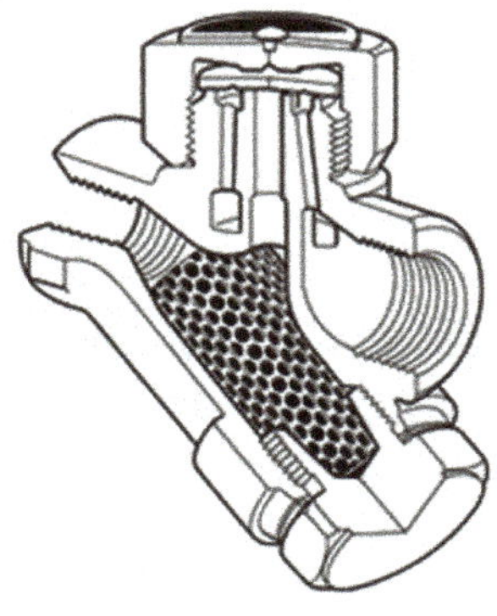

Figure 21.7
Thermodynamic Trap

work well in low differentials. Their presence is felt all along the steam lines with the "phut-phut" discharge noise they make; thus, they are not considered in silent zones.

Thermodynamic steam traps rely partly on the formation of flash steam from condensate. The only moving part is the disc above the flat seating inside the control chamber or cap, which operates with an audible click.

The operation of a thermodynamic steam trap is shown in Figure21.8.

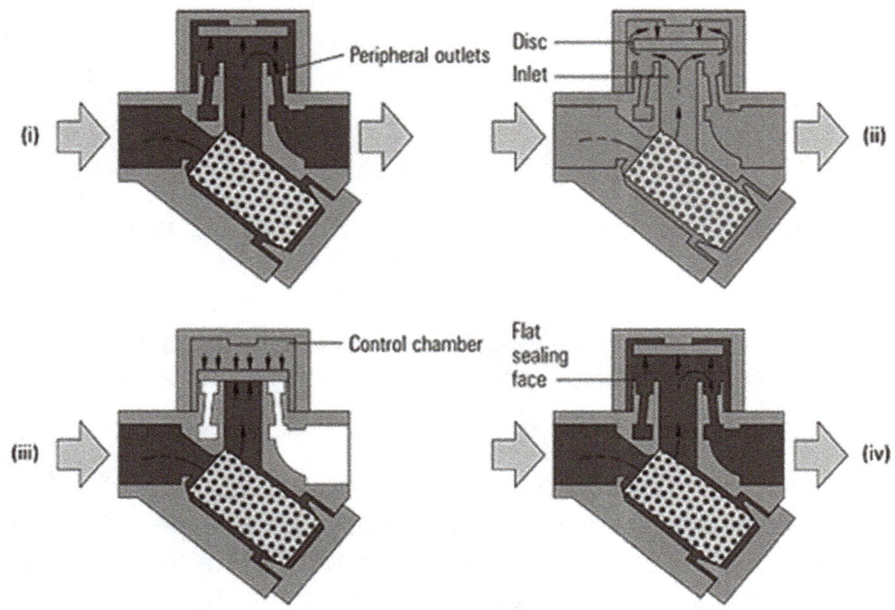

Figure 21.8
Operation of a Thermodynamic Steam Trap

1. At start-up, incoming air and condensate pressure raises the disc off the valve seat, and they are immediately discharged into the outlet line.
2. Hot condensate flashes to steam as it goes through the trap body, which creates a lower pressure area because of its high velocity under the disc.
3. The disc seats with the pressure of the flash steam in the cap .The trap remains closed until the flash steam condenses.
4. System pressure raises the disc off of the valve seat.

MECHANICAL (OPERATED BY CHANGES IN FLUID DENSITY)

Mechanical traps are essentially mechanical in operation by sensing the difference in density between steam and condensate. Ball float traps and inverted bucket traps fall into this category. In the ball float trap, the ball rises in the presence of condensate, opening a valve which passes the denser condensate. With the inverted bucket trap, the inverted bucket floats when steam reaches the trap and rises to shut the valve. They are continuous traps discharging large volumes of condensate.

BALL FLOAT STEAM TRAP A ball inside the valve body floats on the incoming condensate. The ball is linked to the valve seat; as the condensate rises, the valve is opened up, releasing condensate. The valve is always flooded and neither steam nor air will pass through it. Thus, early traps had a manual vent cock to release air (see Figure 21.9) but present traps use a thermostatic air vent, which allows the initial air to pass. A float-thermostatic trap (see Figure 21.10) is the closest to an ideal steam trap; it will discharge condensate as soon as it is formed, regardless of fluctuations in steam pressure and flow. Its capacity is high for its size. Water hammer has a little effect on the trap, but they are susceptible to freezing.

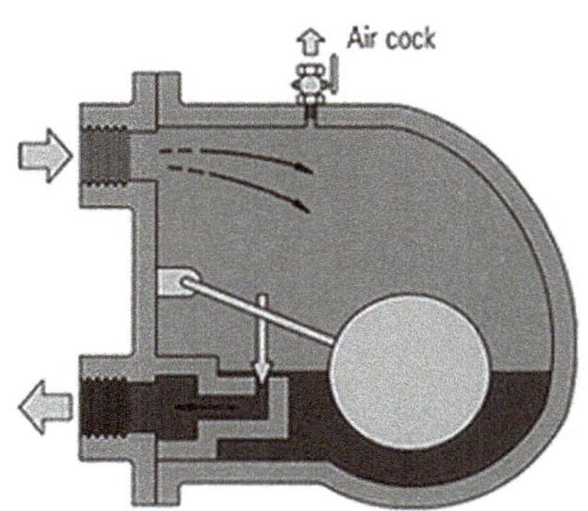

Figure 21.9
Float trap with air cock

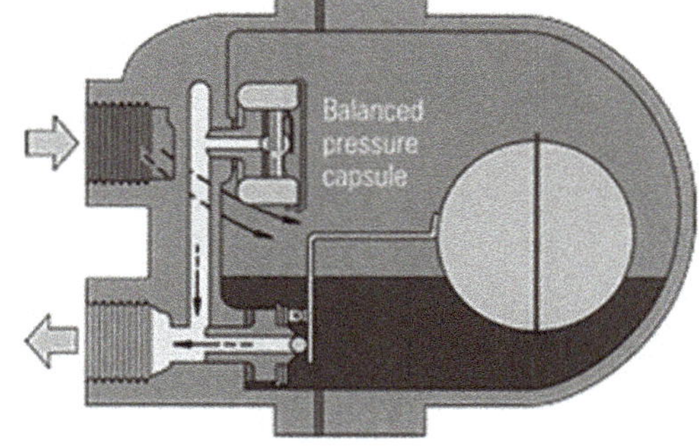

Figure 21.10
Float trap with thermostatic air vent

The automatic air vent uses the same balanced pressure capsule element as a thermostatic steam trap, and is located in the steam space above the condensate level. After releasing the initial air, it remains closed until further accumulation of air or other non-condensable gases.

INVERTED BUCKET STEAM TRAP An inverted bucket steam trap can withstand high pressures and tolerate water hammer. It can be used in conjunction with a check valve on the inlet on superheated steam lines and in places where fluctuations are present. If the water seal is lost in the bucket, the trap can pass steam directly. Air release is slow due to the small orifice because a large orifice would pass steam. The trap should be insulated in freezing conditions.

The trap consists of an inverted bucket attached by a lever to a valve. An essential part of the trap is the small air vent hole in the top of the bucket. Figure 21.11 shows the system of operation.

1. The bucket hangs down, pulling the valve off its seat. Condensate flows under the bottom of the bucket filling the body and flowing away through the outlet.
2. The arrival of steam causes the bucket to become buoyant, it then rises and shuts the outlet.
3. The trap remains shut until the steam in the bucket has condensed or bubbled through the vent hole to the top of the trap body. It will then sink, pulling the main valve off its seat. The cycle repeats.

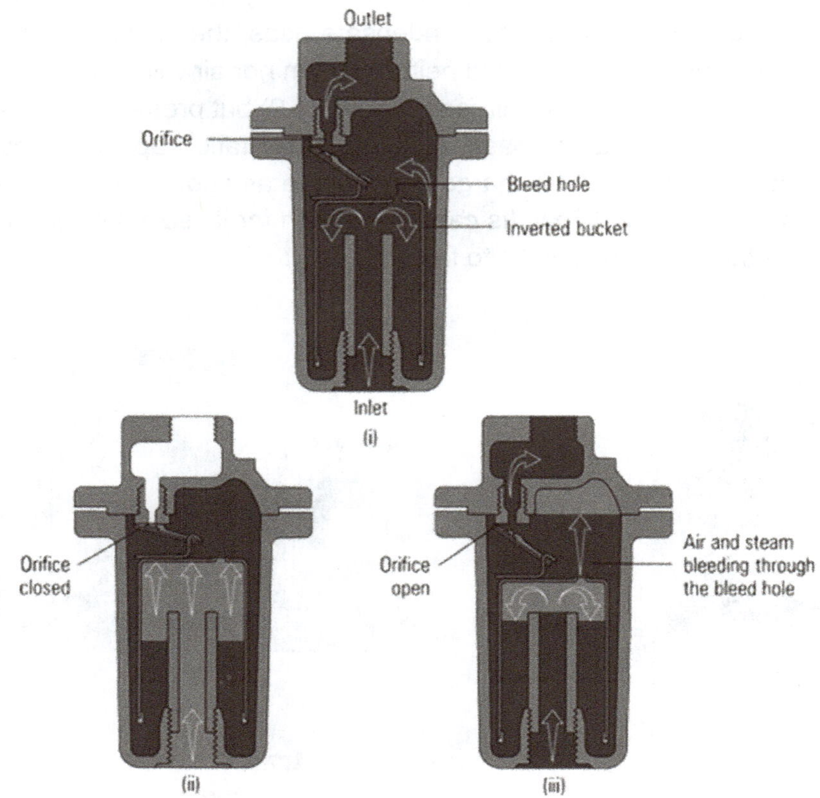

Figure 21.11
Operation Of Inverted Bucket Steam Trap

BALANCED PRESSURE STEAM TRAP There are other variations of the basic liquid expansion trap. The balanced pressure steam trap is small and light in weight; its capacity is large for its size and it's easy to maintain (see Figure 21.12). The valve fully discharges air and other non-condensable gases at start up. Some tolerate fluctuations in steam pressure, but earlier versions were susceptible to water hammer.

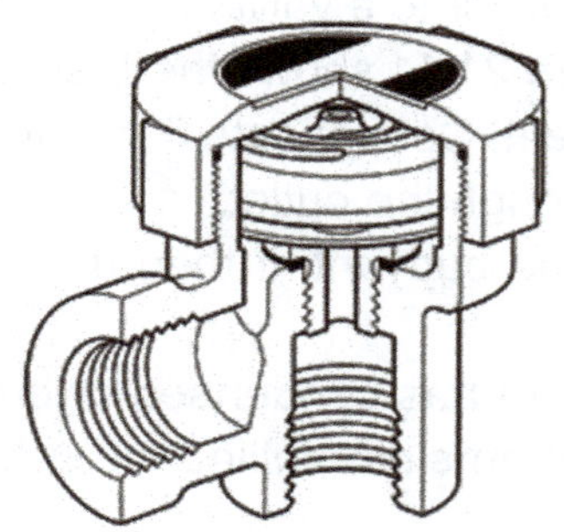

Figure 21.12
Balanced Pressure Steam Trap

A balanced pressure trap contains a capsule with a special liquid and water mixture inside whose boiling point is less than that of water. At start-up, the valve is wide open, removing air. As hot condensate enters the trap, liquid in the capsule vaporizes and closes the valve before steam can escape the trap. But cool condensate in effect contracts the capsule. The valve opens releases the condensate and the cycles repeat (see Figure 21.13).

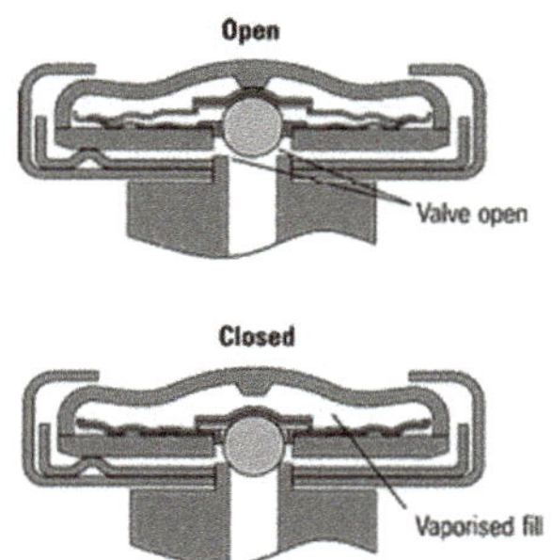

Figure 21.13 Operation Of Balanced Pressure Steam Trap

There are some more variations of the basic trap, such as bimetallic, labyrinth, fixed orifice, and impulse steam trap, with their own advantages and disadvantages.

MAINTENANCE OF STEAM TRAPS In their lifetime, steam traps save much more money than their original cost and repair cost. It is essential that best results be achieved from these silent heroes — thermodynamic traps are not so silent! Generally traps work well, but defective ones may pass steam or block the flow. Passing steam is a tell-tale sign, but blocked flow could be intriguing.

It is difficult to pinpoint a bad trap, but not impossible. Normal or ultrasonic listening devices, sight glasses, and temperature techniques are all adopted to locate leaky, sluggish or not working steam traps, each with its own merits and demerits. Sensors embedded into the traps are ideal but costly. Wrong diagnosis may lead to wastage of money and shutdown time.

Scheduled check and maintenance of the traps is important. A simple cleaning, reuse, or renewal of internals will do wonders for energy conservation. It does not take a high skill to open and assemble back most of the traps. It will be even cheaper to renew an undependable trap.

Comparisons are difficult and each type of trap has its own advantages and disadvantages. Claims of energy and cost savings by various types are often misleading, but proper installation as well as selection of steam traps is important.

INSTALLATION GUIDELINES

Following are the general installation guidelines for steam traps.

- Install at accessible location.
- Install at lowest point of system.
- Use a fine mesh strainer with a blowdown valve ahead of the trap.

- Be sure the trap is the right one for the job. Match the trap type, pressure, and flow suitable to the required conditions to let it operate properly and at peak efficiency.
- Always place float type traps in a vertical position.
- Place lines to trap correctly, that is, inlet-to-inlet and outlet-to-outlet.
- Install shut-off valves on either side to enable isolation for maintenance when required.
- Install unions or flanges on either side of the trap to facilitate removal.
- With most designs, install the trap as close as possible to the equipment being drained.
- Never use pipe smaller than the connection size.
- Pitch horizontal inlet lines to traps to keep them full of condensate and prevent steam binding and water hammer.
- Do not mix condensate if there are different steam pressures as such in the plant.
- Immediately before installing the trap, blow out the inlet lines with full steam pressure to remove dirt, oil, and pipe thread cuttings which may be in the lines.
- Install one trap on each heating coil or each unit of equipment being drained.
- When more than one trap drains into a common return line, install a check valve between each trap and the return to prevent reverse flow, in case one or more units are shut down.
- The return line should be no higher above the trap than the level to which the lowest line pressure can lift condensate.
- The drip trap should be installed for every 100–150 feet of straight piping run, at every change of piping elevation, at risers and ahead of expansion loops. Most often, these are the thermodynamic type of no more than 3/4" (20mm).

Chart 21.1 summarizes the loss due to the leaking steam trap discharge rate (Steam Loss lbs/hr), though it is often repeated at many places.

Chart 21.1
Steam Trap Discharge Rate

Trap Discharge Orifice (inches)	15 psig	100 psig	150 psig	300 psig
1/32	0.85	3.3	4.8	*
1/16	3.4	13.2	18.9	36.2
1/8	13.7	52.8	75.8	145
3/16	30.7	119	170	326
1/4	54.7	211	303	579
3/8	123	475	682	1,303

FLOW MEASUREMENT

Controlling the flow rate of liquids is a key process management mechanism for any chemical plant. There are many different types of devices available to measure flow. The original theories behind fluid flow were published by the Swiss physicist Daniel Bernoulli in *Hydrodynamica* in 1783.

Flow meters are devices that measure the amount of liquid, gas, or vapor that passes through them. They generally consist of a transducer, transmitter, and panel meter. The transducer senses the flow of the fluid and the transmitter generates and transmits a usable flow signal from the measured transducer signal. The panel meter indicates and often records the signal. All these three could be combined and give a filed indication or both field and remote indication, or there can be a just a field indication like the good old rota meter.

There is no single answer flow meter for all services. Flow meters need to be tailored to individual needs. Most of the flow meters are velocity flow meters which measure the velocity (v) of the flow to resolve the volumetric flow. Examples are magnetic, turbine, ultrasonic, and vortex shedding flow meters. Only positive displacement flow meters are volumetric types which measure the volume of fluid passing through the flow meter. Coriolis mass and thermal flow meters are mass flow meters. Differential pressure, target, and variable area flow meters actually deduce the flow from other data parameters and, hence, are known as inferential flow meters.

It is important to select a good flow meter appropriately suited for the application, for which some or all of the following points need to be considered: the fluid measured by the flow meter, chemical compatibility, viscosity of the liquid, minimum and maximum pressure and temperature, MOC, size of the pipe, range required, rate measurement and totalization required, service conditions like clean fluid, and slurries. Whether measurement display is required locally or remote or at both places, and the reliability of the said instrument are also the points to be considered.

Technology and instrumentation are being upgraded at a fast pace — more so with flow meters! Hence, it is also essential to consider the experience and knowledge of the instrument personnel to calibrate or repair the meter. There is no point in procuring a sophisticated meter that only overseas service personnel can calibrate.

In liquid service, the flow meter should be always full of liquid because gas or vapor in the line can alter its accuracy. On the contrary, in gas and vapor service, the flow meter should be always full of gas or liquid because liquid in the line can affect its accuracy. Two-phase flows can adversely affect the precision of many flow meters.

The pipe line should run straight before and after the flow meter. Elbows, valves, and branch offs can adversely affect the flow reading. Similarly, plugging conditions and slugs can affect the flow readings.

TYPES OF FLOW METERS

There are innumerable varieties and variations of flow meters. Here we have a description of some of the more popular flow meters.

ROTAMETERS

The rotameter is probably the most widely-used flow meter because of its low cost, simplicity, low pressure drop, relatively wide range ability, and linear output. Its operation is simple and the reading is straight forward.

Most rotameters/variable-area flow meters consist of a tapered tube and a float in it. The tubes are made up of glass or plastic, and occasionally metal. Fluid flow raises a float in the tapered tube, and increases the passage of the fluid. The float rises higher with higher flows. As the float rises, the area of the annular opening increases. Float is raised by combined forces of the buoyancy and the velocity head of the liquid, whereas only the velocity head raises the float in gases. The float stays stable when the upward force exerted by the flowing fluid equals the weight of the float. This position matches with the particular flow rate depending on the density and viscosity of the fluid. Hence, it is necessary to size the rotameter depending on these parameters.

Be sure to install these flow meters in the vertical position as the float depends on gravity. There should be a straight pipe run of five times the diameter of the pipe. However, there are certain spring-loaded float models that can be installed horizontally. The rotameter has a locally readable linear scale, a relatively long measurement range, and low pressure drop; it is simple to install and maintain. However, typical accuracy is in the order of ±3% full scale.

Water hammer — and for that matter any hammer — can damage this device! This kind of meter is sensitive to changes in flow parameters like viscosity, density, and temperature and is affected by pulsation.

Hence, it is important to use these meters where there is a potential physical damage internal or external. Do not use rotameters for dirty fluids or they can leave a coating on the metering tube or float. Under certain circumstances of high velocity flow, the float can get struck at the top extreme position. Similarly, under extended no-flow conditions, the float can get struck at extreme positions.

In general, most of the rotameters are glass tube types, where the tube is precision made borosilicate glass, and the float is precision machined metal — normally stainless steel, or glass or plastic (see Figures 22.1A and 22.1B). The measuring glass tube is protected by another glass tube which is available in a number of end connections. These tubes are not recommended for water over 194°F/90°C, (its high pH softens glass), wet steam, caustic soda (dissolves glass), and hydrofluoric acid (etches glass).

Glass fixes the temperature limit at 400°F/204°C. However, metal tubes are used for higher pressures and temperatures with magnetic or mechanical linkages for external readout. For certain specific applications like deionized water or corrosive fluids, plastic tube rotameters (generally polycarbonate) are used with the added advantage of lower cost and high impact strength.

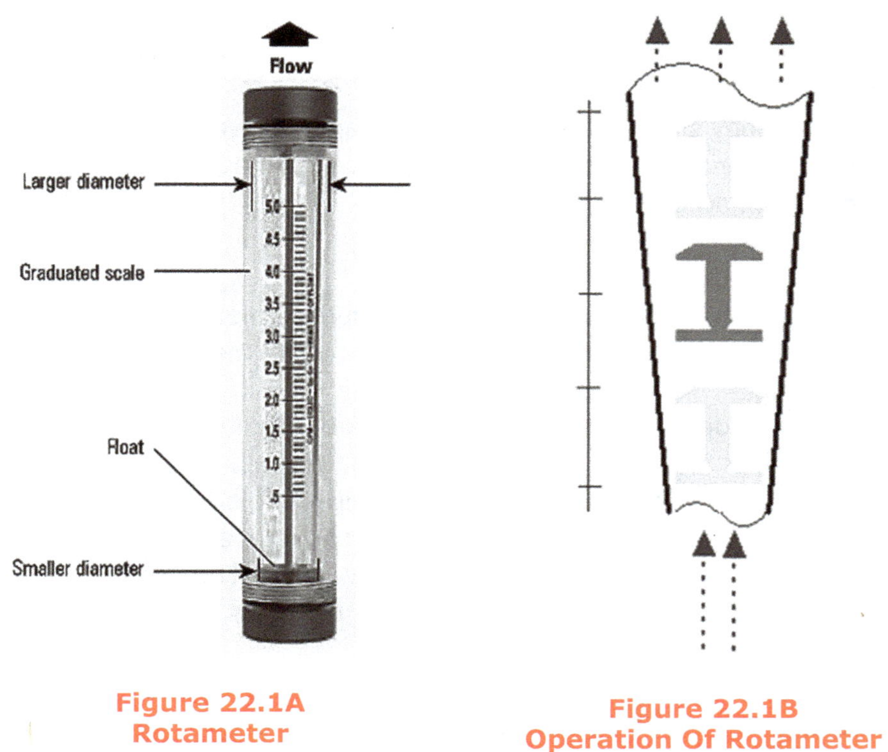

Figure 22.1A
Rotameter

Figure 22.1B
Operation Of Rotameter

The Rotameter was invented by a European inventor Rota, and is named after him. Another version says that the early designs of the rotameters had slots, as a result of which the float rotated or spinned — hence the term rotameter!

VANE STYLE VARIABLE AREA FLOW METERS

Vane-style variable area flow meters have vanes that move in relation to the flow rate. Piston variable area flow meters use pistons that move in relation to the flow rate. The movement is spring loaded which can be coupled to an indicator or transmitter.

ELECTRO MAGNETIC FLOW METERS

The Tobin Meter Company first introduced magnetic flow meters for commercial use in Holland in 1952. Foxboro introduced them in the United States in 1954. Magnetic flow meters use Faraday's Law of electromagnetic induction. According to this principle, when a conductive medium passes through a magnetic field, a voltage is generated. This voltage is directly proportional to the velocity of the conductive medium, the density of the magnetic field, and the length of the conductor. In Faraday's Law, these three values are calculated, along with a constant, to yield the magnitude of the flow. Sensor electrodes placed on the flow tube walls sense the voltage signal.

These flow meters are of two types: full bore and insertion type. These flow meters are used to measure the flow of conductive liquids and slurries, including paper pulp slurries, fertilizer plant slurries, and black liquor. They cannot measure non-conductive fluids like water, acids, caustic, hydrocarbons, and slurries.

For slurry services, they should be operated above a certain velocity (typically 1 ft/sec) where the solids would not settle and obstruct the flow. In the case of abrasive liquids, they should operate at low velocity (typically below 3 ft/sec) to reduce wear.

TURBINE FLOW METERS

Most common and extensively used flow meters are turbine meters. They are used in a number of designs, most simple being the water meter. Domestic water meters are generally rugged and require little maintenance.

They are mechanical devices with a wheel or turbine rotor in the fluid flow. If there is more flow, the wheel turns faster. The wheel is directly or magnetically linked to a dial reading. In the more sophisticated meter, rotor blades are magnetically coupled to an electronic transmitter. With each blade generating a pulse, the processor (micro) integrates and reads the flow. These flow meters are generally accurate, but not so at low flows. Though they are durable, high velocity flows and non-lubricant flow can damage the bearings. The case is similar with corrosive fluids.

Then there are paddle wheel flow meters, which are called "poor man's flow meters." Similar to turbine flow meters, they can be inserted into the pipe by drilling a hole on the pipeline and

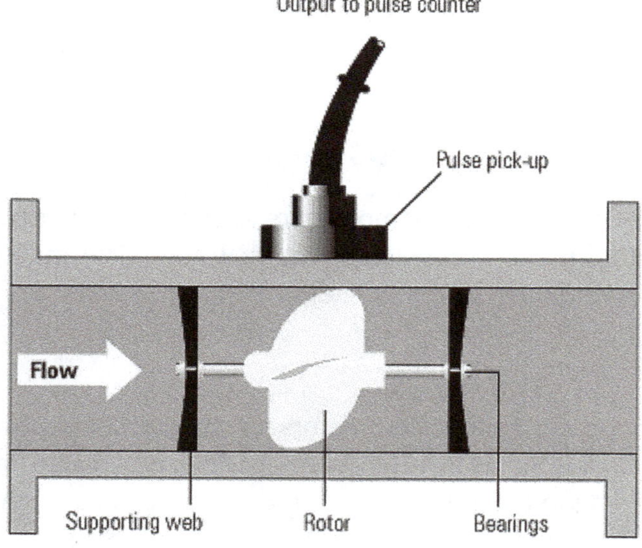

Figure 22.2
Turbine Flow Meter

fixing them. They are fairly accurate and convenient, and require minimal maintenance (see Figure 22.2).

Turbine flow meters have come a long way since the original turbine meter was invented by Reinhard Woltman in the 18th century.

ULTRASONIC FLOW METERS

The Doppler Effect takes its seat in the ultrasonic flow meters that use ultrasonic sound waves to detect the velocity of a fluid flowing in a pipe. If there is no flow, the transmitted and reflected signals are the same. Under flowing conditions, flow is detected by virtue of the Doppler Effect in as much as the frequency of the reflected wave is different from the transmitted frequency. As the fluid travels faster, the frequency shifts upward linearly. The electronic transmitter transmits the ultrasonic signals, receives them, processes them, and determines the flow rate. See Figure 22.3A The Doppler Effect Ultrasonic Flowmeter.

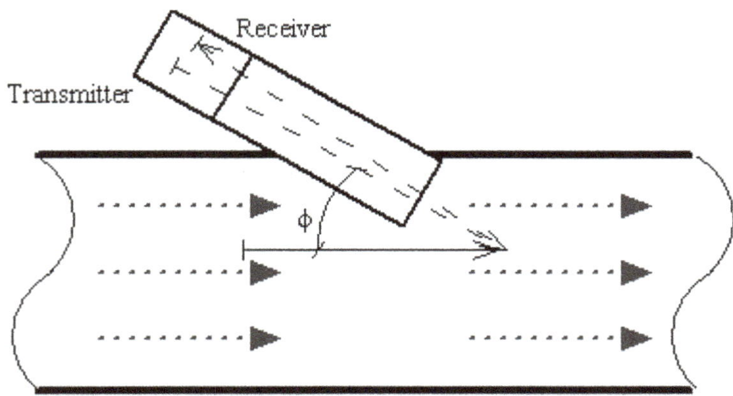

Figure 22.3A
Doppler Effect Ultrasonic Flowmeter

On the other hand, transit time ultrasonic flow meters send and receive ultrasonic waves between transducers in both the upstream and downstream directions in the pipe. Doppler ultrasonic flow meters are more economical. See Figure 22.3B for a transit time or time of flight. Ultrasonic flow meters can be used for any liquid that allows ultrasonic waves to pass, such as water, molten sulfur, cryogenic liquids, and chemicals. They are not effective for slurries and certain other liquids, which cannot conduct and reflect ultrasonic waves. These flow meters do not obstruct flow and do not have any wetted parts so they can safely be used on sanitary, abrasive, and corrosive liquids. Some ultrasonic flow meters use clamp-on transducers that can be mounted externally on the pipe. Clamp-on transducers can be used without disturbing the piping, without the worry of MOC (materials of construction), and are ideal for making temporary but accurate measurements.

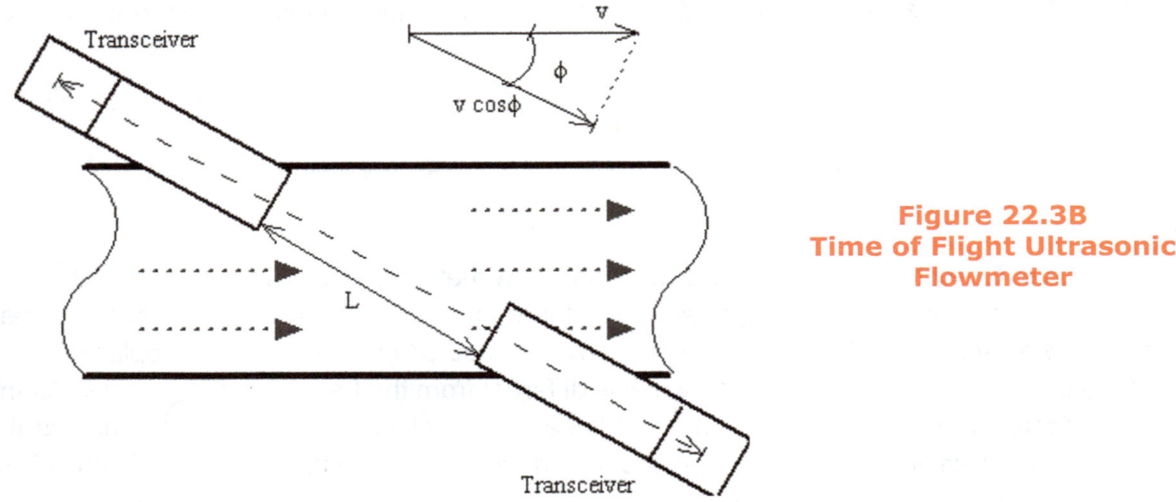

Figure 22.3B
Time of Flight Ultrasonic Flowmeter

The Doppler Effect is named after the Austrian physicist Christian Doppler who discovered the effect in 1842. It was more than a century after the discovery that the effect was used to measure the flow of fluids. Tokyo Keiki first introduced them in Japan for industrial use in 1963.

CORIOLIS MASS FLOW METERS

Coriolis meters are developed to satiate the continuing demand for more accurate flow. They operate on the natural phenomenon called the Coriolis force, hence, the name. Coriolis meters give a direct mass flow measurement, irrespective of density, temperature, and pressure. They are true and direct mass flow meters. These devices are extremely accurate, typically with an error in the order of 0.02 to 0.2 percent of the total flow.

Coriolis mass flow meters measure the mass flow of liquids, such as water, acids, caustic, chemicals, and gases and vapors. They can be used on extremely corrosive services with proper metallurgy and almost all services found in mining, mineral processing, power, pulp and paper, petroleum, chemical, and petrochemical industries.

Coriolis mass flow meters measure the force resulting from the acceleration caused by mass moving toward (or away from) a center of rotation. The effect can be demonstrated by holding with both hands a loop of flexible hose with flowing water. It will be seen that the hose is "swung" back and forth in front of the body, as the water is flowing toward and away from the hands. Opposite forces are generated, which cause it to twist. The swinging is generated by vibrating the tube(s) in which the fluid flows. The amount of twist is proportional to the mass flow rate of fluid passing through the tube(s).

The meter generally consists of a U-shaped flow tube with its sensors placed in a suitable housing (see Figure 22.4). The unit can be fixed into the flow path and the processing electronic can even be placed 500 feet from the sensor. The flow tube is vibrated at its natural frequency by an electromagnetic circuit like a tuning fork. As the liquid takes an upward and downward path in the U tube, it develops twists in the tube in corresponding directions. The amount of twist is directly proportional to the mass flow rate of the liquid flowing through the tube. Pickups on each side of the U tube measure the twists. These signals are then processed to give the mass flow.

These flow meters are not accurate at very low flow rates. Piping vibrations can cause improper measurements. Pressure drop also could be a matter of concern.

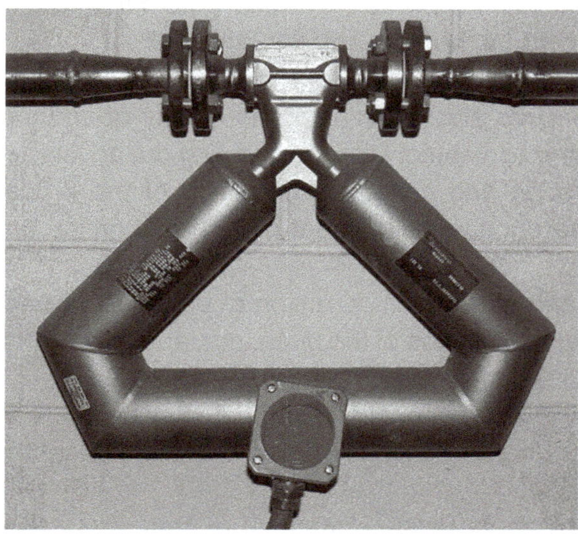

Figure 22.4
Coriolis Mass Flow Meters

Basic theories behind the Coriolis flow meter were discovered by a French civil engineer, Gaspard Coriolis, in 1843. However, the industry had to wait until the 1970s for its practical use, mostly due to the development of sophisticated sensors required.

DIFFERENTIAL PRESSURE FLOW METERS

Differential pressure flow meters make use of the famous Bernoulli's equation to measure the flow of fluid in a pipe. A constriction is introduced in the pipe that creates a pressure drop. The pressure drop increases when the flow increases. Bernoulli's equation states that the pressure drop across the constriction is proportional to the square of the flow rate.

The conventional system uses an orifice plate fixed in between two orifice flanges. An orifice plate is basically a thin plate with a hole in the middle, fitted between flanges in a pipe. Upstream and downstream pressures are measured and transmitted by impulse lines to the transmitter that measures the differential pressure and calculates the fluid flow. Constrictions have different geometries for different requirements, such as the orifice plate, flow nozzle, pipe or full flow taps, Vena Contracta taps, and Venturi tubes. These are inferential flow meters that are used in almost all relatively clean services except in cases of high viscosity (where the Reynolds number is low) and in dirty services. They are almost an industry standard for the measurement of fluid flows. They have no moving parts and can be installed using flanged, welded, or threaded-end fittings.

The orifice plate can be made of any material, although stainless steel is the most common. The thickness of the plate used (1/8–1/2") is a function of the line size, the process temperature, the pressure, and the differential pressure. The traditional orifice is a thin circular plate (with a tab for handling and for name plate details), inserted into the pipeline between the two orifice flanges.

Measuring the differential pressure at a location close to the orifice plate minimizes the effect of pipe roughness because friction has an effect on the fluid and the pipe wall.

These plates must be carefully installed to make sure that no portion of the flange or gasket interferes with the opening. Precision calculations and machining, the quality of and care in the installation, and the condition of the plate are the contributory factors in giving out adverse readings. Other factors to be considered are tap location and condition, condition of the process pipe, adequacy of straight pipe runs, gasket interference, misalignment of pipe and orifice bores, and lead line design.

A dulled edge of the orifice plate, nicks and burrs on it, corrosion or erosion, warpage due to water hammer or excessive pressure for a given thickness of the plate, dirt, and grease or deposits on either orifice surface can cause improper readings.

In measurement of differential pressure, the pipe connections, or taps, may be made at various points with respect to the orifice location. There are three kinds of taps in common use: the flange, the vena contracta, and the pipe or full flow taps.

Taps are not recommended below a 45° angle to the horizontal as they may plug up with foreign matter. The center line of the taps should pass through the center of the pipe. The inside of the pipe wall must be free of burrs around the tap openings.

ORIFICE FLANGES

An orifice plate is basically a thin plate with a hole in the middle, fitted between flanges in a pipe. Flanges have tappings for measuring the pressures. Fluid flows through the pipe, with a certain velocity and pressure. When the fluid reaches the orifice (hole) in the orifice plate, the fluid converges to pass through the hole. The velocity increases and pressure drops to the maximum

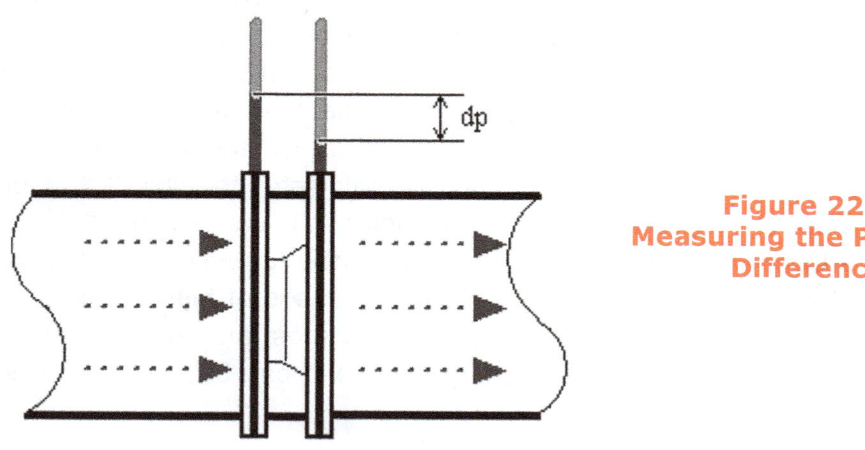

Figure 22.5 Measuring the Pressure Difference

at a point called vena contracta which is a bit downstream of the actual orifice. Beyond this point, the fluid expands and the velocity and pressure change once again. The greater the flow, the greater is the difference in pressure at the nominal pipe diameter and here. With a differential pressure cell, the pressure difference is measured from which actual flow can be calculated (see Figure 22.5).

Flange taps take a special set of flanges, which is a disadvantage. They have the advantage of being standard in design. The meter leads are close together and will be close to the same length. The up-stream face must always be as flat as possible and generally the downstream edge be beveled. The orifice plate is installed between raised face flanges with a gasket on each side. It should be properly centered when fixed between tongue and groove flanges or ring joint flanges. Different types of orifice plates — like concentric (see Figure 22.6A), eccentric (see Figure 22.6B), and segmental (see Figure 22.6C) holes — are used for measuring different processes.

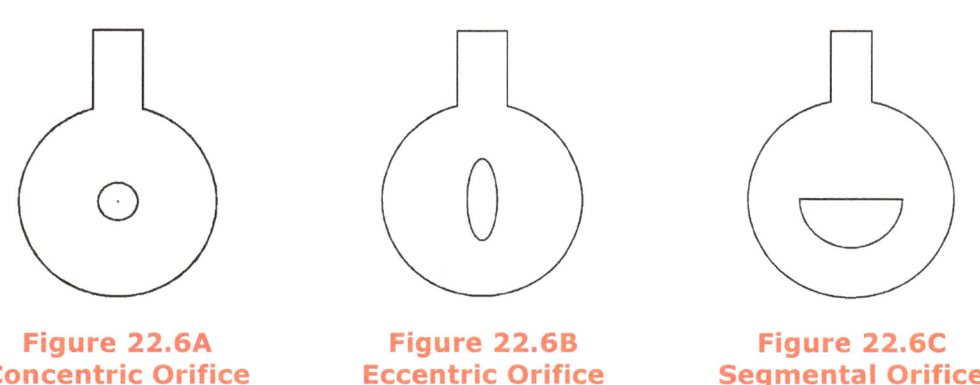

Figure 22.6A Concentric Orifice

Figure 22.6B Eccentric Orifice

Figure 22.6C Segmental Orifice

VENA CONTRACTA TAPS

This type obtains its name from the fact that the low pressure connection is located at the region of minimum pressure (vena contracta). The maximum velocity and minimum static pressure occur at some 0.35 to 0.85 pipe diameters downstream from the orifice plate. The upstream connection is located one pipe diameter before the orifice plate, and the downstream connection at a point giving the greatest differential pressure. This point is designated by the manufacturer and is generally about one-half pipe diameter downstream.

These are theoretically the most desirable. Their accuracy seems to be slightly superior to other forms of taps. Vena contracta taps provide the maximum pressure differential. If the plate is changed, it may require a change in the tap location. Also, in small pipes, the vena contracta might lie under a flange. Vena contracta taps are not recommended for pipe sizes smaller than 4" as there is interference between the flange and the downstream tap.

Sometimes, the manufacturer designates a variation of the vena contracta, called throat taps, where the downstream connection is constant at one half pipe diameters below the orifice and the upstream tap, but is generally around one pipe diameter upstream.

PIPE OR FULL FLOW TAPS

These taps were widely used originally because the first reliable chart tables were made for these connections. These taps are located two and one half pipe diameters upstream and eight pipe diameters downstream from the orifice. However, the room occupied by the meter leads is necessarily greater; being longer, they demand more support than other types. The meter is more sensitive to deviations from the correct specifications such as roughness of the pipe.

FLOW NOZZLE

The flow nozzle is a long radius nozzle that is located in the pipe to measure the flow and is fixed in the pipe between two flanges (see Figure 22.7). The tap location is designated by the manufacture. It is more accurate than a regular orifice plate. It can handle higher velocities; offer some 60% more flow than orifice plates with the same pressure drop. There is less chance of wire drawing. A disadvantage is that the initial cost of the nozzle and its installation is high. Liquids with suspended solids can be metered, but it is not recommended for highly viscous liquids.

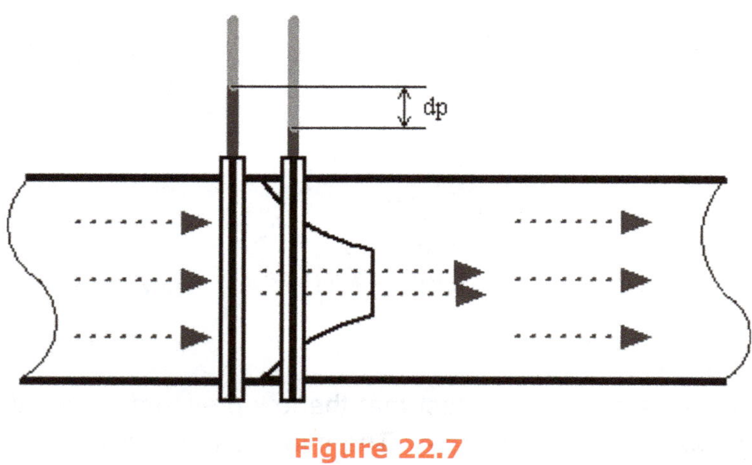

Figure 22.7
Measuring the flow

VENTURI NOZZLE

The venturi nozzle is a complete unit made in a factory with the taps positively fixed by the factory. This is a machined piece of equipment and is highly accurate. It is also quite expensive. A venturi tube is a pipe piece with a tapered entrance and a straight throat. As the liquid passes through the throat, its velocity increases, causing a pressure differential between the inlet and outlet regions (see Figure 22.8). The venturi offers higher throat velocities without unreasonable head loss. Hence, it allows for flow measurement with lower head losses than orifice plates. Venturi tubes are made up of cast iron or machined steel or of welded sheet metal.

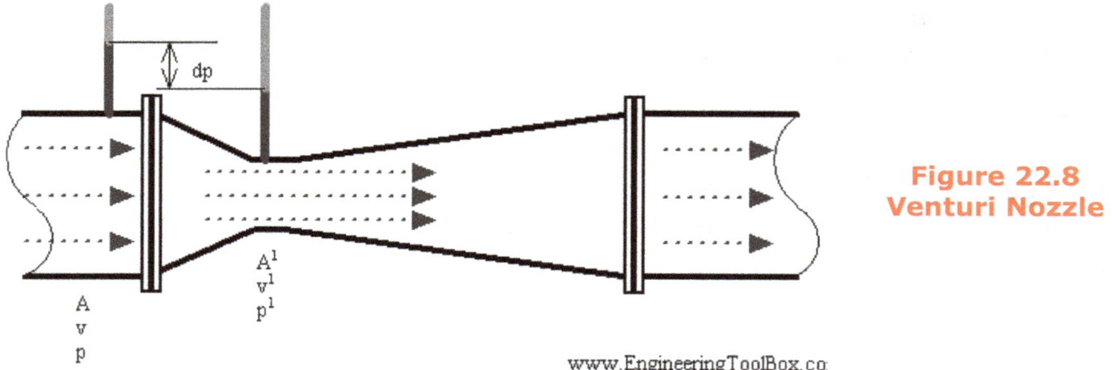

Figure 22.8
Venturi Nozzle

PROBLEMS AND SOLUTIONS

Plugging of the impulse lines is a perennial problem. In some cases, some additional purging may have to be incorporated to keep the lines clean. In liquid service, impulse piping should always contain full liquid without any gas; in gas service, impulse piping should always contain only gas and should not collect any liquid.

The thermo well and the thermocouple used for temperature measurement should be installed at least 10 diameters downstream of the flow element. For the same reason, straight pipe runs should be allowed both upstream and downstream of the differential pressure transmitter (d/p) element.

Sizes of impulse lines vary, including 1/4" for pipes under 2" in diameter, 3/8" for 2" and 3" pipes, 1/2" for 4-to-8", and 3/4" for larger pipes. Both taps should be of the same diameter. The tappings should be square with no roughness, no burrs, and no foreign material.

The d/p transmitter should be located as close to the primary element as possible. Lead lines should be as short as possible and of the same diameter.

THERMAL FLOW METERS

Thermal mass flow meters have traditionally been used for gas measurements, but designs for liquid flow measurements are available. These mass meters also operate irrespective of density, pressure, and viscosity. They make use of the thermal properties of the fluid to measure its flow in a pipe or duct. Thermal meters use an isolated heated sensing element in the fluid flow path (see Figure 22.9). The flow stream conducts heat from the sensing element. Some of this heat is lost to the flowing fluid. As flow increases, more heat is lost. The conducted heat is

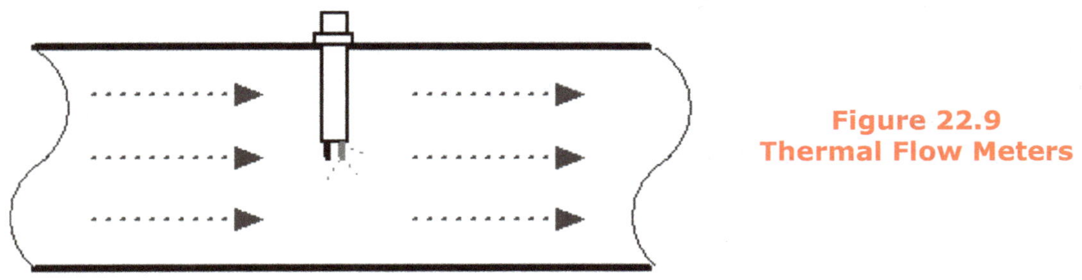

Figure 22.9
Thermal Flow Meters

directly proportional to the mass flow rate. The electronics analyze the signals, process it, and give readout directly proportional to mass flow.

Thermal flow meters are most commonly used to measure pure gases, such as in laboratory experiments, and in semi-conductor production. They can be used for clean, sanitary, and even corrosive gases if the thermal properties of the fluid are known and proper MOC is selected.

Abrasive fluids can damage the sensor; similarly, fluids with a tendency to coat the sensor can give adverse readings. Aerosols and liquid droplets in the gases can give false readings.

VORTEX AND FLUID FLOW METERS

When a fluid passes by an object or obstruction, oscillations can occur. You can see them in nature; swirls produced downstream of a rock in a rapidly flowing river, and the waving of a flag in the wind. But when the river flows calmly, the water flows smoothly around the rock; in a mild breeze, the flag does not wave at all. So these oscillations are indicative of the flow. Eddies or vortices are shed alternately downstream of the object. Vortex meters make use of this natural phenomenon. The frequency of the vortex shedding is directly proportional to the velocity of the liquid flowing through the meter. Oscillations are generated as a result of flow. Increasing flow increases the frequency of oscillation. A sensor detects the oscillations and an electronic transmitter generates a flow measurement signal (see Figure 22.10).

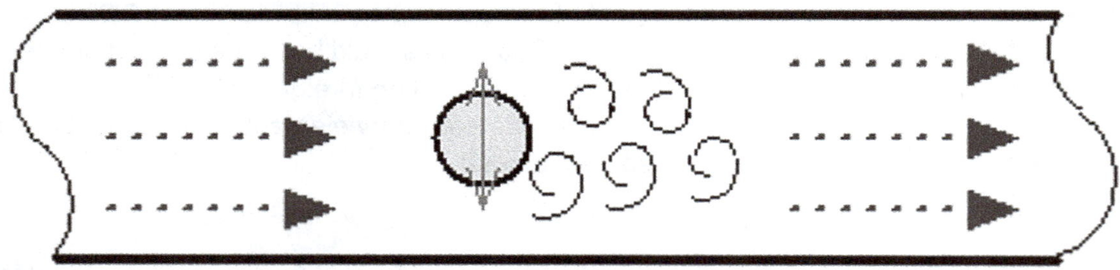

Figure 22.10
Vortex Flow Meter

Now you need a bluff body across the flow meter bore, a sensor to detect the presence of the vortex and processor to do the rest of the job. These meters are used for water, boiler feed water, steam, cryogenic liquids, hydrocarbons, chemicals, air, and industrial gases, but not recommended for slurries or high viscosity liquids.

Vortex precession flow meters have inlet vanes that rotate the fluid to form a vortex center (similar to a cyclone) that rotates around the inside of the pipe. The rotation of the vortex can be related to the flow rate. These flow meters turn off at low flow rates, typically 1 ft/sec for liquids, but the rate could be higher for gases and vapors. For obvious reasons, these meters cannot tolerate piping vibrations.

Basic principles behind the vortex flow meter were discovered by the Hungarian-American aeronautical engineer Theodore von Karman.

PILOT TUBE FLOW METERS

The pilot tube flow meters are frequently used in ventilation and HVAC systems. It is one of the cheapest means to measure fluid flow. The fluid flow velocity is measured by the pilot tube by converting the kinetic energy of the flow into potential energy (see Figure 22.11).

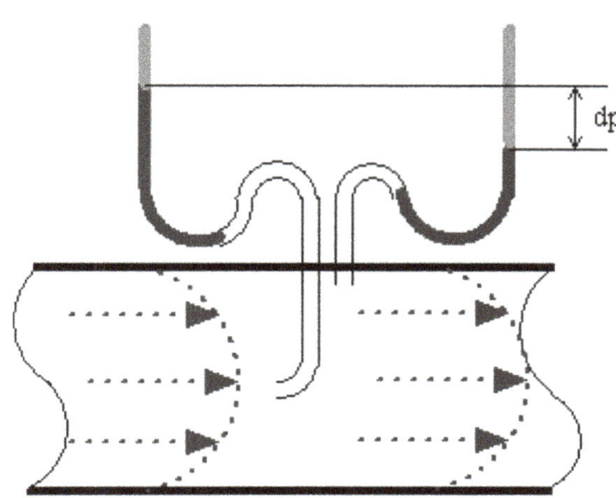

Figure 22.11
Pilot Tube Flow Meters

TARGET METERS

In target meters, a target or a drag disc is suspended in the flow path. The flow makes an impact on the target. The force is sensed and measured, which is a direct indication of the liquid flow rate. Simply speaking, it consists of a hinged, swinging plate that moves outward in the flow path. More sophisticated versions make use of strain gages and electronics to process the signal. They are useful for measuring flows of dirty or corrosive liquids.

ROTARY-VANE METERS

Rotary-vane meters are positive displacement meters. A rotating impeller with two or more blades (now compartments) is housed in the meter. The impeller is in continuous contact with the casing. A fixed volume of liquid swept between each set of blades or one compartment gives out the actual reading of the flow. The number of revolutions is counted and flow measured. Alternate design is helical rotors geared together.

CALIBRATION AND MAINTENANCE

Initially, the manufacturer calibrates the instrument. Depending on the service conditions, quality of manufacture, and maintenance, they may need field calibration. Abrasive, erosive, or corrosive services damage the internal parts requiring recalibration, repair, and replacement. In critical applications, the instrument may need frequent check up. Most of the flow meters, more so of the present day flow meters, require special equipment for calibration. Companies should source them to the best of their need and ability.

All flow meters will eventually need some care and maintenance, with each type having its own peculiar necessities. Unless the meter is properly matched in the first instance, it will call for repeat recalibration, repeat maintenance, and unavoidable downtime. Flow meters without moving parts require less attention though the costly instrumentation, piping, and electronics may require a larger capital cost and higher skill in maintaining it.

As with most other piping devices, improper MOC, bad erection practices would end up in higher maintenance and inaccurate readings. Those flow meters with moving parts need periodic internal inspection and preventive maintenance, more so for the abrasive, dirty, or viscous fluids.

Here are a couple of terms commonly used in flow meters.

CONDUCTIVITY

The electrical conductivity of a liquid is a measure of the ability of the liquid to conduct electricity (mS/cm). Note that water with few impurities (such as de-ionized water) is not very conductive, whereas water with impurities can be highly conductive.

REYNOLDS NUMBER

The Reynolds number is a dimensionless number that describes the flow characteristics of the fluid. Operating a flow meter outside of its Reynolds number constraint can degrade accuracy and make some flow meters turn off.

The Reynolds numbers was proposed by a British mechanical engineer Osborne Reynolds in 1883.

Chart 22.1
Comparing Flow Meters

Device	Liquids	Viscous Liquids	Slurries	Accuracy % Full Scale
Orifice	OK	Limited	No	0.25 -2
Rotameter	OK	Limited	Limited	0.5 -2
Venturi	OK	Limited	OK	0.25 -3
Target	OK	Limited	Limited	0.5 -2
Magnetic	OK	OK	OK	0.25 -1
Vortex	OK	Limited	No	0.5 -2
Turbine	OK	Limited	Limited	0.25 -1

Stress Analysis

Present demands in modern chemical plants are for higher pressures, temperatures, and flows need systems that are designed, built, and operated with intrinsic reliability. Gone are the days when only pressure was considered a primary condition for pipe line design. Nowadays, a number of operating conditions are evaluated, a large number of pipe line stress conditions determined, and worst case scenarios discussed, much before the pipe line is designed and built. Stress analysis is conducted based on these conditions; the pipe line is then designed, built, and operated.

Pipe stress analysis on a piping system is carried out

- To ensure the piping is well supported and does not sag or deflect
- To comply with legislation
- To ensure that the line deflections are within control after applied thermal and flow loads
- To ensure that moments developed by its thermal growth do not impose excessive loading on the connected equipment
- To see that thermal stresses imposed on the pipe line are within limits

CONDITIONS

Properly identifying complex loading conditions calls for not only rich experience in design, building, and operation of pipelines, but also sound knowledge of various engineering disciplines. Some of the important loading conditions are:

- Weight of piping and contents in operating, as well as in test conditions
- Internal / external pressure and cyclic conditions
- Temperature — ambient and operational, range and cyclic conditions
- Climatic loads, traffic, wind, earthquakes, movement of the ground and buildings
- Reaction forces and moments due to anchors, supports, attachments. i.e., vibrating equipment nozzle connections
- Structural changes in overall system geometry, material-related factors
- Dynamic effects like flow variations, water hammer, relief valve pop up, fluctuating and intermittent flows, slugs

- Corrosion, erosion, fatigue, etc.
- Most of the loading can occur simultaneously

The load categories defined can otherwise be classified as follows.

STRESS DUE TO SUSTAINED LOADS

Sustained loads are those which do not vary considerably over the period. Examples are internal and external pressure as well as the weight of the pipeline and the fluid it is carrying. Maximum operating pressure and temperature are the basis for calculation of design parameters. Short-term excursions of pressure and temperature within defined limits are permissible.

STRESS DUE TO OCCASIONAL OR EXCEPTIONAL LOADS

Stress can be caused by factors such as wind, seismic, and snow. Wind load is a transient dynamic load that can impose deflections and vibratory loads. However, wind loads can be ignored in indoor locations. Stresses are also induced on the pipeline due to subsoil conditions and ground settlement. Dead weight is not considered for buried piping, but stresses imposed by the settlement of the earth need to be considered. Hydrostatic tests are predictable intermittent loads, whereas seismic loads are exceptional loads.

Vibrations can be caused on the piping due to fluctuating flow — say from reciprocating compressors — or can be transmitted by the vibrating rotary equipment. Disaster can result if the vibration frequency matches with the natural frequency of the piping.

STRESS DUE TO THERMAL EXPANSION AND CONTRACTION

Piping is fabricated and fixed at an ambient temperature which reaches operating temperatures once it is in service. Temperatures revert if the line is shut down and the line contracts. Loads can be dynamic if temperatures fluctuate or cycle. Rapid changes or cyclic changes in temperature, or expansion due to different piping materials with different expansion coefficients, can set up additional loads on the system.

Piping materials operating at their limits of operability, or high temperature piping operating in the creep range, can load the system sufficiently.

CIRCUMFERENTIAL STRESS DUE TO INTERNAL PRESSURE

Pressure surges created by the water hammer; starts and stops of the pumps, compressors, and valves; and popping of pressure relief valves can cause dynamic loads. Nominal design stress is calculated from the material physical data, then used to determine the allowable stress. The dead weight of the system is taken as the piping and associated component weight. The weight of the system fluid is taken as occasional load by the code.

Considering these requirements, stress analysis is carried out for the following parameters

- **Thermal Analysis**

Analysis for free and restrained thermal growth conditions

- **Deadweight Analysis**

Analysis at ambient temperature with a system of hangers at specific locations to support the weight of the system, for allowable stress and reactions at equipment connections

- **Seismic Analysis**

Either equivalent static or dynamic analysis

- **Wind Load Analysis**

Equivalent static stress analysis

- **Transient Analysis**

For various transient loading conditions such as turbine trip, pipe whip, and safety relief valve pop up

RECOMMENDATIONS

Based on the calculations, recommendations are made to make alterations to the system as required using one or more of the following tools:

RESTRAINTS

A device that resists or limits the free thermal movement of the pipe. Restraints can be directional, rotational, or a combination of both.

ANCHORS

A rigid restraint stops all movement. In ideal conditions, they should not allow any moment, though this is difficult to achieve. A rigid bracket even when properly gusseted may not qualify if the structure to which it is attached is weak.

EXPANSION LOOPS

Expansion loops absorbs thermal growth; usually used in combination with restraints and cold pulls.

NEUTRAL PLANES OF MOVEMENT

A pipe restraint positioned in line with a neutral plane prevents differential expansion forces between the pipe and the machine. This refers to the planes on the three axes of a turbo machine or pump from where expansion of the machine starts, e.g., the fixed end of a turbine casing. This information is normally provided by the equipment manufacturer. If not available from this source, the fixed points of the machine must be determined by inspection and an estimation of the turbine growths calculated.

COLD PULL OR COLD SPRING

This is used to pre-load the piping system in the cold condition in the opposite direction to the expansion, so that the effects of expansion are reduced. Cold pull is usually 50% of the expansion of the pipe run under consideration. Cold pull has no effect on the code stress, but can be used to reduce the nozzle loads on machinery or vessels.

SPRING HANGERS

These are used to support a piping system that is subjected to vertical thermal movements. Commercially available single coil spring units are suitable for most applications.

SOLID VERTICAL SUPPORT

In places where vertical thermal movement does not create undesirable effects, or where vertical movement is intentionally prevented or directed, solid supports in the form of rollers, rods, or slippers are used. It is important that free horizontal movement of the pipe is not hindered unless its restriction is desired. Roller supports must be well designed and lubricated.

Software programs are now available to make the analysis, among the most popular being Caesar II.

EXPANSION JOINTS

Expansion joints are used in pipe lines made to allow for misalignment; movements of the line due to expansion or contraction because of flow conditions or other reasons; and to contain vibrations. Pipelines twist and turn, expand and contract, and also vibrate in all three axes. Expansion or contraction of pipe line occurs due to temperature, pressure, or flow fluctuations. Misalignment can be either angular or parallel, mostly caused during installation, but can happen due to process problems. Vibrations are generally flow induced, but faulty drives also can transmit them. All these can induce cracks, leaks, and failures unless properly attended to.

Expansion joints are generally attached between two fixed assemblies, say, two pipeline flanges, pump delivery lines and piping, etc. They are constructed with bellows of thin section of metal, elastomer, or fabric. Convolutions in the bellows allow for the misalignment, movements of the line, vibrations, and isolation.

Expansion joints are widely used in the petrochemical plants, petroleum industry, nuclear plants, and food and pharmaceutical industries. They are highly engineered products and should be installed, commissioned, and operated as such for safe and reliable operation. They are generally available from full vacuum applications to about 1000 psi or 75 Kg/cm^2 to other applications at temperatures cryogenic to about 1832°F or 1000°C.

Unlike the common piping components, which are controlled by schedules and stresses, expansion joints are made out of relatively thin corrugated metal to allow for the expansions, contractions, misalignments, and vibrations. EJMA Standards, the ANSI Piping Codes, and the Boiler and Pressure Vessel Codes specify various design requirements of expansion joints.

Expansion joints are made in round, square, or rectangular construction and in a number of standard sizes and nonstandard sizes. End connections are made to suit the system requirements; those used in the piping have flanged ends or butt welded ends as per various standards to suit variety of applications. Some of the joints — particularly those with elastomer bellows — are made such that they can be inserted between flanges and bolted.

EXPANSION SCENARIOS

Expansion joints should take care of the axial and lateral expansion and contraction, angular and parallel misalignment, and vibrations in the piping.

Figure 24.1 illustrates the axial movement of expansion joints, whereas Figure 24.2 shows the lateral and angular movement of expansion joints. Figures 24.3 illustrates the axial and angular movement of expansion joints.

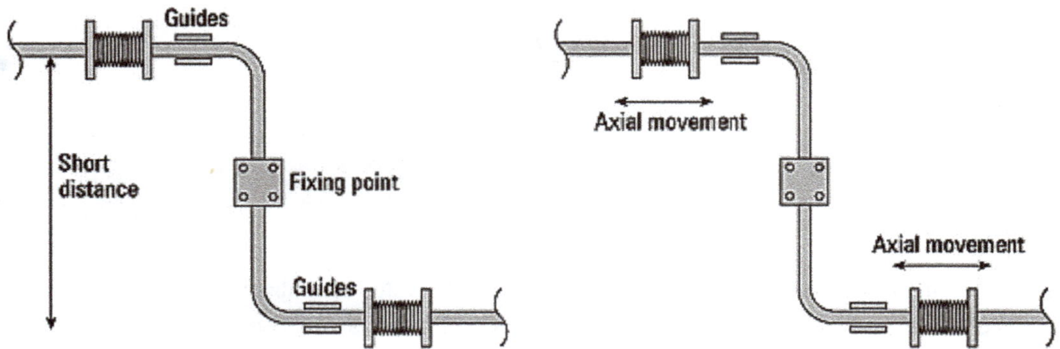

Figure 24.1
Axial Movement of Bellows

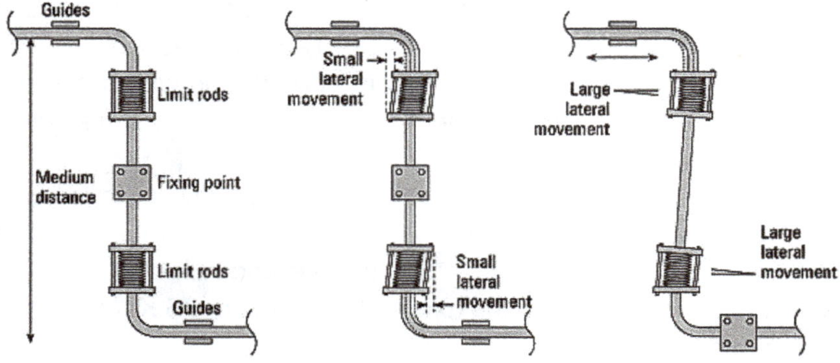

Figure 24.2
Lateral and Angular Movement of Bellows

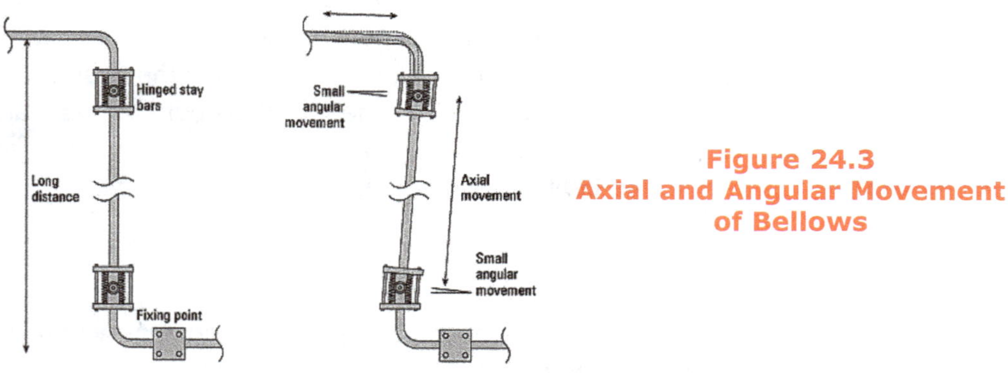

Figure 24.3
Axial and Angular Movement of Bellows

MOVEMENTS AND DEFLECTIONS

AXIAL TRAVEL

Axial travel is the longitudinal movement or axial compression and expansion, not taking into consideration misalignment or angular deviation (see Figure 24.4).

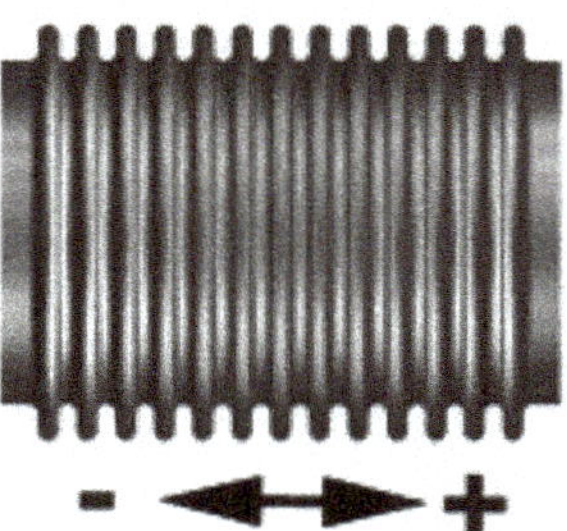

Figure 24.4
Axial Travel

MISALIGNMENT

Misalignment is lateral offset or deviation, not considering axial travel or angular deflection (see Figure 24.5).

Figure 24.5
Misalignment

ANGULAR DEFLECTION

Angular deflection is angular deviation not considering axial travel or lateral offset (see Figure 24.6).

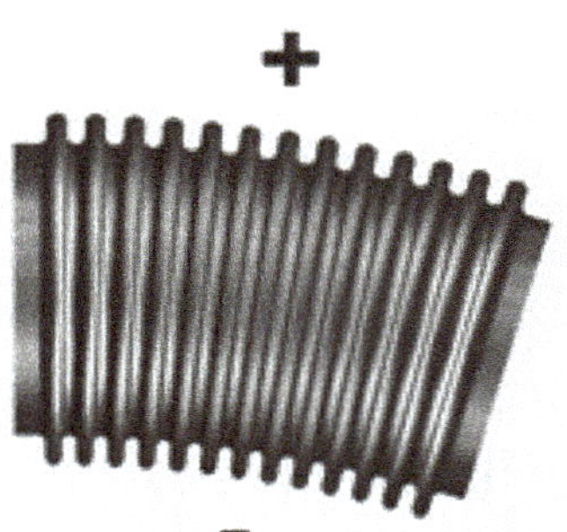

Figure 24.6
Angular Deflection

TORSIONAL MOVEMENT

Torsional movement is the rotation about the axis through the center of a bellows (twisting) (see Figure 24.7).

Figure 24.7
Torsional Movement

Vibrations can be flow induced or mechanically transmitted from the faulty drives of rotating equipment like the pumps, or compressors. If not properly controlled, they can cause cracks, pin holes, and leaks in the piping.

But piping systems can have any or all these deviations. Expansion joints should take care of all the problems.

BELLOW EXPANSION JOINTS

A short description of the major types of bellow expansion joints is given here.

AXIAL EXPANSION JOINT

The axial expansion joint is the simplest of all expansion joints used to take up lateral movements (see Figure 24.8). Sometimes a dual expansion joint is used when the axial movement caused by thermal expansion or contraction is too much for a single joint.

Figure 24.8
Axial Expansion Joint

UNIVERSAL EXPANSION JOINT

The universal expansion joint can take care of lateral, angular, and axial movement in three planes. It makes use of two or more bellows elements (see Figure 24.9). Limiting stops/rods are provided to support the assemblies and prevent destruction.

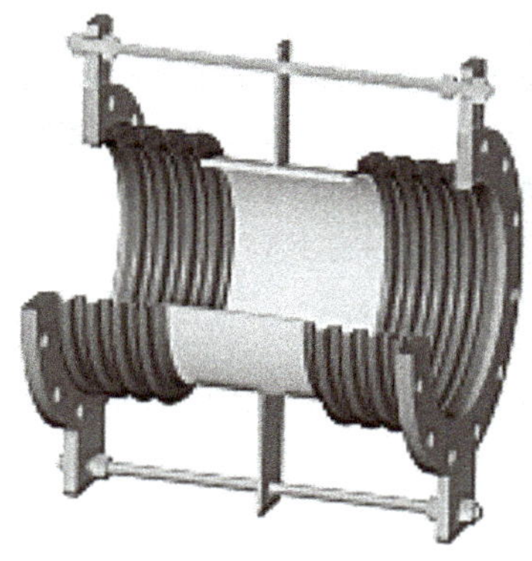

Figure 24.9
Universal Expansion Joint

HINGED EXPANSION JOINT

Hinged expansion joints can take care of angular movements in a single plane (see Figure 24.10).

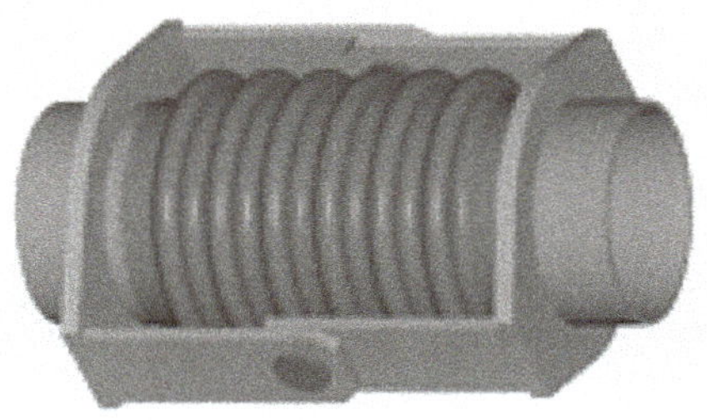

Figure 24.10
Hinged Expansion Joint

PRESSURE BALANCED EXPANSION JOINT

Two balancing bellows produce constant volume which effectively compensates the opposing forces. These have a balancing bellow sealed with a blind flange and an elbow.

GIMBAL EXPANSION JOINT

Gimbal expansion joints are used in piping lines where proper anchoring and guiding are not feasible because they can take care in all planes (see Figure 24.11).

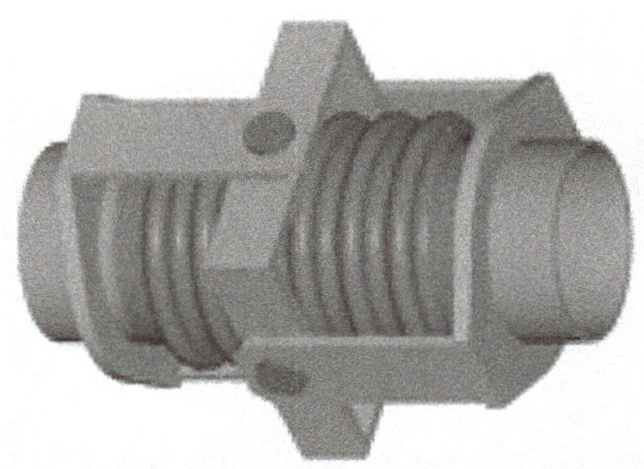

Figure 24.11
Gimbal Expansion Joint

FABRIC EXPANSION JOINTS

Fabric expansion joints, sometimes known as compensators, are used on ductings. Though basically made out of fabric, it is not unusual to find them made out of various elastomers. More often than not, they are used to cover up errors in fabrication and to avoid mechanical shock and vibrations induced from drives (see Figure 24.12).

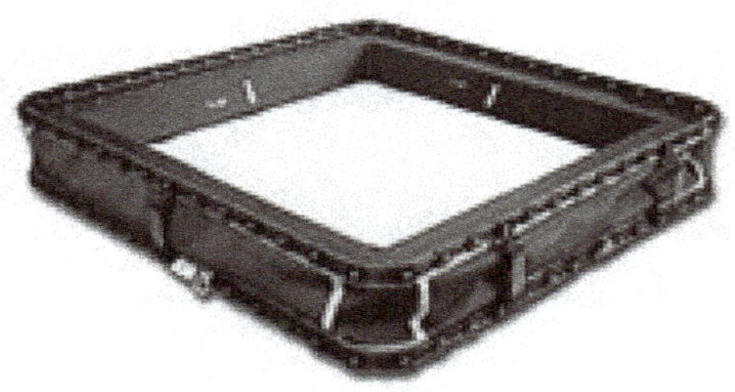

Figure 24.12
Fabric Expansion Joints

Ties rods are attached to the expansion joint assembly to absorb pressure loads and external loads like dead weight (see Figure 24.13).

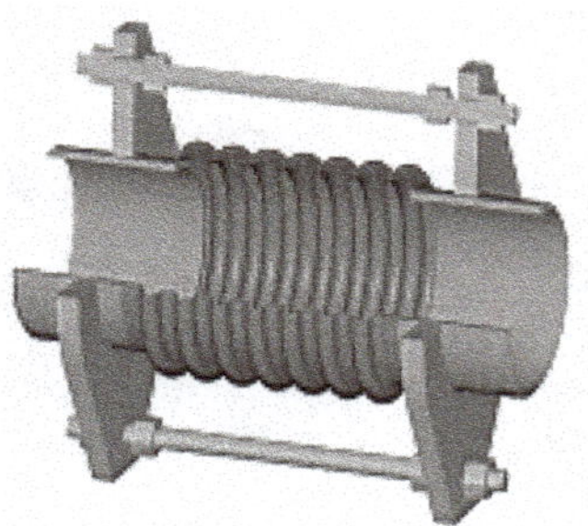

Figure 24.13
Ties Rods

Limit rods are designed to protect the bellows from excessive movements occurring due to plant problems or support failure. They are not meant for the pressure thrust during normal operation (see Figure 24.14).

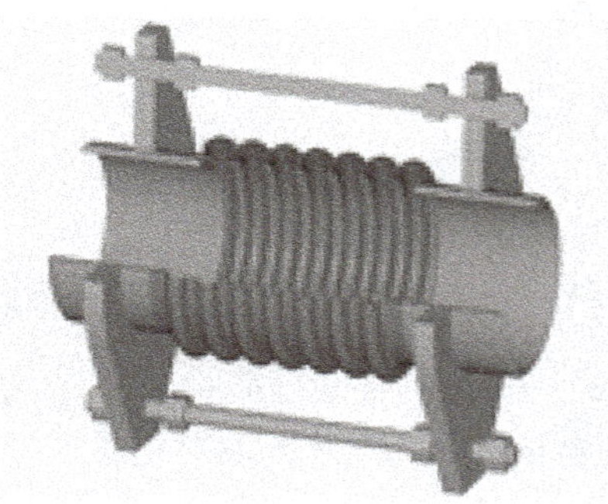

Figure 24.14
Limit Rods

Liners (Internal Sleeves) are used to protect the bellows internally (see Figure 24.15).

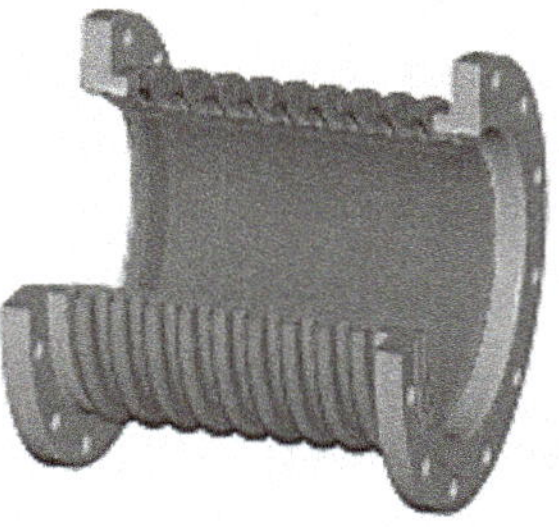

Figure 24.15
Liners

Covers are the external protection for the bellows (see Figure 24.16).

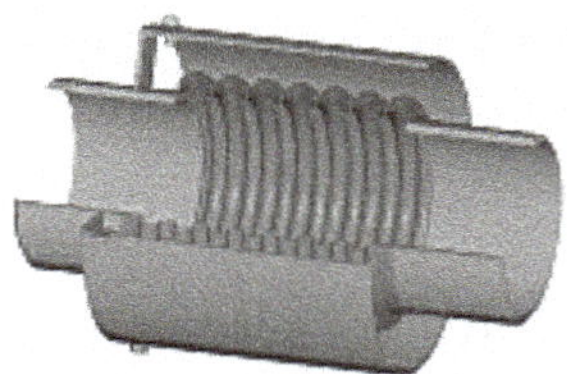

Figure 24.16
Covers

ELASTOMERIC EXPANSION JOINTS

More and more expansion joints are manufactured with elastomers like Teflon, various grades of rubber to cater to the severe corrosion conditions of the process flows. In general, elastomeric joints with only two convolutions can take care of say 1/4"–1/2" lateral movement and with misalignment of 1/8"–1/4" These are full vacuum rated and highest pressure rating in the class. Those with three convolutions can handle more moderate axial movement of 1/2"–1 1/8" and moderate misalignment of 1/4"–9/16" and handle more pressures from full vacuum. Those with five convolutions tolerate most axial travel of 1/2"–1 1/4 and misalignment of 1/2"–5/8", but are rated lowest in the pressure range and least in the vacuum range. See Figure 24.17 for elastomeric expansion joint.

Figure 24.17
Elastomeric Expansion Joint

CARE

Take good care of expansion joints for they will take care of the line. They are delicate pieces of engineering excellence.

- First and foremost, expansion joints, including the bellows, must be resistant to corrosion of the flow medium.
- Material of flow should not clog or build up solids in the convolutions, which will prevent proper operation of the same.
- Piping layout, flow characteristics, and conditions such as reverse flow, turbulence sudden flows, pressure, temperature, flow velocities, and line vibrations must be taken into account before selecting an expansion joint.
- Oversizing an expansion joint is a bad decision as it will involve additional cost, but also additional thickness may induce undue fatigue on the bellows.
- Resonant and induced vibrations are an easy way to destruction. Similarly, torsional stresses should be avoided at all costs.
- Insulation requirements need special attention and design.
- Limiting rods, supports, and guides should not be compromised.
- It is better to use multiple-ply expansion joints with each ply capable of full pressure in case of hazardous fluids.
- Corrosion and erosion are age old enemies of process piping.

INSTALLATION

- Remove the expansion joint from its shipping packages just before the joint is ready to be installed, after all the piping supports and anchors are in place. Improper storage and handling is one of the primary causes of expansion joint failure.
- Remove it from the cartons with extreme care and avoid any damage including small dents and scorings. An innocent looking scratch may result in a premature failure.
- Falling weld slag and gas cutting spatter can easily spoil thin bellows.
- Unlike normal pipe fittings, expansion joints cannot be leveraged into position. They should fit just like that. Pushing them into place, or pulling to match the adjoining flanges, sloppy installation practices, over-tightening the tie rods, or removing one or more of them are some of the killers of these delicate pieces.
- Do not reverse the joint's position as this will severely affect the internal sleeve activity.
- It is a good practice to fix a guard or cover over the expansion joint to prevent any spillage falling on personnel or surrounding equipment.
- Do not hydro test the expansion joint unless it is properly locked in position, unless you want to watch it really expanding and breaking while the test is progressing. Please be careful about this when an expansion joint becomes a part of hydrotesting circuit such as in heat exchanges, piping etc.

After installation, check for even small damages occurred during installation, as well as for pinholes or minute cracks. Check if the limiting rods supports, etc., are properly in place. Inspect the joint regularly and the adjoining piping for anything unusual. If properly designed, installed and commissioned, and maintained, they will easily last their lifetime.

Expansion Loops

All pipe work is installed at ambient temperature but will reach service temperature after commissioning. Hence, the pipe line tends to expand and contract depending on the actual temperatures. The piping should be sufficiently flexible to accommodate the movements of the components as they expand and contract. Fortunately a given length of pipe is flexible by its sheer length. However, this should not be taken for granted. Sufficient design consideration must be provided to take care of these variations. If these lines are not properly taken care of, they may crack and break due to the stresses created in them. Needless to say, different materials expand differently.

EXPANSION IN STEEL PIPES

Chart 25.1 summarizes the thermal expansion of heated or cooled pipes and tubes with commonly-used piping materials. Though it is well known that copper, brass, and aluminum expand more, it can be seen that stainless steel expands much more than carbon steel. This should be kept in mind when designing or laying pipelines or combining pipelines with different metals.

Carbon steel pipes are the most commonly used piping. **Chart 25.2** summarizes graphically the temperature expansion of carbon steel pipes expressed in *inches per 100 feet* per degree Fahrenheit. **Chart 25.3** shows the same information expressed in *mm per m* per degree Celsius.

It can be observed from **Chart 25.2** that 100 feet of carbon steel pipe at an initial temperature of 100°F expands by about 1.6 inches for about 300°F rise. This has a catastrophic effect if not properly taken care of. Most common methods to take care of these expansions are expansion elbows, loops, or Z bends. Expansion loops are a cheaper and easier alternative to expansion bellows. They can be fixed in line with a length of piping made out of the same piping materials. Most often these loops are placed in the horizontal lines rather than on the vertical runs to prevent accumulation of condensates. (This is particularly important on steam piping. Read the saturated steam temperatures from **Chart 25.1**) Loops should be placed at the center point between two anchors.

In practice, however, piping loops are usually installed approximately every 100 feet, depending on the locational availability for pipe anchors, guides, and supports. Expansion loops do offer

Chart 25.1
Temperature Expansion of Pipes per 100 Feet (inches)

Saturated Steam Pressure [1]	Tempera ture (°F)	Cast Iron	Carbon and Carbon Molybdenu m	Wrought Iron	4-6% Cr. Alloy Steel	12% Cr. Stainless Steel	18 Cr. -8Ni Stainless Steel	Copper	Brass	Aluminum 2-S
	-200	-1.058	-1.282	-1.289	-1.25	-1.17	-2.03	-1.955	-2.065	-2.69
	-180	-0.982	-1.176	-1.183	-1.15	-1.07	-1.85	-1.782	-1.89	-2.44
	-160	-0.891	-1.066	-1.073	-1.03	-0.97	-1.67	-1.612	-1.705	-2.18
	-140	-0.797	-0.948	-0.955	-0.97	-0.87	-1.48	-1.428	-1.508	-1.93
	-120	-0.697	-0.826	-0.833	-0.8	-0.75	-1.3	-1.235	-1.308	-1.67
	-100	-0.593	-0.698	-0.705	-0.7	-0.63	-0.09	-1.04	-1.098	-1.4
	-80	-0.481	-0.563	-0.57	-0.55	-0.52	-0.88	-0.835	-0.888	-1.12
	-60	-0.368	-0.428	-0.435	-0.43	-0.4	-0.67	-0.63	-0.673	-0.85
	-40	-0.248	-0.288	-0.295	-0.29	-0.27	-0.45	-0.421	-0.452	-0.58
	-20	-0.127	-0.145	-0.152	-0.145	-0.13	-0.225	-0.21	-0.227	-0.28
	0	0	0	0	0	0	0	0	0	0
	20	0.128	0.148	0.18	0.14	0.14	0.223	0.238	0.233	0.32
	32	0.209	0.23	0.28	0.234	0.234	0.356	0.366	0.373	0.5
29.39	40	0.27	0.3	0.35	0.28	0.28	0.446	0.451	0.466	0.63
	60	0.41	0.448	0.54	0.43	0.43	0.669	0.684	0.69	0.93
28.89	80	0.55	0.58	0.71	0.5	0.55	0.892	0.896	0.92	1.24
27.99	100	0.68	0.753	0.887	0.65	0.69	1.115	1.134	1.15	1.53
26.48	120	0.83	0.91	1.058	0.8	0.82	1.338	1.366	1.39	1.84
24.04	140	0.97	1.064	1.24	0.95	0.96	1.545	1.59	1.625	2.15
20.27	160	1.11	1.2	1.42	1.1	1.09	1.784	1.804	1.865	2.46
14.63	180	1.24	1.36	1.58	1.25	1.23	2	2.051	2.1	2.77
6.45	200	1.39	1.52	1.75	1.4	1.38	2.23	2.296	2.34	3.08
0	212	1.48	1.61	1.87	1.5	1.46	2.361	2.428	2.467	3.28
2.5	220	1.53	1.68	1.94	1.55	1.51	2.46	2.516	2.58	3.41
10.3	240	1.67	1.84	2.12	1.72	1.65	2.68	2.756	2.83	3.73
20.7	260	1.82	2.02	2.3	1.88	1.79	2.92	2.985	3.07	4.07
34.5	280	1.97	2.18	2.47	2.05	1.93	3.15	3.218	3.315	4.4
52.3	300	2.13	2.35	2.67	2.2	2.08	3.39	3.461	3.565	4.74
74.9	320	2.268	2530	2.85	2.37	2.22	3.615	3.696	3.82	5.1
103.3	340	2.43	2.7	3.04	2.53	2.36	3.84	3.941	4.065	5.43
138.3	360	2.59	2.88	3.23	2.7	2.51	4.1	4.176	4.35	5.78
180.9	380	2.75	3.06	3.425	2.86	2.67	4.346	4.424	4.61	6.13
232.4	400	2.91	3.23	3.62	3.01	2.82	4.58	4.666	4.87	6.47
293.7	420	3.09	3.421	3.82	3.18	2.98	4.8	4.914	5.15	6.84
366.1	440	3.25	3.595	4.02	3.35	3.13	5.05	5.154	5.4	7.19
451.3	460	3.41	3.784	4.2	3.53	3.29	5.3	5.408	5.68	7.55
550.3	480	3.57	3.955	4.4	3.7	3.45	5.54	5.651	5.95	7.9
664.3	500	3.73	4.151	4.6	3.86	3.6	5.8	5.906	6.22	8.25
795.3	520	3.9	4.342	4.81	4.04	3.76	6.05	6.148	6.5	8.61
945.3	540	4.08	4.525	5.02	4.2	3.93	6.28	6.41	6.78	8.98

1) Vacuum in degree below 212°F, psi gauge above 212°F

Chart 25.2
Temperature Expansion of Carbon Steel Pipes (Inches/100 ft/Degree F)

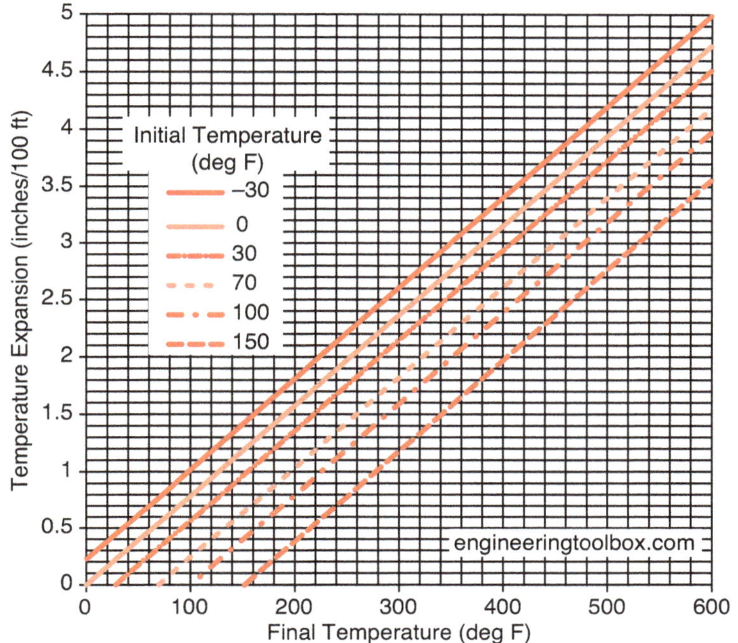

Chart 25.3
Temperature Expansion of Carbon Steel Pipes (m/mm/Degree C)

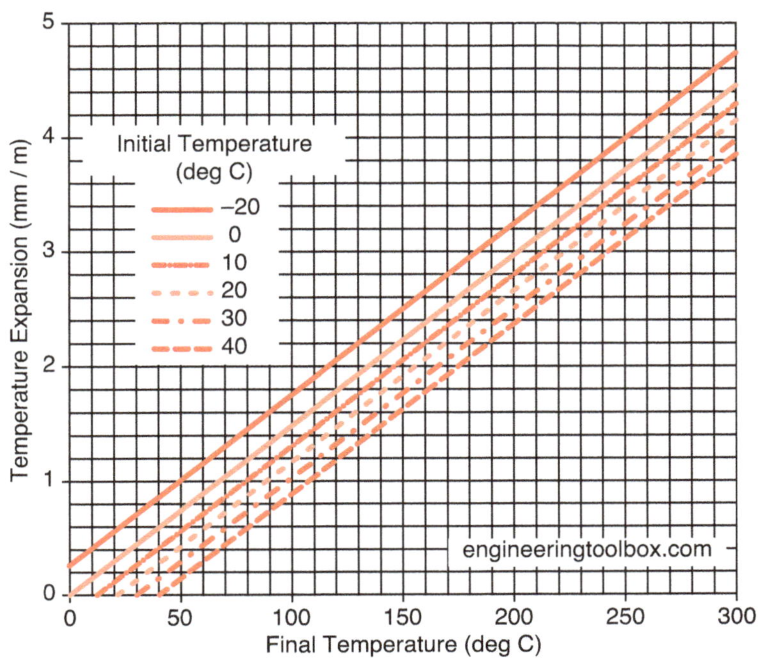

higher pressure drops. If there are a number of them in the piping system, they would be cheaper only as an initial solution, but not in the long run. It is usually best to evaluate the options and make a proper stress analysis in such cases.

TYPES OF LOOPS

There are three kinds of expansion loops most often used in the industry: full loop, lyre loop and — most common — U-loop. A condensate collection system must be incorporated in all cases.

FULL LOOP

Figure 25.1 shows one complete turn of the pipe or a full loop. As can be seen, the downstream line passes below the upstream; it must be specified and fixed in proper direction. It must be fixed on horizontal lines. This loop may exhibit a tendency to unwind which may create undue stress on the flanges. However, these are generally obsolete now, given more easily available bellow expansion joints and loops.

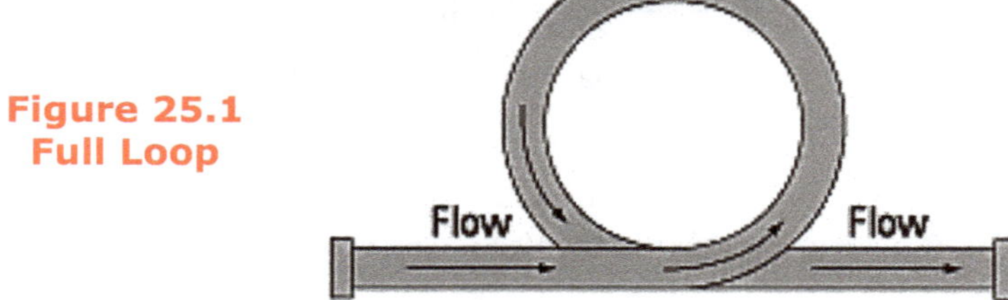

Figure 25.1
Full Loop

HORSESHOE OR LYRE LOOP

When space is not a limitation, a horseshoe or lyre look is used (see Figure 25.2). Though best on horizontal runs, a condensate collection system must be placed if used in vertical lines.

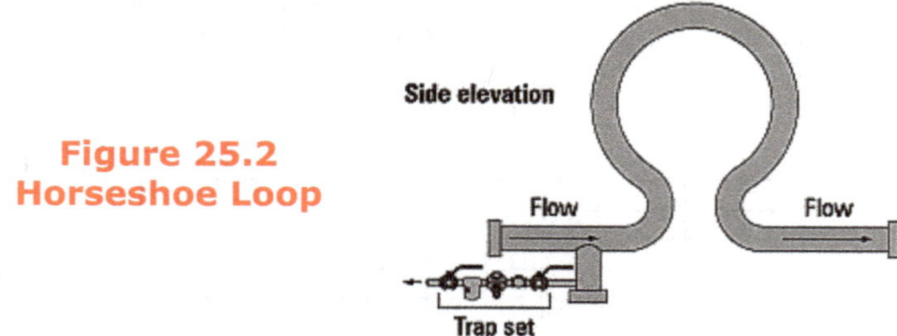

Figure 25.2
Horseshoe Loop

U LOOP

A U loop is the most generally used expansion loop that can be fabricated easily at site using the length of pipe and elbows of the same material (see Figure 25.3). It should be located in the center between two anchors. If dismantling is required, flanges can be fixed. The vertical length of the loop should be twice the width, and the width is determined from the chart below, after calculating the actual expansion taking place from the pipes on either side of the loop. The welded bend radius should be 1.5 times the diameter of the pipe.

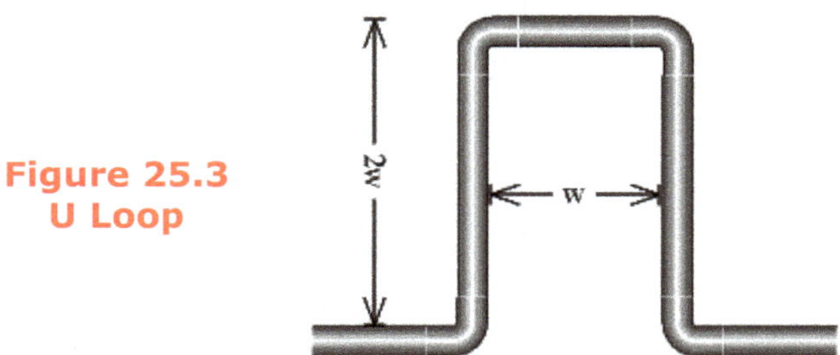

Figure 25.3
U Loop

Here are some useful graphs and charts that help us with computing the expansion and dimensions of the loops and the expansion capacity of the pipes. **Chart 25.4** shows graphically the expansion capacity of pipes from 1" to 4" and the required width of the loop. The details are then given in **Chart 25.5**.

Chart 25.4
Expansion Capacity of Pipes

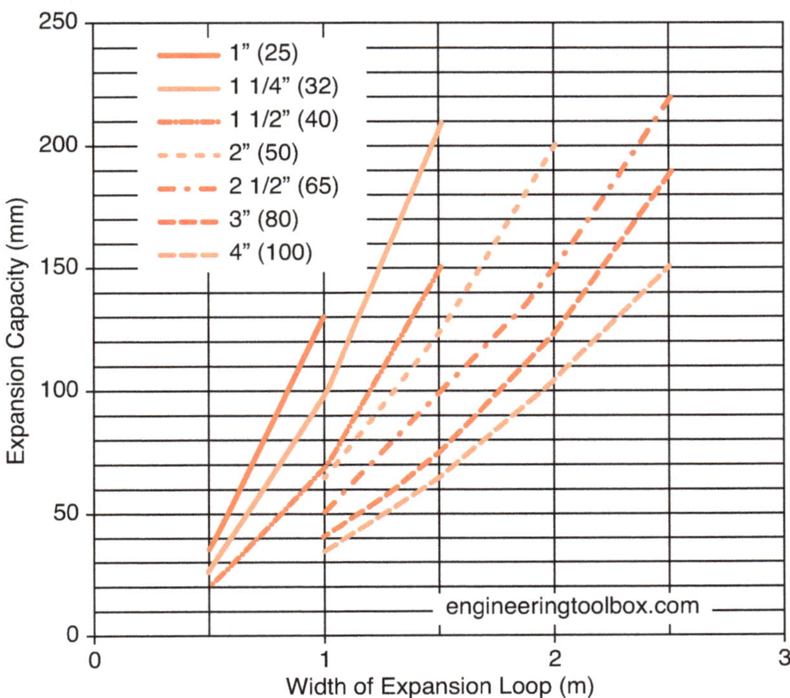

Chart 25.5
Expansion Loop Capacity for Carbon Steel Pipes

Approximate Expansion Capacity (mm)					
Nominal Pipe Size (mm)	Width of Expansion Loop - *w* - (m)				
	0.5	1.0	1.5	2.0	2.5
25	35	130			
32	25	100	210		
40	20	70	150		
50		65	125	200	
65		50	100	150	220
80		40	75	125	190
100		35	65	115	150

SUPPORTS

Roller supports are most generally adopted to support pipes, which at the same time allow them to move in both directions. Figure 25.4 shows chair and roller support; Figure 25.5 shows chair, roller, and saddle support.

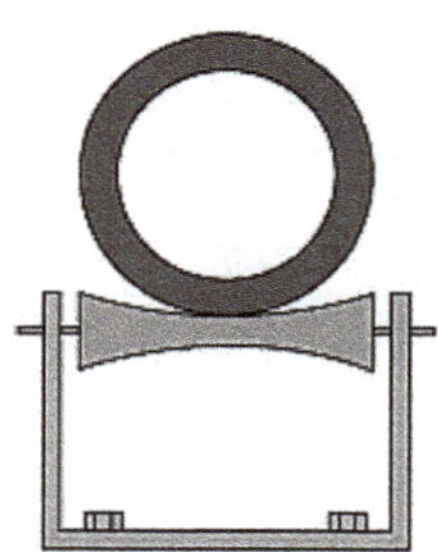

Figure 25.4
Chair and Roller Support

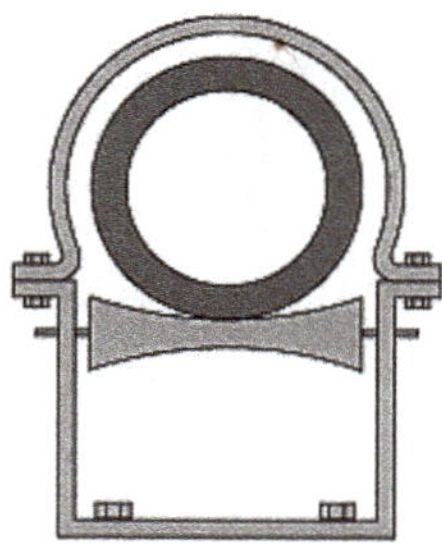

Figure 25.5
Chair, Roller, and Saddle Support

Note that thermoplastic plastic materials exhibit much more expansion compared to steel, and stainless steel expands more than carbon steel. This must be borne in mind while laying the pipelines with these materials. **Chart 25.6** shows graphically the relative expansion of PVC, CPVC, Carbon Steel, Stainless Steel, Fiberglass in English units, whereas **Chart 25.7** shows the same in SI units. Details are given in inches per 100ft in **Chart 25.8**.

Chart 25.6
Thermal Expansion of Some Piping Materials in English Units.

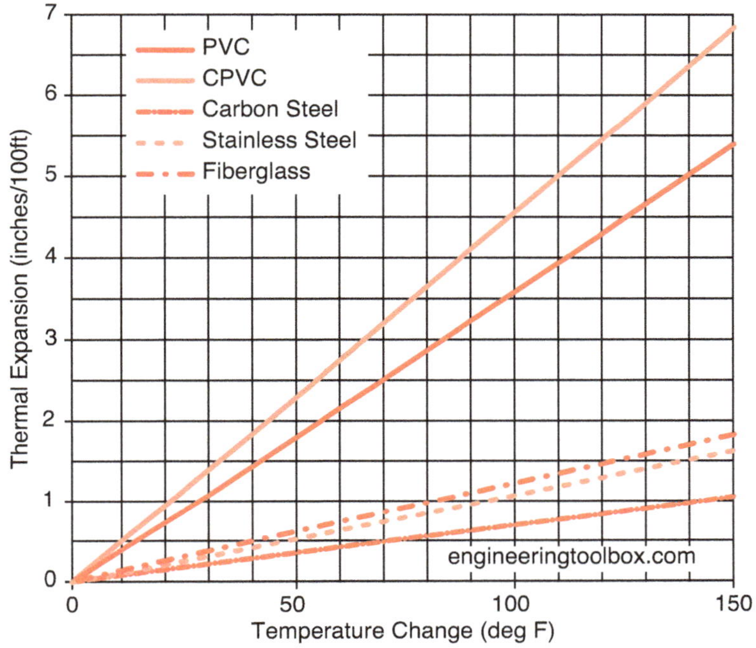

Chart 25.7
Thermal Expansion of Some Piping Materials in SI Units.

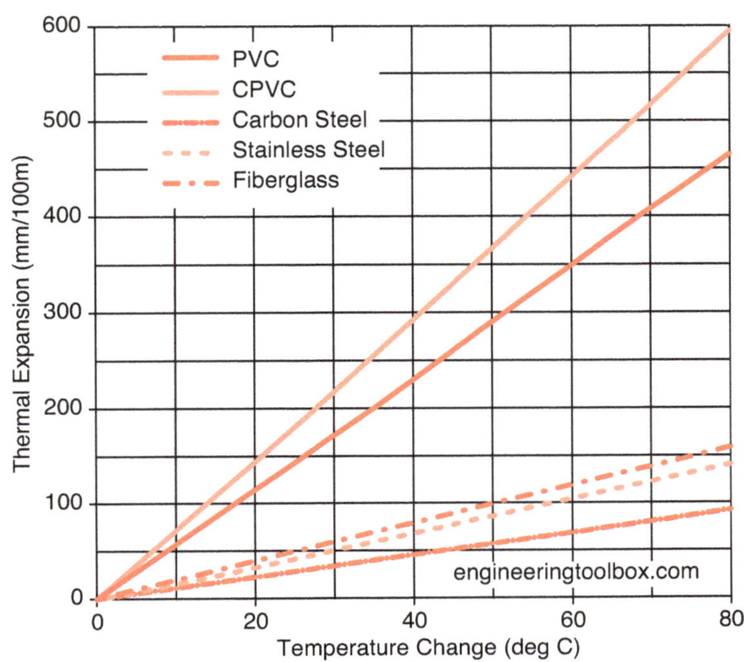

Chart 25.8
Thermal Expansion in Inches Per 100 ft.

Thermal Expansion *(inches/100 feet)*					
Temperature Change *(°F)*	Fiberglass	PVC[1)]	CPVC[1)]	Carbon Steel	Stainless Steel
25	0.31	0.90	1.14	0.18	0.27
50	0.61	1.80	2.28	0.36	0.54
75	0.92	2.70	3.42	0.54	0.82
100	1.23	3.60	4.56	0.72	1.09
150	1.84	5.40	6.84	1.08	1.63

Pipe Supports

It is imperative that piping should be supported adequately. Its own weight and the weight of the fluid impose stresses and strains on the line resulting in sagging of the line; leakage; distorted piping, fittings, valves, and associated equipment; and breakage of the line.

Careful design of piping support systems of above grade piping systems is necessary to prevent failures. Manufacturers Standardization Society of the Valve and Fitting Industry, Inc., has developed several standards for materials, design, and manufacture of the design, selection, and installation of pipe supports.

The locations of piping supports and type are dependent upon the following factors: pipe size and weight, fluid carried with its pressure and temperature, piping configuration, locations of valves and fittings, gradient, and the structure available for support. Provisions must also be made for expansion of the pipe line. Support-type selection and spacing can be affected by the seismic zone.

Weight is defined as the total distributed weight of the piping system plus insulation and the contained fluid. The weight of the pipe increases invariably with the pipe size and schedule, while the amount of fluid also adds to the weight. Span is a function of the weight that the supports must carry. Concentrated loads such as valves, meters, and clusters of fittings need independent supports.

Spacing is a function of the size and MOC of the pipe, the fluid in the piping, its temperature and the ambient temperature, and whether the pipe is horizontal or vertical. The maximum span between supports is based on the maximum deflection the pipeline may encounter. Typically, a deflection of 2.5 mm (0.1 in) is allowed. Improper spacing of supports makes the pipe sag; fluids then tend to collect in the sag of the pipe. The elevation of the down-slope pipe support should be lower than the elevation of the lowest point of the sag in the pipe, such that the piping drains properly.

As a rule-of-thumb, piping should be supported at about 10-foot intervals for the ordinary installation. Generally, piping below 1 1/2" is and should be supported from adjacent larger lines. Special consideration must be made for tubing and thin-walled piping of less than 3/4" size to give adequate support and flexibility to contain vibrations. The available structure often dictates location and selection of pipe supports, but it should be able to accommodate the load from the support and pipe. However, if two or more pipes are supported on a common bracket, the spacing between the supports should be governed by the smallest pipe.

The configuration of the piping system affects the location of pipe supports. Where practical, a support should be located adjacent to directional changes of piping. But they should not restrain or impose additional loads on the developed length of expansion loops, offsets, etc. Piping supports should restrain lateral movement and should direct axial movement into the corresponding compensators.

Anchors give full fixed support and allow very limited movement. Guides permit movement along the axis, but prevent lateral movement. Restraints control, limit, and redirect thermal movement, thereby reducing thermal stress and loads on equipment connections. They absorb imposed loads from wind, earthquake, flow slugs, flow induced-vibration, and water hammer.

Vertical pipes should always be adequately supported at the base, to withstand the total weight of the vertical pipe and the fluid within it. Additional supports may be given as deemed fit.

Chart 26.1
Hanger Spacing and Rod Size for Horizontal Pipes

Nominal Diameter Pipe NPS (inches)	Recommended maximum space between Hangers (feet)			Recommended Rod Size (inches)	
	Standard Steel Pipe		Copper Tube		
	Water	Steam	Water	Copper	Stainless Steel
1/2	7	8	5	3/8	3/8
3/4	7	9	5	3/8	3/8
1	7	9	6	3/8	3/8
1 1/2	9	12	8	3/8	3/8
2	10	13	8	3/8	3/8
2 1/2	11	14	9	1/2	1/2
3	12	15	10	1/2	1/2
4	14	17	12	1/2	5/8
6	17	21	14	5/8	3/4
8	19	24	16	3/4	3/4
10	22	26	18	3/4	7/8
12	23	30	19	3/4	7/8
14	25	32			1
16	27	35			1
18	28	37			1
20	30	39			1-1/4
24	32	42			1-1/4

Chart 26.2
Calculating the Distance Between Pipe Supports for Steel and Copper Pipe Work

Nominal pipe size (mm)		Interval of horizontal run (metre)		Interval of vertical run (metre)	
Steel bore	Copper outside diameter	Mild steel	Copper	Mild steel	Copper
	15		1.2	2.4	1.8
15		1.8		3.0	
20	22	2.4	1.2	3.0	1.8
25	28	2.4	0.5	3.0	2.4
32	35	2.4	1.8	3.7	3.0
40	42	2.4	1.8	3.7	3.0
50	54	2.4	1.8	4.6	3.0
65	67	3.0	2.4	4.6	3.7
80	76	3.0	2.4	4.6	3.7
100	108	3.0	2.4	5.5	3.7
125	133	3.7	3.0	5.5	3.7
150	159	4.5	3.7	5.5	
200		6.0		8.5	
250		6.5		9.0	
300		7.0		10.0	

Tee joints from the vertical pipes should not be treated as supports because this would impose additional stress on the line.

HANGER SPACING

Charts 26.1 and **26.2** provide some recommended maximum span between hangers as well as rod sizes for straight horizontal pipes — not for concentrated loads which should be supports on both sides as a common rule of thumb. The information, though authentic, is general. Specific applications must be evaluated based on engineering calculations.

Large seismic activity can disturb piping systems with catastrophic results. Piping is connected to the process equipment such as pumps, compressors, or vessels. Displacements or disturbances of the equipment (nozzles being the most susceptible lot) can alter the piping systems. The pipe itself is very ductile and can withstand quite a deal of twist and shake; however, piping connections to equipment are the most prone areas during earthquakes. Differential motion of the soil and the attached structures can cause piping failure. Small branch lines taking off from the headers, bolted joints are most prone. Hence, the piping should be capable of holding up in case of such an eventuality. Pipe supports must be braced and piping itself must be restrained to take up seismic loads.

TYPES OF PIPE SUPPORTS

Selection of the proper type of support is all the more important to the design of the piping system. But then pipe supports should not damage the pipe material or impart other stresses on the pipe system! The basic type of support is dictated by the expected movement at each support location. Vibrations transmitted by the adjoining equipment must invariably be prevented from reaching the pipe lines.

A very large number of pipe supports and configurations are developed to suit an equally large number of applications. Pipe supports should be specifically designed, fabricated, and fixed to suit the outside diameter of the pipe concerned. The use of oversized pipe brackets, shimming is not good practice.

Here are the most popular supports.

PIPE CLIPS

Pipe clips are more commonly used to support smaller, often domestic lines (see Figure 26.1A and B). Similar clamps made out of heavier sections of metal are used for supporting and guiding bigger lines (see Figure 26.1C).

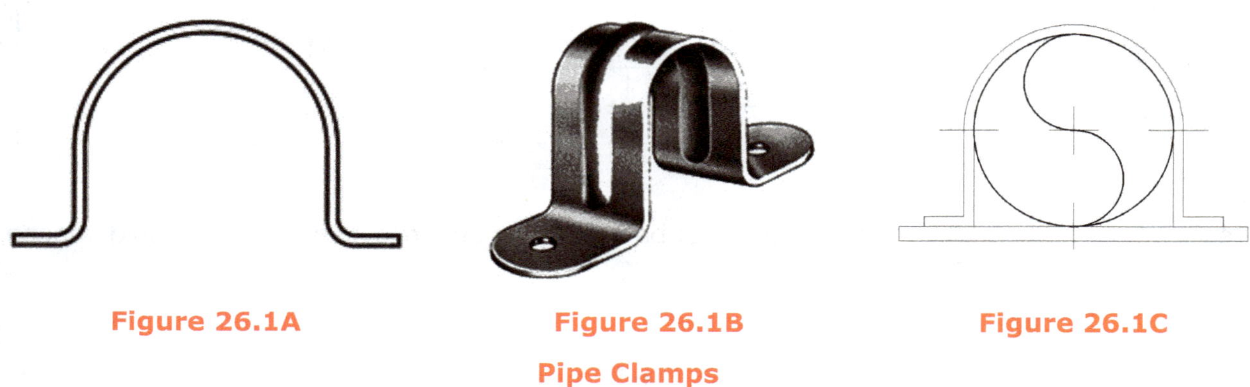

Figure 26.1A Figure 26.1B Figure 26.1C

Pipe Clamps

U CLAMP BOLTS

The standard and most widely used support is the U clamp bolt (see Figure26.2). It helps secure pipes to standard structures such as I-beams, angles, and channels.

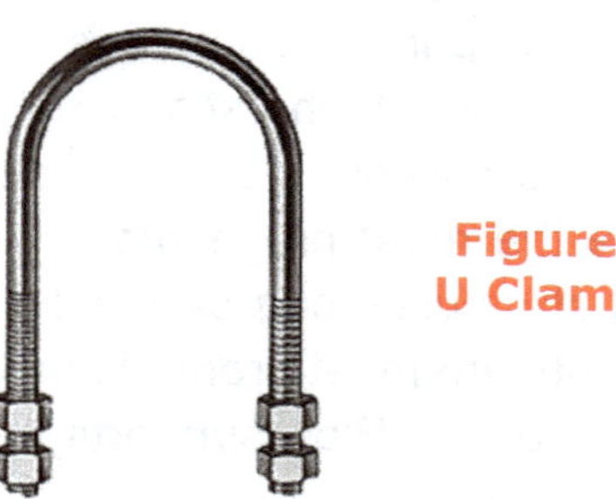

Figure 26.2
U Clamp Bolt

HALF SADDLE CLAMPS

Half saddle clamps are used when pipe is supported between two rolled plates, or a single plate rolled into shape. One of the plates is welded to a structure (see Figure 26.3A). These supports are used to support the pipe from the ground or bottom structure. Some models are adjustable to give proper loading to the pipe (see Figure 26.3B).

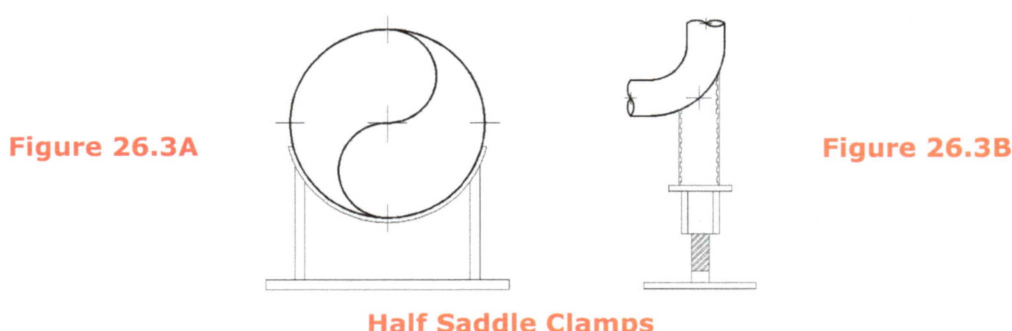

Figure 26.3A **Figure 26.3B**

Half Saddle Clamps

FULL SADDLE CLAMPS

Full saddle clamps are similar to the half saddle clamps, but both plates are rolled into shape to suit the pipe (see Figure26.4A). This support is generally fixed to the beam with a center beam clamp, as shown in Figure 26.4B. The clamp is bolted to the nearby structure or welded. It is also used to support the pipe from the bottom or the ground, if necessary, with adjusting bolts for proper loading.

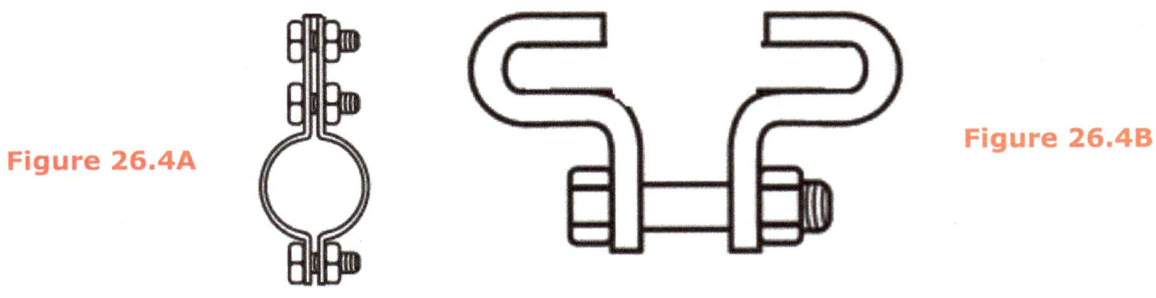

Figure 26.4A **Figure 26.4B**

Full Saddle Clamps

WELDED SUPPORT

Generally half a rail is welded to the pipe. The other piece is placed between two angles, where the rail is free to move, thereby the pipe (see Figure 26.5A). It is common practice to support insulated pipes in this manner. This is often a better choice as less corrosion takes place at the interface. This support acts as a guide. But it is not always required to have a kind of sliding support. Then the rail is directly welded to the structure, making it rigid (see Figure 26.5B). Care should be taken while welding a long seam on the pipe.

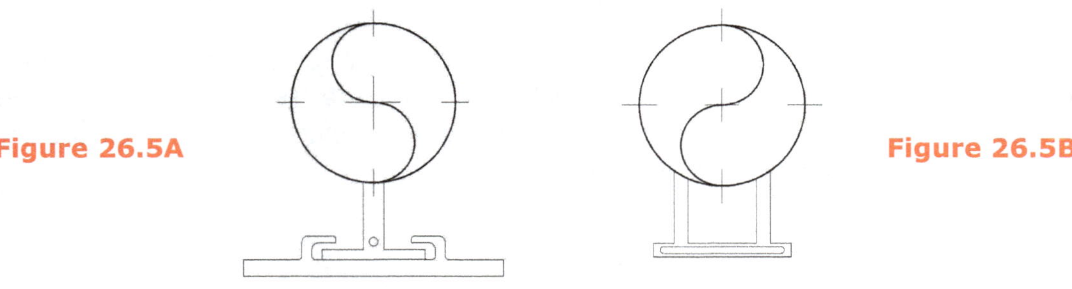

Figure 26.5A

Figure 26.5B

Welded Support

PLASTIC PIPE SUPPORT

The plastic pipe support shown in Figure26.6 offers a large bearing area for plastic pipes.

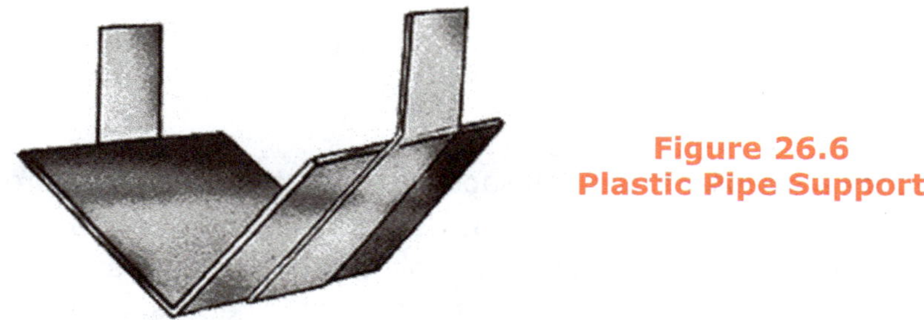

Figure 26.6
Plastic Pipe Support

A large variety of pipe supports are engineered and manufactured to suit individual needs like valve support, flange support, and elbow support. There are variations in each type tailored to individual needs. An adjustable pipe hanger is shown in Figure 26.7. There are variations to the basic designs to suit individual needs.

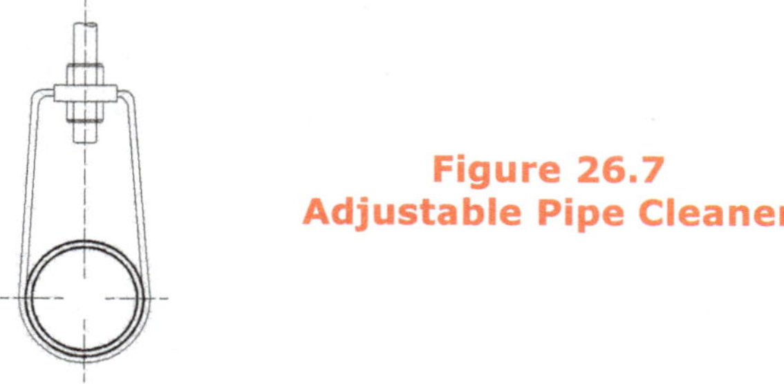

Figure 26.7
Adjustable Pipe Cleaner

ROLLER SUPPORTS

Roller supports are most often adopted to support pipes; at the same time, they allow pipes to move in both directions. When an appreciable expansion and contraction can occur, roller type supports are recommended. See Figure 26.8 for chair and roller support and Figure 26.9 for chair, roller, and saddle support.

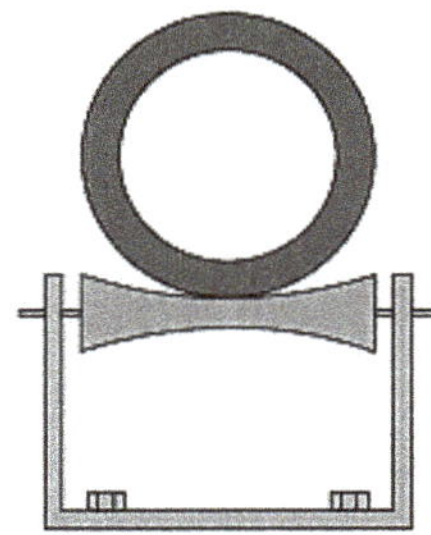

Figure 26.8
Chair and Roller Support

Figure 26.9
Chair, Roller and Saddle Support

SPRING SUPPORTS

If a pipe needs to have freedom of axial movement due to thermal expansion and contraction or other axial movement, a roller type support is selected. If minor axial and transverse (and minimal vertical) movements are expected, a hanger allowing the pipe to swing is selected. If vertical movement is required, supports with springs or hydraulic dampers are required.

Pipes expand and contract as they carry hot or cold fluids, with differences in operating and ambient temperatures. Pipe supports must support the pipe in all conditions and permit the degree of movement. They should also be able to cater to the needs of long vertical runs and those with relative movement, such as those running in and out of a building. They should also withstand earthquake conditions. Spring-loaded supports (constant effort or variable effort supports) are designed and extensively used all over the industry to meet specific conditions. See Figure 26.10 for an adjustable spring support hanger.

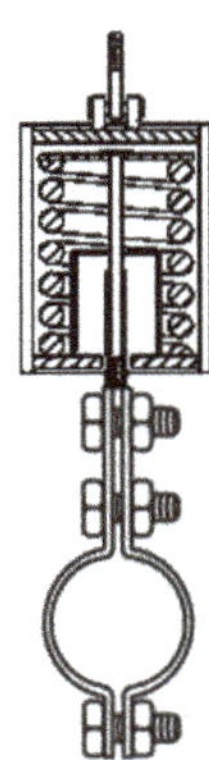

Figure 26.10
Adjustable Spring Support Hanger

Spring-loaded supports are just that! They are additional attachments over the regular pipe supports. They are compact and self contained with springs, adjustment studs, nuts, and locks. Springs are designed to take care of the calculated and possible loads. They allow for a certain over-travel and have limit stops.

Normally it is not difficult to fix the spring loaded support in position. It is lifted into position and secured to the hanger rod, to the beam attachment with a suspension bolt or clevis pin, or to the support beams of the original pipe support. The spring is loaded or tensioned by rotating the turnbuckle or by adjusting the nut. It is locked into position. Although the supports are preset to the customer's specified load and travel, they can generally be adjusted within limits of plus or minus 20% of the average load.

Find out the required supporting effort and possible pipe movement (up or down) or travel range. Take the pipe load, including the weights of associated beams and pipe clamps. Make sure that the selected spring can accommodate the operating travel within the working range.

They do not need any maintenance except that a scheduled inspection for visual damage, corrosion, and debris collected in the support should not obstruct the movement of the spring.

Fluctuating flow and unsteady temperatures produce dynamic loads on the piping systems. Consequential failures and unforeseen conditions like water hammer and earthquake can contribute to this dynamic loading. Supports discussed so far can only accommodate static loading; they are not suitable for dynamic loading. Under such dynamic loading conditions, rigid struts — which act in both directions instead of hanger rods — should be fixed to the piping systems. Snubbers are used where dynamic restraint is required, but the pipe needs a movement due to thermal conditions.

Thermoplastic pipes need more closure spacing of the pipe supports because the plastic pipe has a tendency to deflect more than metallic pipe. Valves, meters, and fittings should invariably be supported independently or the pipe is sure to break.

CORROSION AT SUPPORTS

It has been observed and is a peculiar problem of process piping that corrosion takes place at these supports. More often than not, one of the main reasons is that it is neglected area, a kind of fit and forget situation. It is a situation where pipeline is installed, supported well but soon forgotten, without giving a cursory look at the physical condition of the piping. One of the main reasons is that it is difficult to paint this area or even properly inspect it. Water gets trapped at the interface and does the damage. Crevice corrosion takes place. Bolts often get corroded. The pipe eventually fails and even breaks. Rubber pads or fiber glass pads are used habitually at the interface in an attempt to remove the metal-to-metal contact, but the problem persists. Water still accumulates. Improper spacing of supports causes excessive deflection in the line. This deflection, in turn, disturbs the structural integrity of the piping system, resulting in the corrosion and structural failure, typically at joints and fittings of the piping system.

Stainless steel or coated clamps and bolts seem to solve problem, but they may create galvanic corrosion. The welded support seems to be a viable solution. However, it adds significant cost to a typical project both in terms of construction and inspection. In some situations, it would be undesirable to make so many external longitudinal welds to a pressured piping system.

A detailed discussion on the subject is available in Chapter 27, Corrosion.

Spring-loaded piping supports and hangers for high-temperature service or low-temperature service must be properly designed and frequently checked. Maintenance such as blinding, replacement, painting, and insulation requires adequate space. Failure to provide for it would be loss of time at a later date.

UNDERGROUND PIPING

It is advisable to keep the underground piping to a minimum to avoid unknown corrosion and unnecessary maintenance. Some of the fire water mains are located underground. Sewer lines would naturally stay underground. In some cases, cooling water lines are buried underground to avoid frosting. Still some process piping may occasionally be laid underground due to peculiar plant lay-out conditions.

- Underground lines should be at least 300 mm away from the foundations.
- Similarly, if flanged connections are made, they should at least be 300 mm above ground to facilitate maintenance.
- Valves should be located in a box-like concrete structure with adequate clearance for maintenance and ample room for working.
- Ease of maintenance in all underground piping should be the primary concern.
- It should be protected with a corrosion-resistant coating and wrapping.

27

CORROSION

One of the major nightmares of maintenance engineers and production personnel is the loss of piping or related equipment due to corrosion. Corrosion is an electrochemical reaction in the metals that leads to mechanical failure due to loss of metal and consequent loss of structural strength. Metals corrode because they tend to return to their naturally occurring state. Only copper and a few other precious metals like platinum, gold, and silver occur in their metallic state.

In the process plants, corrosion is constantly in action. It deteriorates equipment, interrupts production, and causes accidents. Corrosion attacks in many ways, such as general loss of metal, pitting, grooving, and cracking, as well as other kinds of selective attack (see Figure 27.1). Attacks may be greatly influenced by minor constituents in the metal or by mechanical, electrical, chemical, or biological factors in the environment. To prevent severe cases of corrosion, either more metal is used or more corrosion-resistant metal is selected for service.

Figure 27.1
Result of Corrosion

Corrosion can create hazards in otherwise safe equipment, injury to the personnel, and loss of both precious product and production time. Corrosion may contaminate the fluid passing through the pipe line, reduce the value, and impose additional cost in refurbishing or renewal of the pipeline or equipment.

However, the corrosion did as such induce an extensive research into the metals, metallurgies, and their behavior, yielding whole range and a wide variety of exotic metals which now suit almost every purpose. Still, corrosion continues to play havoc on industrial production.

CORROSION MODES

Corrosion can be classified into two broad categories: uniform corrosion and localized corrosion.

UNIFORM CORROSION

General corrosion accounts for about 70% of industrial piping failures. Uniform or general corrosion affects the entire surface at more or less the same rate. When the natural protective layer is damaged either completely or partially, or if it weakens, the parent metal is directly exposed to corrosive environment (see Figure 27.2). It succumbs to the conditions, often at a considerable pace.

Figure 27.2
Uniform Corrosion

ATMOSPHERIC CORROSION Metals can corrode in general atmospheric conditions, which become more active in the presence of moisture, oxygen, and their cyclic onslaught (see Figure 27.3). The situation is further aggravated by pollutants like chlorides, sulphur compounds, etc. Stainless steels (i.e. steels containing at least 12% chromium) resist atmospheric corrosion. 18-8 stainless steels (austenitic) are still more resistant when an addition of 2–3% molybdenum makes them almost impervious to attack.

LOCALIZED CORROSION

Localized corrosion is much more severe as it can occur suddenly due to one or a combination of several corrosive conditions. Though it contributes about 30% of industrial piping failures, these failures are more catastrophic and sudden. Local corrosion can further be classified as

Figure 27.3
Atmospheric Corrosion

- Local — pitting corrosion, crevice corrosion, intergranular corrosion
- Environment assisted cracking — stress corrosion cracking, hydrogen embrittlement, corrosion fatigue
- Galvanic corrosion — flow enhanced corrosion, microbiologically influenced corrosion.

PITTING CORROSION When the protective film on the metal breaks down, the metal is exposed to localized corrosion. This results in pits on the metal's surface (see Figure 27.4). Though it is a localized corrosion, pits can initiate stress corrosion cracking. They are the product of accumulation of corrosive products, often accelerated in the presence of chlorides.

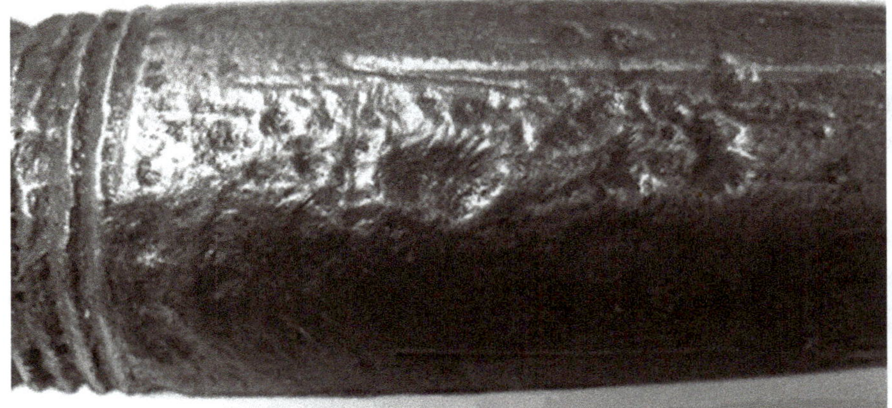

Figure 27.4
Pitting Corrosion

Stainless steels with their passive films resist corrosion. Clean surfaces, protective coatings, inhibitors, and cathodic protection are used to combat this type of corrosion.

CREVICE CORROSION Crevice corrosion is localized, usually associated with a stagnant solution at the micro-level. It selects crevices for attack under gaskets, washers, bolt heads and nuts, surface deposits, broken coatings, and clamps.

INTERGRANULAR CORROSION Microstructure of metals and alloys is made up of grains, separated by grain boundaries. Intergranular corrosion is a preferential attack of the grain boundaries of the metal, leaving the grains intact. Microscopic analysis shows the granular structure of most alloys, with clearly defined and chemically different grain boundaries.

A classic example is the sensitization of stainless steels in the heat affected zone (HAZ). Chromium gets depleted in the grain boundaries due to multi-pass welding exposing the vulnerable areas to attack. Similarly in austenitic stainless steels, either titanium or niobium combine with carbon-forming carbides in HAZ, resulting in a knife-line attack (a specific type of intergranular corrosion).

Another variation in the intergranular corrosion is the exfoliation corrosion which is more common in aluminum. Here the surface grains of metal are lifted up by the corrosion products.

STRESS CORROSION CRACKING Stress corrosion cracking (SCC) is one of the major distressing factors of pipeline operators. It is the consequence of the combined action of mechanical tensile stresses and corrosion, resulting in cracks and eventual failure (see Figure 27.5). The stress can be internal stress due to a manufacturing defect, or external stress due to the mechanical work carried out during bending, welding; improper mechanical loading, or line distortion due to soil deformities and combination of any or all of these. Cyclic loading only worsens the situation.

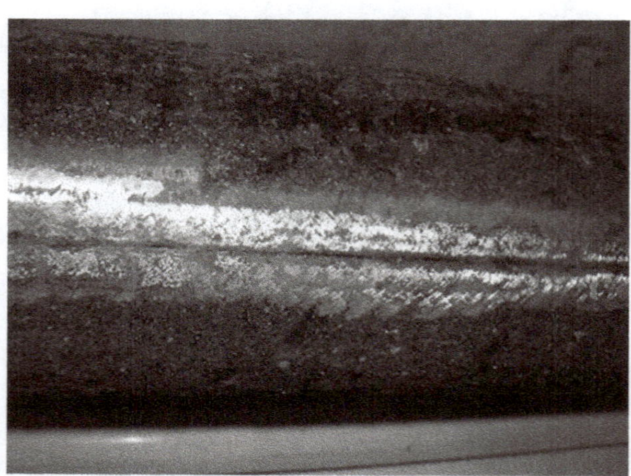

Figure 27.5
Stress Corrosion Cracking

Design and fabrication with the right materials properly loaded and supported, relieving stresses created during mechanical work, improving the environmental conditions, and avoiding stagnant areas are some of the ways to control SCC. High-alloy steels are more susceptible than low-alloy steels, though no metal is immune from it, particularly in a cyclic stressing that occurs in a corrosive environment. Buried pipe lines are more susceptible because they are exposed to environmental abuse, external damage, and loss of coating, manufacturing defects, and unstable soil.

HYDROGEN DAMAGE Hydrogen embrittlement occurs in a critical combination of localized hydrogen atom concentration, stress, and a susceptible microstructure activated by high hardness, temperature, and corrosive stock. The embrittlement of metal or alloy takes place by the ingress of atomic hydrogen into it, reducing the ductility and strength, and causing cracks much below the projected limits. It is more of a problem in the high strength steels, titanium, etc.

CORROSION FATIGUE Cyclic stress conditions in a corrosive environment can lead to a corrosion-fatigue, resulting in failures even at lower loads and in short time. Corrosion fatigue can be reduced by the reduction of cyclic stresses. No metal is immune from it when loaded by corrosive stresses in a corrosive environment.

GALVANIC CORROSION Galvanic corrosion is the damage caused when two dissimilar materials are coupled in a corrosive electrolyte (see Figure 27.6). A galvanic series was developed for metals and alloys in seawater, which shows their relative nobility. When two metals are submerged in an electrolyte, the less noble will experience galvanic corrosion. When a galvanic couple forms, a metal with lower electrochemical potential will be the cathode and will remain unchanged. A metal with higher potential will be the anode and will corrode.

Figure 27.6
Galvanic Corrosion

Chart 27.1 indicates the galvanic relationship between commonly-used piping materials. The metals listed higher will resist the metals listed lower. The greater that the distance is between two metals, the greater the speed of corrosion will be.

Chart 27.1
Galvanic Series in Sea Water

Noble (Least Active)
Platinum
Gold
Graphite
Silver
18-8-3 Stainless steel, type 316 (passive)
18-8 Stainless steel, type 304 (passive)
Titanium
13 percent chromium stainless steel, type 410 (passive)
7NI-33Cu alloy
75NI-16Cr-7Fe alloy (passive)
Nickel (passive)
Silver solder
M-Bronze
G-Bronze
70-30 cupro-nickel
Silicon bronze
Copper
Red brass
Aluminum bronze
Admiralty brass
Yellow brass
76NI-16Cr-7Fe alloy (active)
Nickel (active)
Naval brass
Manganese bronze
Muntz metal
Tin
Lead
18-8-3 Stainless steel, type 316 (active)
18-8 Stainless steel, type 304 (active)
13 percent chromium stainless steel, type 410 (active)
Cast iron
Mild steel
Aluminum 2024
Cadmium
Alclad

Chart 27.1 Continued

Aluminum 6053
Galvanized steel
Zinc
Magnesium alloys
Magnesium
Anodic (Most Active)

THERMOGALVANIC CORROSION Temperature changes can alter the corrosion rate of a material. A good rule of thumb is that a 10°C rise doubles the corrosion rate. The thermal gradient can result in local corrosion somewhere between the maximum and minimum temperature zone. Similar condition pertains in buried pipelines where substantial temperature difference exists between the pipeline and the surrounding sub-soil.

EROSION CORROSION Erosion corrosion is related to the flow of corrosive fluid in the pipeline. A high level of turbulence, abrasive particles in the stream, multi-phase flows, and faulty workmanship add to the corrosion rate. Flow and turbulent flow have an important impact on corrosion rates. High velocity fluids in a corrosive environment can corrode a stationary object in the vicinity, as well as the rotating element. A case in point is high erosion on the pump casing and also the impeller. Harder alloys or harder facings and more corrosion resistant alloys can control this type of corrosion.

MICROBIAL CORROSION Microbiologically-influenced corrosion can occur in natural waters, sea water, and soils, and by degradation of materials by bacteria and their by-products (see Figure 27.7). Microbiological growth forms bio-films, thereby depleting oxygen. The protective film is then destroyed. It produces slimes and deposits which harbor crevice corrosion. Acting through their byproducts, microbiological growth creates corrosive atmosphere (often acidic).

Figure 27.7
Microbial Corrosion

They even destroy corrosion inhibitors added to the system, and may start corrosion reactions. Both aerobic and anaerobic bacteria are capable of causing corrosion. In addition to the impact on corrosion, microbial growths cause fouling. Ozonation and ultraviolet (UV) exposure and biocides are some of the treatments.

FILIFORM CORROSION When moisture permeates the coating or plating of the surfaces, filiform corrosion takes place. Long filaments staring out from the original pit extend and tend to corrode the rest of the protected metal (see Figure 27.8).

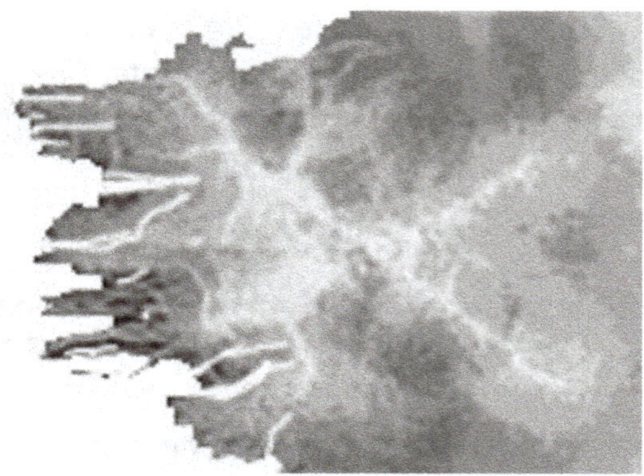

Figure 27.8
Filiform Corrosion

FRETTING CORROSION Metal contact surfaces such as bearing interfaces are subjected to this type of corrosion. Very good lubrication and protection from direct contact of air can exclude this problem.

DEALLOYING Dealloying or dezinking is fairly rare; it occurs in copper alloys, gray cast iron, etc. An alloy may lose an active and corrosion-resisting metal and get exposed to the vagaries of a corrosion environment (see Figure 27.9).

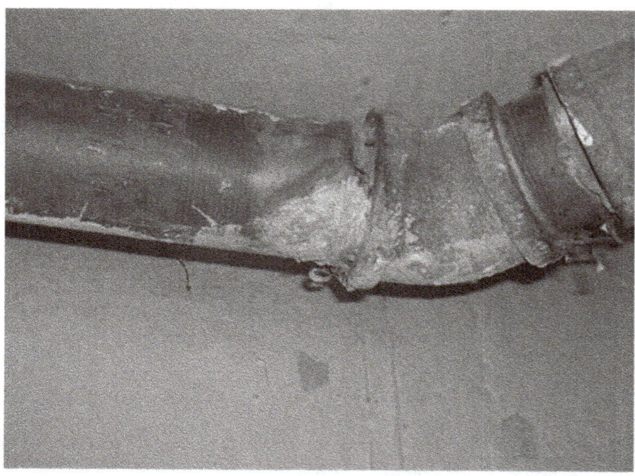

Figure 27.9
Dealloying

PREVENTION AND PROTECTION

Corrosion is an electro-chemical reaction that happens in the combination of material, environment, and component geometry. Corrosion can be controlled by breaking or interfering with the chemical or electrical processes, or altering the component geometry. Corrosion can be controlled by conditioning the metal or the environment.

CONDITIONING THE CORROSIVE ENVIRONMENT

REMOVAL OF OXYGEN By the removal of oxygen from water systems in the pH range 6.5–8.5, one of the components required for corrosion would be absent.

CORROSION INHIBITORS A corrosion inhibitor is a chemical additive which, when added to a corrosive aqueous environment, reduces the rate of metal wastage. These are added in small concentrations to reduce the corrosion rate.

CONDITIONING THE METAL

ELECTROCHEMICAL CONTROL Because corrosion is an electrochemical process, the rate of corrosion reactions may be controlled by passing anodic or cathodic currents into the metal.

CATHODIC PROTECTION Cathodic protection and protective coatings should invariably be provided for the buried steel piping, regardless of soil or water conductivity, such as natural gas, liquid fuel, oxygen, fire water, or chilled water. An external anode is connected to the metal to be protected; electrical DC current is passed so that all areas of the metal surface become cathodic and, therefore, do not corrode. Cathodic protection can be achieved in two ways:

- Galvanic (sacrificial) anodes
- Impressed current

Pipe is connected with an external anode — a galvanic anode otherwise known as a sacrificial anode such as aluminum, zinc, and magnesium. Natural potential difference between the anode and the steel, as indicated by their relative positions in the electrochemical series, makes a positive current flow in the electrolyte, from the anode to the steel. Thus, the whole surface of the steel becomes more negatively charged and becomes the cathode.

In impressed-current systems, an external source of DC power is used to flow current from an external anode onto the cathode surface. In buried piping, this system is used in addition to the coatings. It continues to protect even if the coating is broken or lost. The system can be continuously monitored.

Soil conditions can also worsen the effectiveness of the cathodic protection. Factors like soil type, drainage, temperature, CO_2 concentration, and electrical conductivity all contribute to the corrosive environment around the pipe.

COATINGS

Painting or coating is the mainstay of industrial corrosion protection. A wide variety of coatings and paintings (wrappings, temporary protective materials, paint films, surface coatings of metals, polymers or vitreous enamels, and conversion coatings) are available now to protect the metal basically from atmospheric corrosion, in other words, to protect it from oxygen, moisture, and other ambient corrosive agents.

PAINTS Paints are commonly used to protect the pipe lines. They are versatile organic coatings containing some amount of corrosion inhibitors. However, they are porous and they are liable to mechanical damage. Paints should be applied after thorough cleaning, preferably with abrasives if the surface is already corroded.

POLYMER COATINGS A number of commercial processes are available for applying polymeric materials from either sheet or powder, and comparatively thick, robust coatings. Buried pipe is coated to offer protection from the surrounding environment. A breakdown in the coating will result in pipeline metal being exposed. Better materials are continuously being developed to protect pipelines from underground corrosion. Presently polyethylene tape and extruded polyethylene jacket material are used.

VITREOUS ENAMELS Comparatively thick layers of borosilicate glass can give excellent protection against corrosion and oxidation up to about 900°C. Again, mechanical damage leads to failure of protection against corrosion.

CONVERSION COATINGS Chemical reaction to produce layers of corrosion-resisting scales (particularly phosphate and chromate) can produce a wide range of coatings, suitable both for enhancing the corrosion resistance of the metal and as a preparation for painting.

METAL FILMS In principle, electrodeposits of metals might provide completely impervious barriers: gold, silver, nickel, and chromium deposits are used for both decorative and protective purposes. Additionally, corrosion can be resisted by

- Galvanizing, which coats the metal with another metal. Zinc, cadmium, and aluminum can protect steel by sacrificial action. These metals build up a surface layer that limits corrosion and prolongs the life of the coating.
- A protective film gained from the parent metal itself, e.g., aluminum, stainless steel.

RATE OF CORROSION

The rate of corrosion is measured by

- Weight loss per unit area per unit time, usually mdd (milligrams per square decimeter per day)
- A rate of penetration, i.e. the thickness of metal lost in mpy (mils per year, a mil = thousandth of an inch) or in metric units, mmpy (millimeters per year).

APART FROM THEORY, IN PRACTICE

For a minute, let us forget the galvanic corrosion, crevice corrosion, and stress corrosion cracking. Let us see how corrosion attacks in practical terms. It is easy to understand that it attacks where we fail.

UNDER DEPOSIT CORROSION

Under deposit corrosion is created under the deposits of the corrosion products, flow products, and particulates that settle at various nooks and corners of the pipelines (literally) (see Figure 27.10). Deposits settle more at low flow velocities, mostly in horizontal lines, particularly in lower floors. Interior pipe deposits accelerate corrosion that leads to extensive secondary damage. However, the corrosion can be ultrasonically detected because some areas show near-original specification, and adjacent areas of high wall loss.

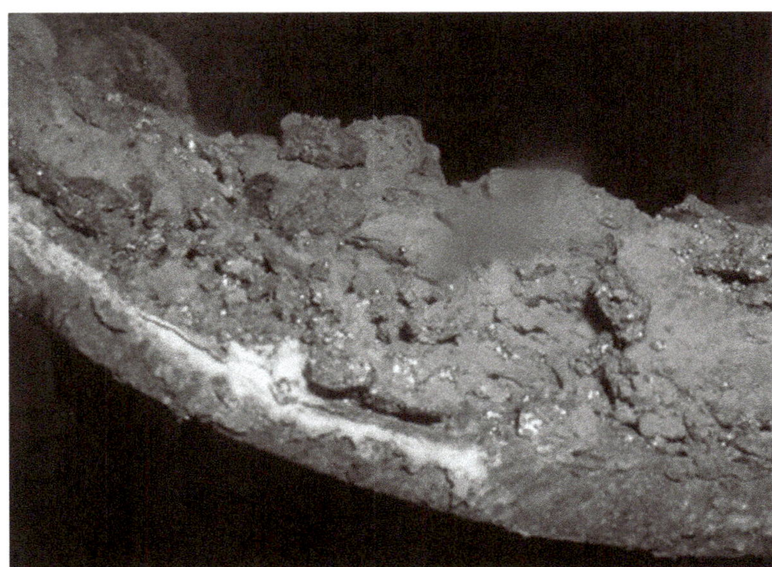

Figure 27.10
Under Deposit Corrosion

THREAD FAILURE

One of the major failure points in the industry are the threaded joints. Slightly leaky and ignored threads are the most vulnerable areas, where a failure of a single threaded joint may demand renewal of the entire length. Interestingly, most of the pipe wall thickness (as much as 65%) is lost in the threading itself. Gaps in the threads harbor particulates and microbiological growths force failure of the joint. Postponing attention of an innocuous leaky joint is just counter-productive. Small leaks are often ignored due to production constraints or due to a view that they may slowly stop. Some of them do stop leaking after some time, but how? Deposits slowly accumulate and stop the leak.

But then...Read the section above, *Under Deposit Corrosion*, to know what happens. See what happens in Figure 27.11.

Figure 27.11
Thread Failure

GALVANIC INDUCED FAILURE

Much has already written about galvanic corrosion, but it is a common problem where brass or copper fittings are joined to steel (see Figure 27.12). Occasionally it also occurs where old steel is coupled to new steel pipe, where different potentials are available.

Figure 27.12
Galvanic Induced Failure

INSULATION PROBLEMS

Though insulation protects the fluid in the piping, more corrosion problems are encountered at insulated piping — bad insulation, that is. When moisture penetrates the insulation, it has all the time in the world to make a thorough damage. It deteriorates the pipe locally and will be difficult to locate before the actual failure occurs. Corrosion is observed only when the insulation is removed for other purposes. It is more common in outside locations where rain, snow, or ambient moisture enters the insulation. Insulation itself is not properly insulated from the atmosphere. Lines carrying brine, ammonia and similar colder fluids are more susceptible to this kind of damage by corrosion under insulation (CUI).

Some insulation is damaged when the piping is fixed in a hurry in a slipshod manner (see Figure 27.13). In other cases, it is removed for some purpose or another (see Figure 27.14) or is left out altogether (see Figure 27.15). Voids left in the insulation and improper bondings are other causes for concern, particularly in cold insulations (see Figure 27.16). Insulation weeps, moisture settles at the voids, and corrodes. The problem is known only when the fluid leaks!!

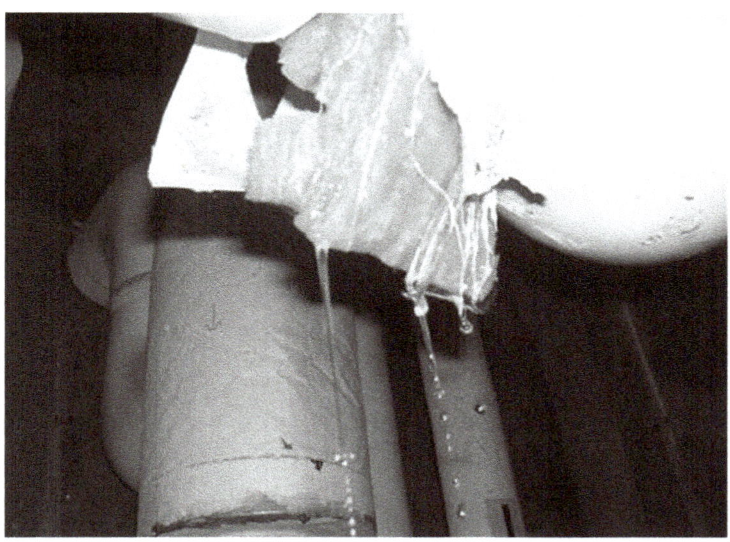

Figure 27.13
Careless Insulation

Results of Bad Insulation Repair

Insulation Partly Removed
Figure 27.14

Insulation Forgotten
Figure 27.15

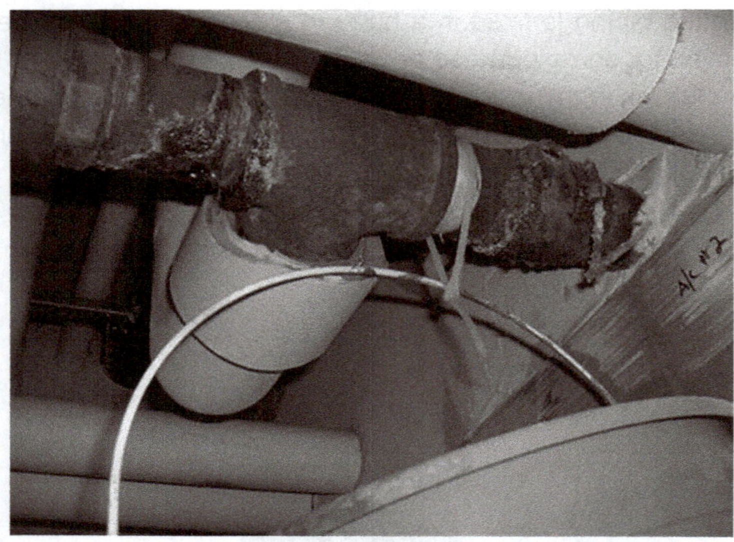

Figure 27.16
Taking Care of Leaks-we do have Better Methods!!

INSULATION FAILURE

Bad insulation can create all these damages, as can be seen in Figure 27.17.

Figure 27.17
Result of Bad Insulation

MICROBIOLOGICALLY-INFLUENCED CORROSION

Microbiologically-influenced corrosion is a peculiar problem (see Figure 27.18). The microorganisms eat away iron as an energy source (often as an alternative to oxygen); in the process they produce corrosive byproducts such as sulfuric acid which further deteriorate the pipeline. Carbon steel is not the only victim in this microbiological attack, which has been found to damage copper, brass, and stainless steel pipe.

Figure 27.18
Microbiologically-influenced Corrosion

Poor water treatment, irregular system maintenance activities like blow downs, and gaps at the pipe fittings offer adequate advantage to the microbiological growth.

OUTER SURFACE PITTING

Pitting is probably the most common form of corrosion, found almost everywhere (see Figure 27.19). It is the resultant of most other forms of corrosion such as crevice, under deposit attack, erosion corrosion, and concentration-cell corrosion. Where protective films such as painting and galvanizing are lost or inadequate, pitting results due to atmospheric attack. The consequence of damaged or insufficient insulation ends up the same. In case of accumulation of deposits whether organic, or iron or plain dirt, the effect would be the same. If neglected or if appropriate action is not taken soon enough, pitting extends all over the surface and you then have the pinhole leak!!!

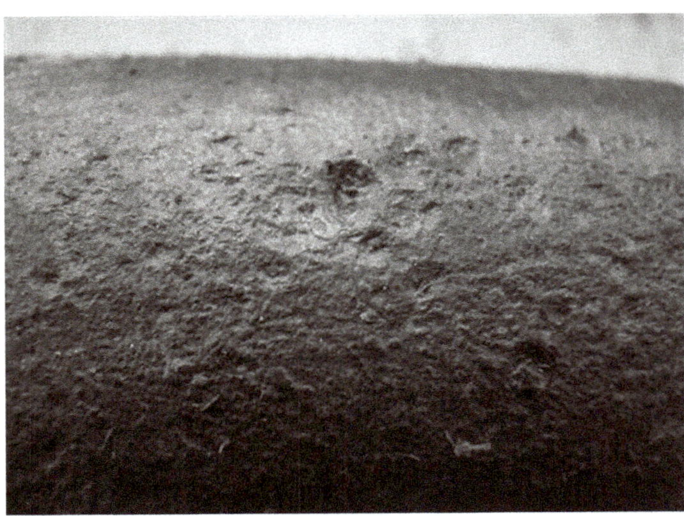

Figure 27.19
Outer Surface Pitting

WEATHERING DAMAGE

Though proper initial care is taken, most often the piping systems are left to the wind. The weather takes its toll (see Figure 27.20).

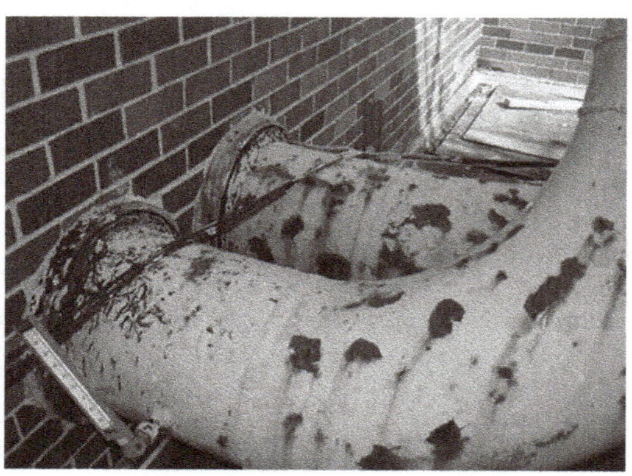

Figure 27.20
Weathering Damage

PIPE REPAIR

Emergency repairs are carried out on pipe lines where leaks are sealed (see Figure 27.21). These are carried out in trying circumstances, without shutting down the plant or process (see Chapter 32, *Emergency Repairs*). The circumstances under which the repairs are carried out often preclude absolutely leak proof joints. (This statement does not in any way imply that all emergency repair jobs are slipshod jobs). Whether it is due to the fact the pipe must be in bad state when it leaks, a combination of old and new pipes, or small left over leaks, the propensity of the metal for higher corrosion is observed at these points.

Figure 27.21
Bad Pipe Repair

DE-ALLOYING OR BRASS DEZINCIFICATION

The active element of an alloy can withdraw itself from the alloy, exposing the rest of it to the vagaries of corrosion. A typical problem occurs in brass, where zinc is leached out. Deep pitting occurs, and the pipe becomes porous. Figure 27.22 shows the surface deposits due to dezincification of brass pipe.

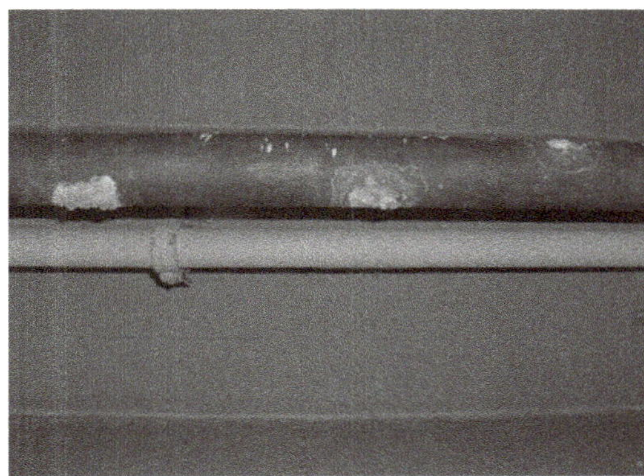

Figure 27.22
Brass Dezincification

CORROSION AT PIPE SUPPORTS

To put it lightly, corrosion prefers pipe supports. (see Figure 27.23). Areas around pipe supports are among the neglected areas of pipe protection, often accentuated by the lack of proper painting into the crevices, nooks, and corners of pipe supports. Moisture or water is trapped there and does the damage. Crevice corrosion often comes into play, though poor surface protection, micro-seepage under painting (filiform corrosion) and galvanic corrosion are strong factors as well.

Figure 27.23
Corrosion at Pipe Supports

EROSION CORROSION

This kind of corrosion is often found in the places where fluid changes direction, such as elbows or tees, where the line sharply reduces in size, and, worst, if the pipeline takes a bend there. Apart from high flow rate, entrapped gas or air bubbles, and solids accelerate the process of erosion corrosion. The top rows of Figure 27.24 show where erosion of the tubes can be seen clearly, particularly in the inlet pass.

ENGINEERING DESIGN AND MAINTENANCE PRACTICE

All methods of corrosion protection must be considered during the design process itself by taking care into the widespread prongs of corrosion enemy. Currently a wide selection

Figure 27.24
Erosion Corrosion

of metallurgies is available to combat specific problems. Carbon steel continues to be the bulwark of industry; it is inexpensive and readily adopts itself to a variety of engineering practices. Before more expensive alloys are considered, carbon steel lends itself to a very long life, when adequately protected. It is now possible to transfer virtually any fluid because a very large number of stainless steel alloys are developed, but they do have their limitations. Aluminum, copper alloys, and titanium are all here but each with its limitations. Painting and protection need not be overstressed again, but do not forget the nooks and corners.

Vertical lines have fewer tendencies to corrode for obvious reasons compared to horizontal runs. Corrosion products naturally tend to settle at the lowest points. In general it is better to keep the pipe filled, rather than drained, to keep corrosion products at bay. Pressure and temperature play a significant role in initiating and accelerating the corrosion even when the metallurgies are designed to resist corrosion. Systems need to be carefully analyzed and designed as higher velocities may induce erosion corrosion whereas lower velocities may create under deposits, and initiate biological growths. Look for problem areas where larger lines branch off into smaller lines.

Failure at any point must be carefully studied because even just a grain of rust may be only the tip of the iceberg.

CORROSION ALLOWANCES

Do not forget the corrosion allowance — a little excess metal to withstand the anticipated load for a given application. Safety factor varies between 20 and 300%, depending on the philosophy of supply and demand.

Luigi Galvani was first to discover the principles behind the galvanic corrosion (hence the name!). The first electric battery or cell was developed by Alessandro Volta in 1800, putting the principles to practical application. Sir Humphry Davy and Michael Faraday applied these principles into engineering practices in corrosion protection.

The first practical use of cathodic protection is generally credited to Sir Humphrey Davy in the 1820s. The Royal Navy sought Davy's advice in investigating the corrosion of copper sheeting used for cladding the hulls of naval vessels. Davy found that he could preserve copper in sea water by the attachment of small quantities of iron or zinc; as Davy put it, the copper became "cathodically protected."

Miscellaneous Practices

PIPELINE SPACERS

FILLERS AND SPACERS

Fillers or spacers are flat, cylindrical, metal discs that are used to fill the unusual gaps in the pipe lines. These are used for temporary installations, to reduce pipe strains, and to fill in where there are excessive gaps between flanges. This problem occurs quite often in cast iron pipes and lined piping where the piping is joined by standard lengths. Finally where the gap is short, it is not possible to get a flanged spool of a very short size. If ever available, it will be difficult to accommodate flanges, gaskets, bolts, and nuts.

Spaces are wafer-style pipe spools bolted between two mating flanges. The outer diameters of the spacers correspond with the inside of the bolt circle diameters and so they fit snugly between the bolts. They are fixed with gaskets on both sides. As a matter of fact, plants do keep these spacers of different diameters, often in lengths of one foot, then machine out the required size and use them.

DUTCHMEN OR TAPER SPACERS

Pipe lines go out of shape due to poor fabrication, support, and process conditions; they have flange gaps that are not parallel. If flanges are tightened in this condition, severe pipe strain will result and, in extreme situation, the line may crack at some point. Dutchmen or taper spacers are used when piping flanges are not parallel.

Therefore, take measurements carefully at the largest and smallest space between the two flange faces. Machine a spacer accordingly with the same materials as the pipe line. Fix it in place between the flanges with long bolts. These are used to reduce pipe strain in lines and on pumps, etc. All kinds of spacers are best avoided in high pressure and high temperature lines.

PADDLE AND FIGURE EIGHT BLINDS

Blinds are one of the main means used to gain a complete shutoff in piping. A paddle blind, or slip blind, is a piece of metal thick enough to withstand the required pressure. The paddle blind is inserted between two flanges, with a gasket on each side, and tightened. The second type of blind, the figure eight blind, is designed to be installed in the piping permanently. On one end of the blind is an opening to be used as a spacer between two flanges when the blind portion is not in use. The figure eight blind has an advantage in that it is always at the location of use, whereas the paddle blind, when not in use, can be misplaced. You can read more on this topic in Chapter 15, *Blinding and Normalizing Lines*.

SPECIAL PIPING

JACKETED AND GUTTED PIPING

There are some fluids like liquid sulfur, crude oil, LSHS, and tar which do not flow at ambient temperature, but flow when they are heated to different degrees. The temperature needs to be maintained throughout their flow. To keep the fluid inside the pipe at the higher temperature, the pipe is jacketed, gutted, or traced. Both jacketed and gutted piping have two concentric (one inside the other) pipes. In jacketed piping, the fluid is conveyed through the inner pipe and a heating medium is conveyed through the jacket (the outer pipe). Gutted piping is the reverse; the fluid is conveyed through the outer pipe and the heating medium is conveyed through the inner pipe. In some special cases, three concentric pipes are used. Innermost and outermost pipes carry the heating medium while the middle pipe carries the stock. See Figure 28.1 for typical jacketed piping.

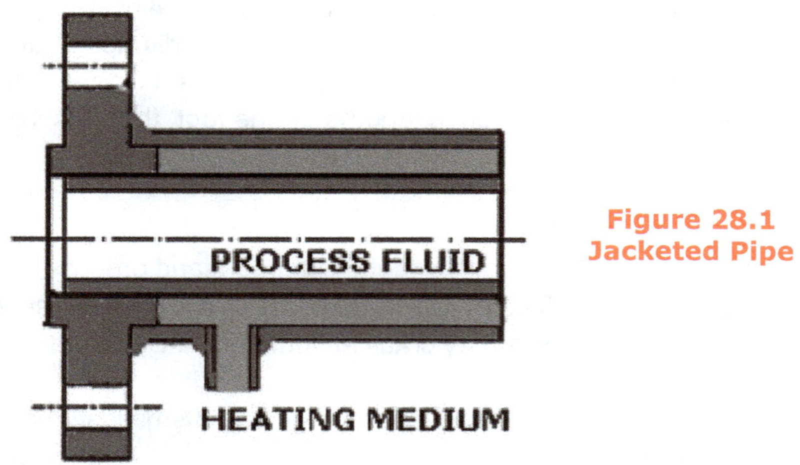

Figure 28.1
Jacketed Pipe

Jacketed piping offers the most uniform heating of the core medium. The jacket pipe or the outside pipe is welded to the back of the flanges, which makes the heating fluid to heat the flanges as well as the process piping. The flanges are oversized and matched to the outside pipe. In effect, they are reducing flanges, where the ID of the flange suits the OD of the inside pipe. Jacket pipe requires tapping — as near as possible to the flange — a takeoff socket for connection and continuation of the heating medium. Needless to say that the companion flanges of the equipment or valves, etc should have the same oversize. Connected valves and fittings, etc., also should have jacketing for effective heating and avoiding cold pockets.

TRACED PIPING

Steam tracing includes wrapping or coiling either copper or stainless tubing around the process pipe and covering it with heat transfer cement or insulation. Hot oil tracing systems are used when the process fluid is hotter than the plant's steam system. Low-pressure steam or hot oil is passed through this tubing during operation, which keeps the main line reasonably hot.

Heat loss will occur if there is no insulation. Therefore, the traced pipe should be insulated for personal protection and to prevent heat loss. A negative aspect of steam tracing occurs when the tubing sweats or leaks under the insulation, trapping moisture next to the piping.

Flanges, valves, pipe supports, etc., also need to be traced. If they are not traced, the fluid inside will solidify and plug the line. So the traced tubing should also take its route along the flanges and valves. Similarly, tracing needs to be more rigorous near the pipe supports in order to compensate for the heat loss due to conduction into them.

In some cases, electrical tape or cable is wrapped around the pipe.

DOUBLE BLOCK AND BLEED

A double block and bleed system is frequently used to stop flow positively. The double block and bleed consists of two valves in series with a small spool in between. A smaller valve, say 3/4'', is fixed in the line between these valves; it opens to the atmosphere to bleed off the built-up fluid between the valves. The system is used where the line should shut off definitely and positively.

This form of positive isolation allows personnel entry into a confined space. The two main valves are closed and the bleeder is opened to remove any trapped built up due to passing of the upstream valve. Thus the downstream piping is protected from undue flow.

Integrated double block and bleed valves have been developed and are available.

PRESSURE AND TEMPERATURE MEASUREMENT

PRESSURE GAGES

Pressure gages and switches are among the common instruments in industry. "Familiarity breeds contempt!" is an old saying that aptly suits pressure gages in a plant. There are so many gages in a plant that people often tend to take them for granted. However, they are some of the basic instruments in the process control which always need special attention.

U-tube manometers were probably the first pressure indicators (see Figure 28.2). These tubes are made of U-shaped glass with scales attached. The tube is filled with some liquid, typically oil, water, or mercury. One end of the U tube is open to the atmosphere, whereas the other is connected to the test point. U-tube manometers are inexpensive instruments for measuring positive, negative, or differential pressures with a high degree of accuracy. They are still used widely for measuring pressure drops across heat exchangers and flow meters, and in laboratories. They are simple but a little cumbersome, and cannot be easily integrated into automatic controls.

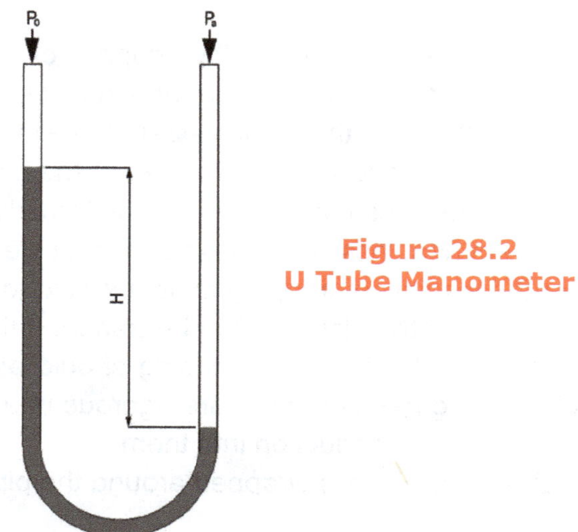

Figure 28.2
U Tube Manometer

The most popular pressure gages are the Bourdon type (see Figure 28.3). A Bourdon tube is C-shaped with an oval cross-section. One end of the tube is connected to the process pressure and the other end is sealed. As the pressure changes, the tube coils or uncoils. The motion is transmitted to the pointer on a dial through a linkage of gears and pinions. In some models, diaphragm, or bellows are used.

Figure 28.3
Pressure Gage (inside view)

The most common reasons for failure of pressure gages are pipe line vibration, dust, and water condensation. Presently most of the gages are filled with a viscous silicone fluid, which protect the parts to a certain extent but also dampens the pointer vibrations.

THERMOWELLS

Thermowells are used in the industry on the piping as well as vessels to measure the temperature of the piping fluid while keeping the sensor (often a thermocouple or the thermometer) isolated from it. Figure 28.4A shows a screwed thermowell and Figure 28.4B shows a flanged version.

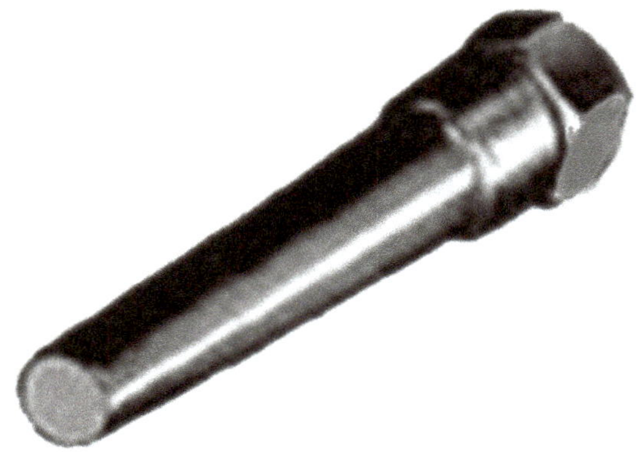

Figure 28.4A
Screwed Thermowell

Figure 28.4B
Flanged Thrmowell

Thermowells are simple — and the most forgotten pieces of instrumentation interface between the piping and automation. It is an isolated and blinded piece of piping fixed into the pipe line by threading, welding, or a flange joint. The sensor itself can now be fixed, removed, and maintained without touching the fluid in contact. However, there may be certain measurement errors due to heat loss down the length of the thermowell and the sensor may respond slowly to temperature changes. Thermowells are fixed within and across the flow; hence, they are subjected to static and dynamic fluid forces. But then that is the price paid for the measurement needs. Thermowells are industry standard.

Insulation

More industries take insulation seriously today than did a few decades ago because the savings on energy, corrosion, safety, and process efficiency outweigh the cost of insulation. Although the principle purpose of insulation is to conserve energy and cost by reducing heat loss or gain as required, insulation also maintains required process temperatures within the operating and regulatory limits, saves personnel from unsafe temperatures, increases their efficiency and comfort, protects the pipe from dangerous temperatures, avoids condensation, and prevents corrosion.

To achieve these benefits, good insulation material should possess low thermal conductivity, low moisture pick up, low corrosivity, and low cost. Insulation should have good thermal, chemical, and mechanical properties.

Based on temperature, insulation has three categories.

Low-temperature insulation covers from about 60°F/15.6°C through –100°F/–73°C; insulation below this range is *cryogenic*.

Medium-temperature insulation covers from about 61°F/16°C through 600°F/316°C.

High-temperature insulation covers from about 600°F/316°C through 1500°F/816°C. Insulation above this range is known as *refractory*.

INSULATION MATERIALS

Based on materials, the three basic types of insulation are fibrous, cellular, and granular.

Fibrous insulation consists of small diameter fibers (glass fiber and mineral wool).

Cellular insulation contains small individual cells separated from each other (glass or foamed plastic such as polystyrene (closed cell), polyurethane, and elastomeric.)

Granular insulation is composed of small nodules that contain voids or hollow spaces (calcium silicate, expanded vermiculite, pearlite, cellulose, and expanded polystyrene).

Insulation material is available as blankets, sheets, preformed shapes, and foam. It is also available as cement for final cover and cementing. Some of the more common insulating materials are listed below.

Calcium silicate is a favored material for piping insulation with its light weight, low thermal conductivity, high temperature, and chemical resistance. It is granular made of lime and silica, reinforced with organic and inorganic fibers. Insulating temperature range is about 100°F/37.8°C to 1200°F/649°C. It absorbs water, but can be dried up.

Mineral fiber (rock and slag wool) has an upper temperature limit of 1900°F/1038°C. It is chemically neutral, noncombustible, and good sound absorbent.

Refractory fibers are non-combustible and have a very high temperature limit of about 3000°F/1649°C with very high thermal shock resistance. These insulations are mineral or ceramic fibers, including alumina and silica, bound with extremely high temperature binders and manufactured in blanket or rigid form. The design and use of refractory insulation borders on engineering marvel.

Expanded silica, or perlite, is volcanic rock, mixed with water and heated; it expands as the water vaporizes. It is used in the intermediate and high temperature ranges.

Glass fiber is resistant to many chemicals, is non combustible, has good sound absorbent qualities, and is available in fibrous and cellular form. Fibrous glass can handle –40°F/–40°C to 100°F/37.8°C whereas cellular glass can take –450°F/–267.8°C to 900°F/482°C.

Elastomeric insulations have a low upper-temperature limit of 220°F/104°C. Though cost effective for low temperature, their fire retardancy is questionable.

Plastics are generally used in the range of –295°F/–183°C to 315°F/149°C. Though they have excellent moisture resistance, their fire retardancy is questionable. Polyurethane, polystyrene, and polyisocyanurate insulations fall into this category.

Insulating cements make a top covering for the main insulation to protect it from mechanical damage, corrosion, etc.

TEMPERATURE LIMITS

Chart 29.1 summarizes the temperature limits of some common insulating materials.

Chart 29.1
Temperature Limits of Common Insulation Materials

Insulation Material	Low Temperature Range		High Temperature Range	
	(°C)	(°F)	(°C)	(°F)
Calcium Silicate	−18	0	650	1200
Cellular Glass	−260	−450	480	900
Elastomeric foam	−55	−70	120	250
Fiberglass	−30	−20	540	1000
Mineral Wool	0	32	1000	1800
Phenolic foam			150	300
Polyisocyanurate or polyiso	−180	−290	150	300
Polystyrene	−50	−60	75	165
Polyurethane	−210	−350	120	250

PROPERTIES OF INSULATION

When selecting insulation, its thermal properties must be considered. They include:

- **Temperature limits:** Limiting temperatures beyond which the material is ineffective
- **Thermal conductance C:** The rate of heat flow for the actual thickness of a material
- **Thermal conductivity K:** The rate of heat flow based on 25 mm (one inch) thickness
- **Emissivity E:** Signifying the surface temperature of the insulation
- **Thermal transmittance U:** The overall conductance of heat flow through a system
- **Thermal resistance R:** R-value refers to insulation's resistance to heat flow, not to its thickness

When selecting insulation for a given service, it is important to consider chemical properties such as Ph value, chemical reaction and resistance, fire retardancy and combustibility, moisture pick up and capillarity, resistance to ultraviolet light and fungal growth, and toxicity. Also consider mechanical properties such as breaking load, coefficient of expansion, compressive strength, density and dimensional stability, and sound absorption characteristics and appearance.

RECOMMENDED THICKNESS

Proper material with sufficient thickness must be used for the insulation to be effective. **Chart 29.2** summarizes recommended insulation thickness for different sizes of pipes within various temperature ranges.

Chart 29.2

Recommended Minimum Thickness of Insulation *(inches)**				
Nominal Pipe Size NPS *(inches)*	Temperature Range (°C)			
	50–90	90–120	120–150	150–230
	Temperature Range (°F)			
	120–200	201–250	251–305	306–450
	Hot Water	Low Pressure Steam	Medium Pressure Steam	High Pressure Steam
< 1″	1.0	1.5	2.0	2.5
1 1/4″–2″	1.0	1.5	2.5	2.5
2 1/2″–4″	1.5	2.0	2.5	3.0
5″–6″	1.5	2.0	3.0	3.5
> 8″	1.5	2.0	3.0	3.5

*based on insulation with thermal resistivity in the range *4–4.6* ft^2 *hr °F/Btu in*

PROBLEM AREAS

Insulation that is wrongly selected, improperly laid, and poorly maintained does more damage than any savings it can generate. It can be a harbinger of severe corrosion under insulation. Lack of sufficient thickness and density, voids in the insulation, wrong selection, chemical environment, and water seepage are some of the problem areas (see Figure 29.1). Seepage of water or chemical reactants into the insulation and further onto the metal surface is the key problem, a threat that stays under the wraps of insulation. Water can enter directly or through capillary action.

Figure 29.1
Result of Voids in Insulation

Another area of serious concern is where the insulation is removed for some purpose or the other, then left off or replaced sloppily (see Figure 29.2).

If chlorides are present in the insulation materials, results can be catastrophic (see Figure 29.3). If insulation is fixed without attention to its density or thickness, it does not serve any useful purpose.

Over-compressing the material will remove all the tiny air pockets, the very basis of insulation. Over-expanding it would surely create voids. Good insulation can save energy, but bad insulation can *sap* our energy!!

Figure 29.2
Insulation Removed and Forgotten

Figure 29.3
Wrong Insulation

Pipe Layout

We live in a piped world. Pipes handle almost every imaginable fluid through them for transport, process, production, and use. Hence engineers and fitters involved in these processes should be familiar with piping drawings and piping symbols. Pipe sketching is an important aspect of fabricating, fixing, and using the pipe.

PIPE SKETCHES AND SYMBOLS

A variety of drawings are developed and used for process piping.

PROCESS FLOW DIAGRAM

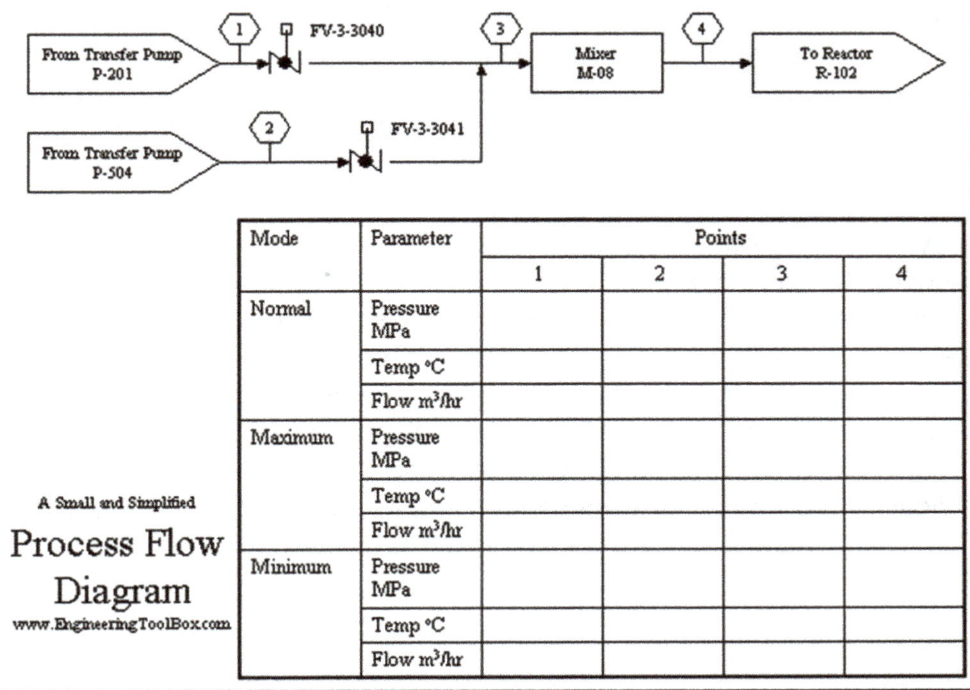

Mode	Parameter	Points			
		1	2	3	4
Normal	Pressure MPa				
	Temp °C				
	Flow m³/hr				
Maximum	Pressure MPa				
	Temp °C				
	Flow m³/hr				
Minimum	Pressure MPa				
	Temp °C				
	Flow m³/hr				

Figure 30.1 Process Flow Diagram

The process flow diagram depicts all the important points and parts of the process and its equipment in detail. It shows the fluid flow to and from the equipment items and the connected process piping with all the data on the process parameters and requirements. All the production plants have unique identification numbers for the equipment, valve, instrument, and line. A PFD should include this data wherever required. It should show interconnections with other systems and bypass and recycle lines. It provides a snapshot of the process. It is the precursor of Piping and Instrument Diagrams (P&IDs) and is not made to scale. A PFD does not show minor components, piping systems, piping ratings, and designations. A sample of PFD is shown in Figure 30.1.

PIPING AND INSTRUMENT DIAGRAMS

Commonly referred to as P&I diagrams, these engineering line diagrams are clear and comprehensive diagrams showing all of the equipment, piping, valves, and instrumentation. They show the progress of the process flow. All the production plants have unique identification numbers for each equipment valve, instrument, and line. The P&ID includes these details and certain process parameters like pressure, temperature, flows, piping and fitting sizes, flow direction, instruments, main interlocks, and power and control interfaces. A legend of identification symbols is included. It is the primary source of information. P&ID should be similar to, as far as practicable, that of the process flow diagram. These diagrams are not made to scale.

Piping components and pipe lines are represented by various symbols. Piping symbols are not really standardized as much as the other piping systems. Various companies take liberties with basic symbols, make variations, and use in-house signs. This lack of standardization goes well *as long as* the diagram is circulated internally or a legend is attached to it.

Symbols are used to denote common process equipment. Figures 30.2 through 30.7 show some of the more commonly used process piping, components, and equipment. These figures are not exhaustive but only representative, granted that the process equipment and components are vast and variations even within each group are enormous. Pipe line symbols are given in Figure 30.2. Symbols for steam traps, gages, and orifices are shown in Figure 30.3. The symbolic representation for valve operations, whether manual, hydraulic, or pneumatic, is shown in Figure 30.4. The representation for a gate valve only is shown; it is generally adopted for other valves as well. Other valves such as globe valves, needle valves, and plug valves, are shown in Figure 30.5. Symbols for major equipment like heaters, condensers, boilers, pumps, compressors, and turbines are shown in Figure 30.6. Symbols for indicators, transmitters, and controllers of the various process parameters are normally shown with abbreviations like PI for pressure indication, TC for temperature control, and FT for flow transmitter inside a circle, as shown in Figure 30.7. Figure 30.8 shows a typical P&I diagram for a process flow of cooling water.

Main Process Line	
Minor Process Line	
Underground Line	U/G
Insulated Line	
Capped Line	
Pneumatic Line	
Electric Line	
Reducing Line	
Screwed Joint	
Flanged Joint	

Figure 30.2
Pipeline Symbols

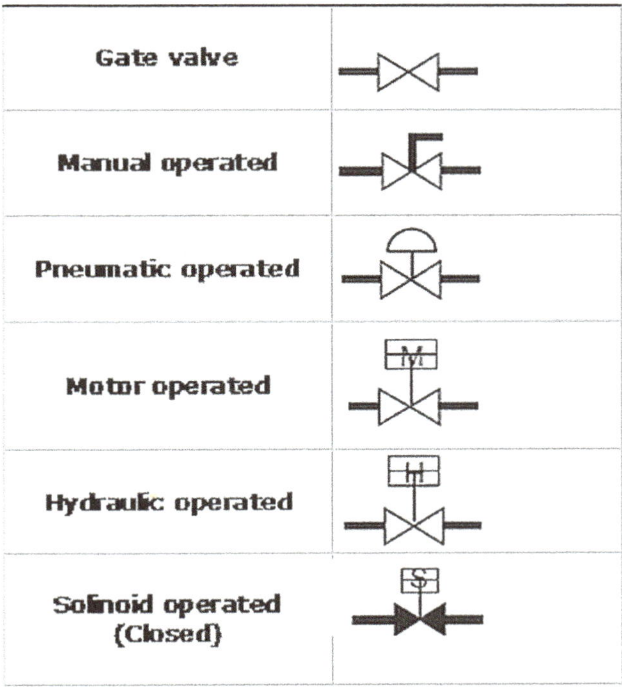

Figure 30.4
Symbols for Valve Operation

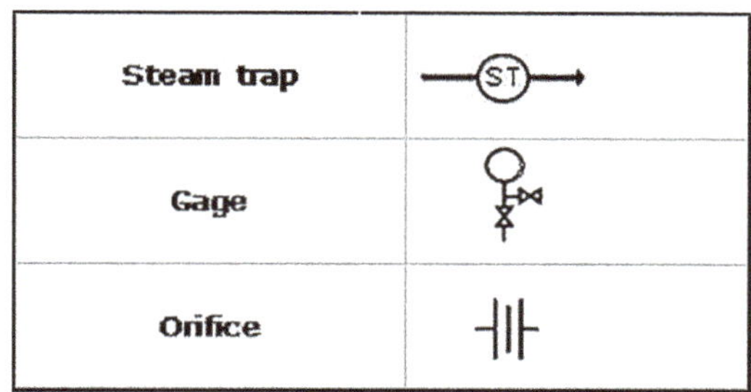

Figure 30.3
Symbols for Trap, Gage, Orifice

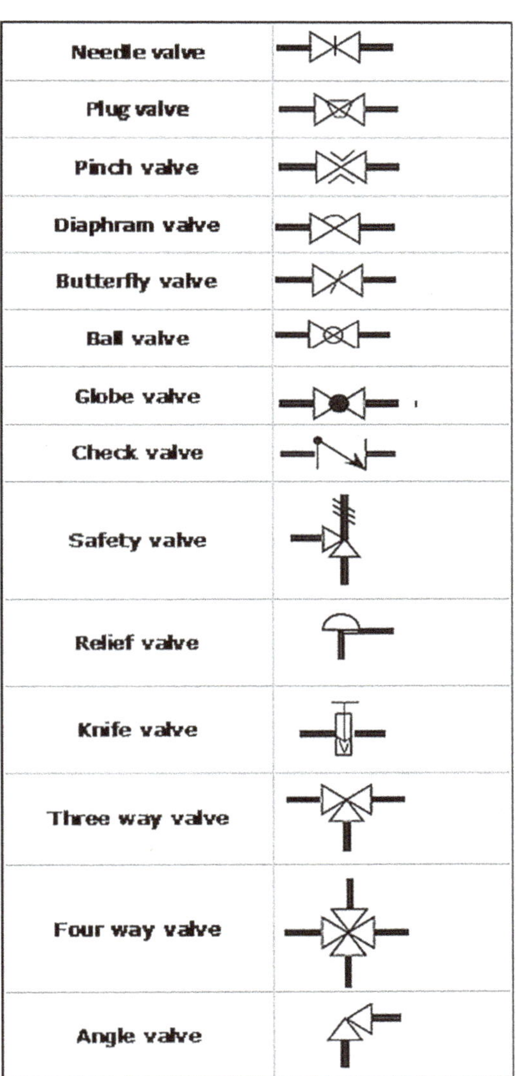

Figure 30.5
Symbols for Valves

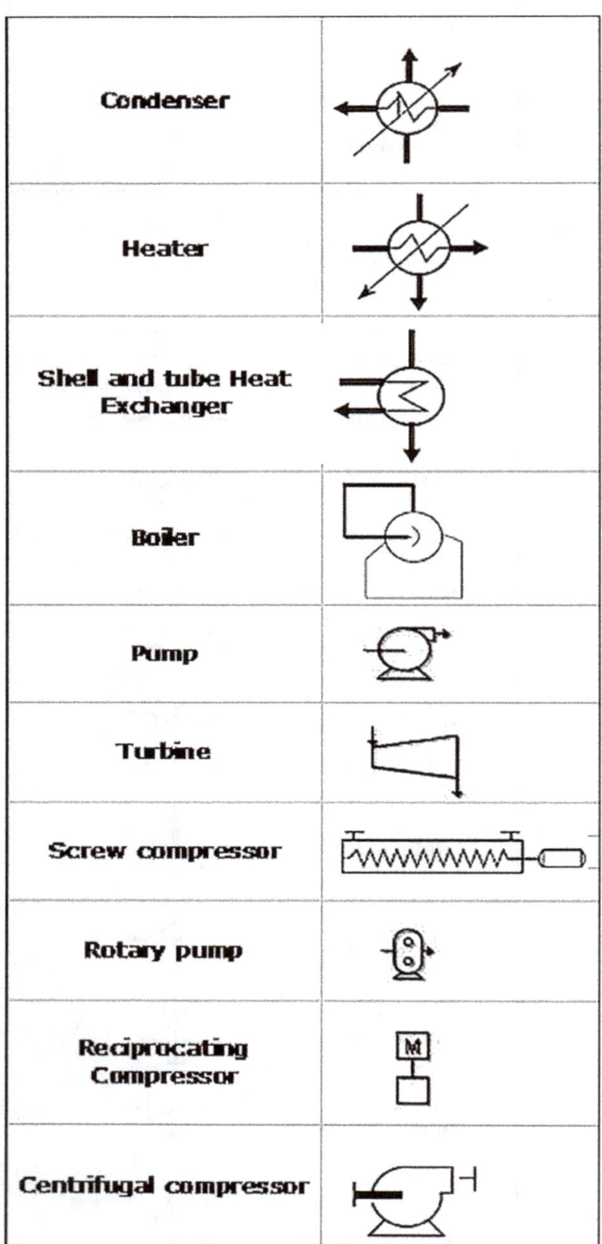

Equipment	Symbol
Condenser	
Heater	
Shell and tube Heat Exchanger	
Boiler	
Pump	
Turbine	
Screw compressor	
Rotary pump	
Reciprocating Compressor	
Centrifugal compressor	

Figure 30.6
Symbols for Some Equipment

Indicator	Symbol
Temperature indicator	TI
Pressure Indicator	PI
Flow Indicator	FI
Level Indicator	LI
Temperature Transmitter	TT
Pressure Transmitter	PT
Flow Transmitter	FT
Level Transmitter	LT

Figure 30.7
Symbols for Indicators

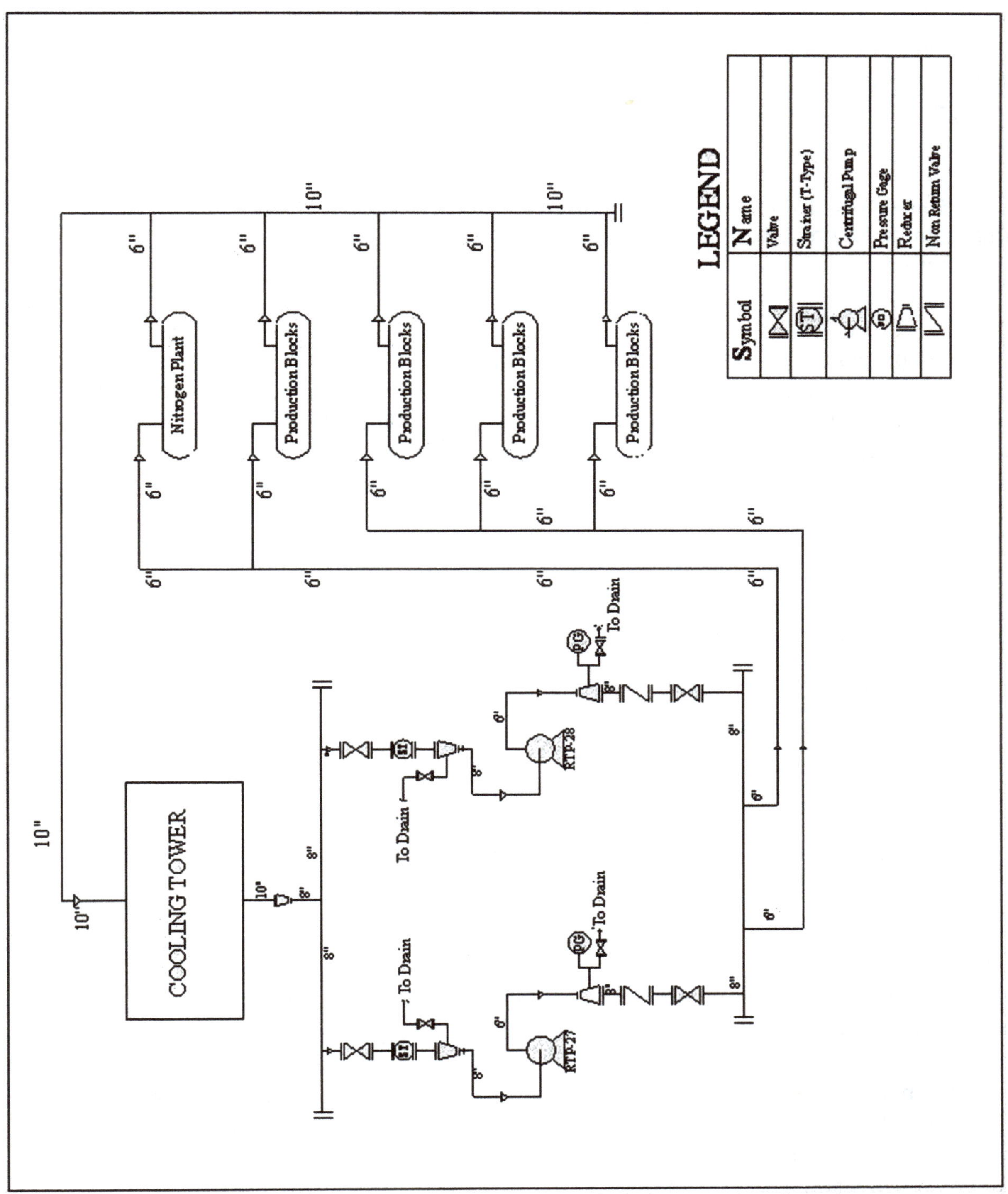

Figure 30.8
Pipeline and Instrument Diagram

PIPING ARRANGEMENT DRAWINGS

Piping arrangement drawings give in detail the pipe runs and the interfaces with the equipment, sizes and shapes, metallurgies, insulation, and support details. They are either isometric, orthographic, or a combination of the two.

INSTALLATION DRAWINGS

Installation drawings are the next level, with instructions for installation along with general arrangement for equipment and the piping arrangement.

There are two methods of piping drawings: orthographic and isometric.

ORTHOGRAPHIC PIPE DRAWINGS

Orthographic pipe drawings show items only in plane either straight or bent. Orthographic pipe drawings may be single-line or double-line drawings. A thick line is drawn for the pipe and symbols for valves and fittings are added as per standard nomenclature. The procedure is speedy and useful for the engineers. Piping is drawn in double line for use, say, in the publicity material where visual appearance is essential. Figure 30.9 shows a double-line orthographic piping drawing.

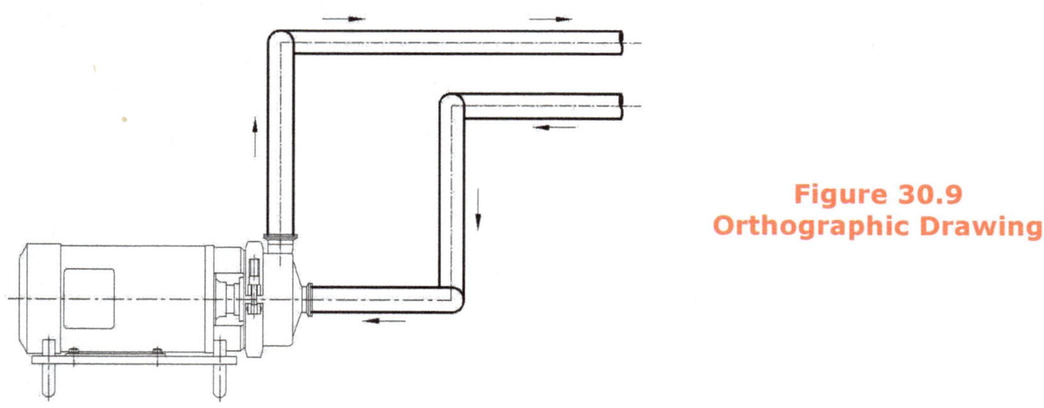

Figure 30.9
Orthographic Drawing

ISOMETRIC PIPE DRAWINGS

Isometric pipe drawings are a more popular version often used in the industry, with the single line version being more preferred. They are used for all pipes bent in more than one plane. As almost all of the piping and fitting details are standardized, drawings made in isometric view are easier to make and understand. In isometric drawings, all vertical lines are drawn vertical and all horizontal lines are drawn 30° off of the horizontal. Drawings need not be to scale but should include lengths, fittings, sizes and schedule, metallurgy, welding, and other relevant information like data on location, parts list, etc.

If the joint is detachable — say, with a flange or union — it is shown with a line across; if it is fixed — say, by welding — the joint is shown with a heavy dot. When pipes cross each other without connection, lines are drawn without interruption. If it is necessary to show which line is behind, the pipe line farthest from the viewer is interrupted. Sometimes lines are shown with crossover when they do not connect.

PIPE LAYOUT AND MEASURE UP

The first step in making a pipe line is to look over the job and make a sketch of the route of the pipe. Then take accurate measurements and record them on the sketch. Overall lengths should be marked on the sketch; also consider various fitting lengths. It will help to include as much necessary information as possible on the sketch.

TOOLS NEEDED

- Tape measure — 10ft/3mtr
- Tape measure — 100 ft/30mtrs
- Two-foot level
- Two-foot square
- Straight edge — 8' or 12' length
- Plumb bob
- Center finder and protractor

In general, all English measurements under two feet are taken in inches and fractions (e.g., 6", 9 3/4", 15"); all measurements two feet or over shall be given as feet and inches (e.g., two feet and eight inches as 2' 8", not 32"). Similarly all metric measurements are given as mm for up to 3 meters, say1800 mm instead of 1.8 mm; those at 3 meters and above are given in meters, say 5.4 m.

Not all of the above tools are required for each and every job, but they should be available so the technicians can use them when needed.

It is important to understand the job completely after thoroughly discussing it with those concerned. The purpose of the line, layout, MOC, available material, and expected material should be known. Go to the place where the piping should be installed and discuss with the concerned to locate valves, other instruments, and flanges for the convenience of the operators and technicians.

Visualize the job before attempting to measure it. Foresee the obstacles that can be overcome when installing the pipe. Take note of the obstacles that cannot be overcome, where the pipe line needs to be detoured. Clearance for flanges, insulation if needed, vents, bleeders, other lines, and equipment should be considered. Stagger the flanges on adjacent lines.

Make a sketch of the job and take measurements. Establish reference points from which to work, such as other pipe lines, buildings, firewalls, and established equipment.

The use of flanges reduces down times later. If the line requires opening at frequent intervals, then connect it with flanges at suitable places. If the product has tendency to cake up like acids or some solvents, make a provision for end blinds so that only they can be easily opened and line cleaned or flushed.

Take note of supports. Small lines can be supported from adjacent lines.

Consider the necessity of a replacement or renewal at a later date of some parts like valves and instruments. Make a note of line expansions due to process or ambient variations. Plastics need special care in this regard. Measurements taken early in the morning will go wrong by afternoon.

If possible, have all lines run on the vertical, horizontal, and square, one with another. Nothing is harder to repair than lines which were not originally put in the correct position.

IMPORTANT POINTS WHILE LAYING PIPING FOR PROCESS EQUIPMENT

Routing of pipe should be based on safety, operation, maintenance, and of course economics. The layout should have a well-organized look. Pipe line should preferably run on overhead pipe ways. Lines should be accessible for periodic inspection and maintenance. If they are buried, they should be adequately protected.

FLANGE BOLT HOLE ORIENTATION

All flange bolt holes shall straddle the natural centerlines. The flange bolt hole orientation rule is as follows:

Vertical flange face (the flange face in vertical and the line is horizontal): The bolt holes shall be oriented to straddle the vertical and horizontal centerlines.

Horizontal flange face (the flange face is horizontal and the line is vertical up or down): The bolt holes shall be oriented to straddle the (plant) north and south centerlines.

Two of the flange holes must straddle the natural centerline. The rule is invariably followed by all conscientious equipment manufacturers and pipe fabricators. If not, there will be problem in mating the flanges in the field, particularly to the equipment.

PUMPS

Suction lines should be short and have minimum bends as far as possible. Piping should give access to removal of the pump and the driver where isolating valves are in place. For instance, one should be able to remove the impeller and cover in end suction pumps, while the suction and discharge valves are in place. Incorporate additional pipe spools to accomplish this. But do make provisions to support the disconnected piping. Valve hand wheels or handles should be readily operable from grade and should not interfere with the pump maintenance. Piping spools must be incorporated into the suction lines to facilitate easy removal and repair of suction strainers. Piping should offer minimum dead load on the pumps and should not misalign it by pulling.

COMPRESSORS

Routing pipes to compressors must follow the basic principles behind the pump lines. The same is true when connecting any process equipment. Piping should be adequately and properly supported to take care of vibrations from the reciprocating compressors. Cylinder valves should be accessible without disturbing the piping.

INSTRUMENTS

The requirement for straight runs of piping, distance of immediate bends in the control valves, and flow meters should be according to the corresponding manufacturers and should be strictly adhered to. These instruments must be located for easy accessibility and visibility. Control valves and any related flow, level, pressure, and temperature instruments or indicators should have a handshake distance such that the process variables are visible from the control valve.

Provide bleeding points with valves at level switches, level controllers, control valve stations, and gage glasses per job standard.

SAMPLE CONNECTIONS

All sample connections must be easily accessible. Provide funnels over the drains where liquid samples are taken for free flow of liquid before sampling and to avoid spillages. Provide a cooler for hot samples.

ORIFICE FLANGES

Placing orifices in vertical runs is not recommended. Instead, place them in horizontal lines with adequate straight run of the piping, as recommended. A top vertical centerline is recommended for orifice taps in air and gas lines, and horizontal centerline for steam and liquid lines.

RELIEF VALVES

Relief valves should be located in a vertical position. If they are discharging to a closed loop, the common header should be lower than the valves. If they are discharging into the atmosphere, the tail pipe should extend at least 8 feet from the nearest platform.

EXCHANGERS

Locate the exchanger piping such that the internals can be easily removed without so much dismantling of its piping. If any section of piping needs removal to open the end covers, the block valves should still be in position.

FURNACE PIPING

Fuel valves should be located such that they can be operated while watching the flame. But the snuffing steam manifolds and fuel gas shutoff valves should be horizontally away from the furnace by at least 50 feet or 15 m.

FLANGES, VALVES, STEAM TRAPS, AND BLINDS

Adjacent flanges should be staggered. Lines must be adequately spaced for maintenance of valves. Lines need to be blinded or the flanges with spectacle blinds should be accessible. Locate all steam traps at low points, long runs of piping, and dead ends of steam headers. Steam traps should be accessible for maintenance. Provide a funnel where open outlets from steam traps or process drains are routed to drains.

Valve handles should be within easy reach, but wheels stems should not protrude into platforms. If the valve is located about seven feet or above 2.1 Mts, a suitable chain drive may be provided. A knuckle clearance of about 3" or 80 mm should be maintained between the outside of the hand wheel and any obstruction.

Provide hydrostatic vents and drains at the high and low points of piping. They are preferably fixed with plugs or caps rather than with valves.

UTILITY STATIONS

Utility stations like water, steam, or air should be reachable with a single 50-foot/20m hose. Plant air connection should invariably be near heat exchangers when air-operated tools are used. Similarly, water connections should be available near pumps or equipment to be washed and in areas of product spills or steam connections to those places requiring steam purging. In general, all the three outlets are placed strategically to provide for easy use all over the plant.

TESTING OF EQUIPMENT

Pressure tests are carried out to determine that the equipment is safe for operation. Normally hydrostatic tests are carried out

- Before it is initially put in service
- At scheduled intervals as per statutory controls
- After repairs or alterations
- When it is considered necessary by the concerned authorities

Hydrostatic testing is a standard procedure adopted to verify the equipment worthiness of piping systems and pressure vessels, such as heat exchangers, cylinders, boilers, etc. These tests are also conducted to check for any manufacturing defects that may result in leaks. The system or vessel is filled with an incompressible fluid, normally water or, when water is detrimental to the process, another liquid like oil. Then it is pressurized typically to 150% of the maximum operating pressure. Next, the system is checked for pressure drop and for any leaks. Color dyes are also used to aid in the detection of leaks.

Some of the terms related to testing of vessels should be considered now.

Design Pressure is the maximum pressure to which the system is subjected at the most severe condition of coincident internal and external pressure and temperature. The most severe condition is that which results in the greatest required component thickness and the highest component rating.

Design temperature of each component in the piping system is the temperature at which under the coincident pressure, the greatest thickness or highest component rating is required.

TYPES OF TESTS

ASME specifies the codes for testing of the piping systems. The code specifies the procedure, requirements, and limitations for the various testing methods used for testing for metallic piping, non metallic piping, process piping, and pressure piping, etc. Code also specifies the compensation for room temperature tests. This is apart from the non-destructive examination carried out on the piping, details of which can be found in Chapter 33, *Inspection*. The normally used such test procedures are as follows.

TESTS WITH LIQUIDS (HYDROSTATIC TEST)

Water is the most general testing medium, wheras oil or other liquid can be used if water is detrimental to the equipment under test. Test pressure is not less than 1.5 times the design pressure.

The test liquid should be at a minimum temperature of 15°C and maximum of 49°C. However, in cases, where the test liquid temperature is higher than this, special precautions must be taken to prevent scalding of personnel in case of a leak.

TESTS WITH AIR, STEAM, OR INERT GAS (PNEUMATIC TEST)

Certain limitations must be understood before attempting air tests. Tests made with air or inert gas shall be limited as per the code. *Do not use oxygen, nitrogen, or carbon dioxide for this test.* Test pressure should be 110% the design pressure.

HYDROSTATIC PNEUMATIC TEST

A combined hydrostatic–pneumatic test is also specified under specific conditions.

ALTERNATE LEAK TEST

Under certain conditions where a hydrostatic test is contraindicated due to typical process conditions, and a pneumatic test is also ruled out due to inherent problems, the code specifies alternate leak testing procedure.

TIGHTNESS TESTS (INITIAL SERVICE TESTS)

Tightness tests are applied before starting up to determine that the equipment is tight for service. The test pressure for tightness need not be higher than the normal working pressure. The test fluid is the service fluid.

SENSITIVE LEAK TEST

The test shall be in accordance with the Gas and Bubble test method. The test pressure shall be lesser of 15 psi or 25% of design pressure. A tracer gas such as helium is introduced into the system and leakage is detected.

REPAIRS OR ADDITIONS AFTER LEAK TESTING

If repairs or additions are made after the leak test is over, the affected piping may be retested. However if the repairs and additions are minor, the test may be waived subject to the fact that good engineering practices are taken up while repairing.

There are other tests followed in some plants.

STRENGTH TESTS

Strength tests are applied to new and existing equipment that has undergone major repairs or alterations to determine that it is strong enough for the intended service.

REDUCED PRESSURE TESTS

During construction and maintenance work it is sometimes desirable to test assembled piping systems at some pressure less than the full test pressure for the pipe or fittings. Assembled process piping systems that include vessels or other equipment may be tested at the pressure for the weakest part of the system.

BUBBLE TESTING

Air or gas is introduced into the system at a low pressure and bubble solution is applied at the welds and joints. The points are then visually inspected for leaks as bubbles would show them up.

TEST PROCEDURES

Provide vent valves to release air from the vessel while filling with water and a bottom valve to drain water after the testing is done.

The pressure vessel is blinded from all connected and associated lines to make sure that only the vessel under test is subjected to the test pressure. All the associated piping should be blinded using slip blinds. When the line clearances do not permit this, one may have to use full-face blinds. Valves cannot be relied upon for test pressure.

All pressure gage readings shall be corrected for elevation.

The test pump should be located in a clear area. Piping or tubing from the test pump to the vessel should be capable of withstanding the test pressure.

A calibrated test pressure gage should be used directly onto the system under test; it should be visible to the pump operator at all times. A calibrated and well-documented gage should be used. It is common experience during any typical hydrotest that frequent failure of gages is observed. Always replace with a calibrated gage.

The indication on the gage must be bold. The range should not be less than 1.5 and not more than 4 times the test pressure.

Fill the system with water and remove the air from one or more vent points. Close the vent valves after making sure that all the air is removed from the vessel. It will be extremely difficult to reach proper test pressures with air inside the vessel.

It is a good practice to stop the pump at about half the test pressure and inspect for leaks and other signs of weakness. After this preliminary inspection is OK, go ahead for the full test pressure.

Follow the safety precautions.

Indications of leaks or failures should be addressed immediately.

When the test pressure is reached, it is held for a statutory or accepted time limit (normally 30 minutes). Pressure should hold steady and not change significantly during the test. Neither water leaks nor drips should be observed.

Again, it is a good practice to drop the pressure to 2/3 before making a final close inspection or if there any leaks.

To release the test pressure, *the pipeline or the system under test should be vented first.* Do not attempt to drain the pipeline before it is properly vented out because there are genuine chances of vacuum pulling in and damaging the equipment.

All defects should be reported and repaired. Repeat the test after repairs.

Test readings must be recorded and well documented for future reference.

Avoid testing in wet and rainy times as leak detection could be difficult and deceiving.

SAFETY FIRST

Equipment under test is energy under closet. A sudden release of such energy can be devastating. Hence, the employees witnessing and making these hydraulic or pneumatic tests must take all safety precautions.

Employees operating the test pump must stay at the pump at all times during the test with the test gage visible to them.

The concerned must make sure that the system under test — and its associated connections subjected to test such as expansion joints, gages, gage glasses, and testing lines — are safe and can handle the pressures subjected.

If the pipeline or vessel under test has any expansion joints or bellows, they should be properly secured such that they do not stretch dangerously under test pressure.

Keep everyone else away from the test area. Nobody should be allowed to go up on the vessel for a close inspection during the course of full test pressure. They should be allowed only after the pressure has dropped to 2/3 of the test pressure. However, in some cases, the vent valves are located on top of the vessels or such elevated places; in these cases, extreme care should be exercised while mounting the vessels or reaching the places.

Please note if the system fails catastrophically, there is a likelihood of flying objects hitting the personnel within the vicinity. Take every care to avoid the eventuality.

Extra precautions should be taken while making high test pressures and where a large amount of liquid is handled, as with spheres and large vessels.

Because the test pressure is much higher than normal operating pressure, bolts and other joining materials are overstressed, however temporarily they may be. Hence, do not compromise on the number, size, or quality of the bolts and other materials. The normal tendency to use half the bolts — because the test is, after all, a temporary event — should be avoided.

When the pressure of the testing medium (plant air, water, or steam) is more than the required test pressure, it is necessary to install a relief valve with full capacity discharge set at 110% of the equipment under test. Such a relief valve should also be used when the medium is gas from cylinders. Use a pressure regulator set at no more than the test pressure.

PNEUMATIC PRESSURE TEST

- A pneumatic pressure test should only be considered if a hydrostatic test is contra-indicated. A pneumatic pressure test is fraught with danger.
- ASME B31.3 refers to the dangers of performing this test and provides for safety considerations in the standard.
- Because a large amount of energy is stored in compressed gas and there is a potential hazard of a sudden release of this energy, pneumatic testing should be avoided as far as possible.
- Water is practically incompressible. Hence, even a little amount of energy increases the pressure considerably. On the other hand, air (like all gases) is compressible; hence, much more energy is stored in it for a given pressure. As much as 200 times more energy is stored in compressed gas compared to water at the same pressure and volume. If any component should fail under pneumatic test pressure, the stored energy released can be deadly!
- A responsible person should be in charge of pneumatic test operations and should oversee it at all times during its course. A written record also must be made. The person in charge must also make very sure that the system is depressurized after the test is over.
- Before the actual pneumatic test, a leak test should be carried out with a maximum pressure of 0.5 bar. Check for any leaks, depressurize the system, and rectify them. Then start the pneumatic test all over.
- All flexible connections from the pressure source to the equipment under test should be positively secured. At one time or another, everyone in the industry must have watched the snake like whipping of detached pneumatic hoses.
- The compressed air pressure must preferably be controlled away from the test area. However, if it cannot be done as such, the piping must be well protected by barricades or by sandbags!
- For a pneumatic pressure test, a pressure relief valve must be incorporated in the system which should be set at 50 psi or 10% of the test pressure, whichever is lower.
- Air is normally used, but a nonflammable and non-toxic gas may be used if required. The test pressure is 110% of the design pressure.
- To perform the pneumatic test, start increasing the pressure up to a gage pressure of 25 psi. Stop at this point and make a preliminary check for leaks.
- If everything is fine, increase the pressure in steps while carefully watching the pipe line condition.
- Once the test pressure is reached, check for leaks as soon as possible and reduce the pressure to the operating level.
- Make sure that the system is depressurized.

EMERGENCY PIPE REPAIRS

A small leak in a piping system could shut down any factory. However, if the leak could be repaired without shutting down the system, operations could continue. Part of the maintenance personnel job is to do all they can — in a safe manner — to keep the plant operating. Emergency pipe repairs save the plant from being shut down, and save the cost of new pipe and labor. Repairs can be generally left in place for a considerable time or until the plant itself goes down for a normal shutdown.

There are many different piping materials used in any industry with different fluids flowing through them, some hot, some cold, some corrosive, some hazardous. For this reason, many different methods and metallurgies must be adopted to make emergency pipe repairs. Now the pipe repair industry is as big as the piping industry, with a number of ingenious devices developed for the job. Let us have a look at some of them.

CLAMPS AND CURES

A C-clamp is probably the most handy and most wanted pipe repair clamp. C-clamps come just in time to seal a pinhole leak or small crack in the pipe line. Suffice it to say that they are available in a number of sizes that can be selected depending on the pipe size. It has a pressure pad fixed on a ball joint, which can be moved in or out by the screw (see Figure 32.1).

Figure 32.1
C Clamp

Check the pipe from where the leak is emanating. If there is one hole in the pipe, you can assume the rest of the pipe is also very thin and so ready to fail. Take an appropriate clamp. For larger sizes and for peculiar positions, C-clamps can be modified or even fabricated to fit various conditions. However, leaks from the holes larger than the c-clamp pads cannot be fixed. Release the screw just enough to fit on the pipe. Take a small piece of suitable rubber of say 3–6 mm thick and fix it to the fixed pressure plate. Now place this pad directly on the leak and start tightening the screw. Tighten the screw just enough to stop the leak. Wise technicians will use common sense to determine just exactly how tight it should be. If it is tightened more, the pressure pad will invariably punch through the pipe because it is highly possible that the surrounding areas of the pipe are just as weak.

Certain precautions must be taken with C-clamps because they are used while the pipe line is active.

- Always use a perpendicular clamping force. C-clamps can slip off easily, particularly when used on round pipe lines.
- Engage the pressure pad fully; otherwise it can slip.
- Do not use a C-clamp to pull, lift, or carry something just because it looks handy for that purpose.
- Use safety goggles because C-clamps are used on filled or partially filled-in piping lines. Serious eye injury is possible.
- Do not use C-clamps without pressure pads. They will slip all the more.
- Do not use C-clamps with bent screws or that are otherwise damaged.
- C clamps on pipe lines are generally used for emergency jobs.

While attending any emergency repair — whether it is with this clamp or by any other method — it t is extremely necessary that proper physical and mental balance is maintained. More often than not, the line is not depressurized and any simple mistake can result in a serious accident.

Sometimes jubilee clamps are used (see Figure 32.2). Take a clamp suitable to the pipe. Unwind it and open it up. Take a piece of rubber and place it on the leak. Place the jubilee clamp and tighten it just enough to stop the leak. These steps are often ideal to stop the leaks on small water lines and air lines, etc.

Figure 32.2 Leak Fixing with Jubilee Clamps

A threaded tapered steel plug is sometimes used for an emergency pipe repair. The plug is screwed into the hole until the leak is stopped. The steel plug is of harder material than the pipe, so the hole or leak is made to conform to the shape of the plug. The threads, then, make the seal.

Tapered wooden plugs, made from soft wood, can be used to stop leaks in water piping where the pipe line pressure is not too high. The plug is carefully driven into the hole with a hammer until the leak is stopped; then the portion of the plug remaining outside the pipe is carefully cut off with a hack saw, and a pipe clamp with a piece of rubber is installed.

Then there are a number of elastomer-based, ceramic–based, and epoxy-based compounds that come in handy at the time of crisis. Holes and cracks in sections of pipe work, couplings, threads, and welding seams can be repaired. But to attend the job, the line must be depressurized and the fluid must be taken out.

The damaged surface should be dry. The area must be scrupulously cleaned with a wire brush and slightly roughened with a coarse emery paper to improve the bonding. Most of the compounds are two-part mixtures. Follow the manufacturer's instructions in mixing and curing. They often come with a tape or bandage which should be wound tightly over the pipe; then the mixture is applied. Some of them are amazingly fast curing and permanent.

D I Y CLAMP AND OTHER CLAMPS

Problem is the father of innovation!!

A two-part bolted pipe clamp used with a piece of rubber is probably the most generally used emergency repair. This pipe clamp is used for repairing pin holes, and cracks in both steel and cast iron pipe lines. With proper care, the clamp can be effectively used under extreme pressure or temperature services. The clamp can even be used on plastic pipes. See Figures 32.3A and 32.3B to get an idea of the intended clamp.

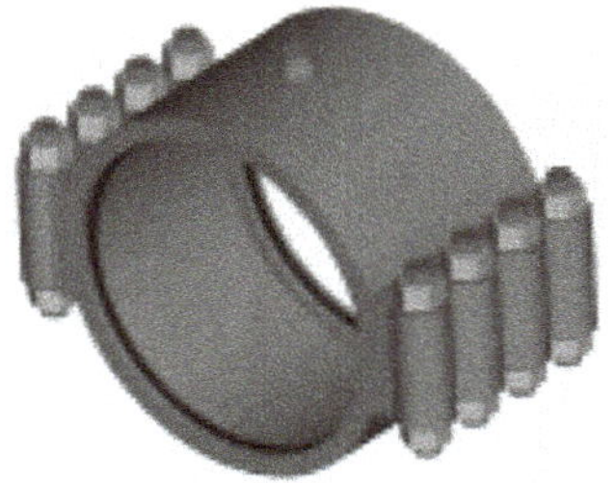

Figure 32.3A Two Piece Pipe Clamp

Figure 32.3B Pipe Clamp with a View

These pipe clamps can be fabricated and modified to suit welded connections, flanges, or most any other piece of equipment. Take a piece of pipe suitable to the outside diameter of the leaking pipe. A heavy schedule pipe does not clamp well. Take one with less thickness of sufficient length to cover the leak or crack. Cut it along the length into two pieces, trim it, and if necessary bend a little to suit the leaky pipe. Take two strips of steel of just enough width and length and weld them as flanges along the length of pipe pieces on both sides. Make holes for fixing the bolts. Now take a suitable and sufficient piece of rubber and fix it on the insides of this pipe clamp. It is better to cement it with a rubber adhesive compound. Place one part of the clamp and align it on the leak. Place the studs now, at least a few of them. Now slide the other part of this clamp on the pipe. As soon as the holes are matched, fix the nuts and hand-tighten them. Fix the rest of the bolts and tighten them.

Start from the center and tighten them all. Tighten the bolts *just enough* to stop the leak. Any time a clamp is used to stop a leak, wise technicians will use common sense to determine just exactly how tight the bolts should be. If not, the leaky pipe will dictate what is in store for you. It will simply break. More often than not, emergency leaks are attended when the line is full and operating. This should be kept in mind always. All safety precautions apply.

There is pressure inside (due to system fluid) and pressure outside (from the bosses that be). Do not panic.

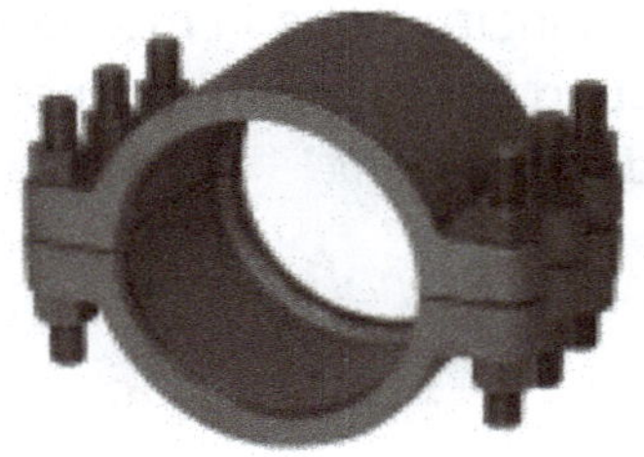

Figure 32.4A
Ready Made Pipe Clamps

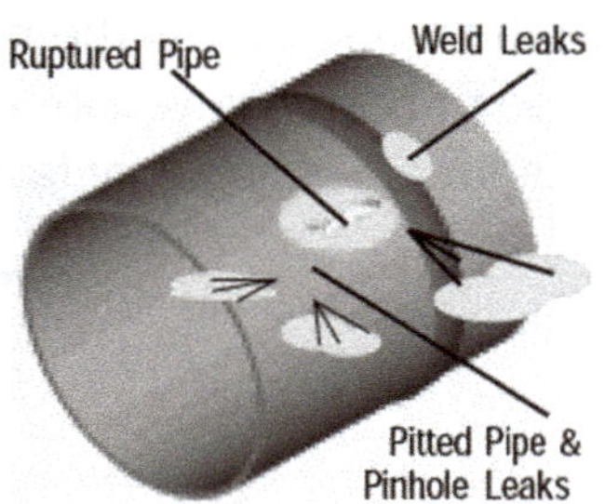

Figure 32.4B
Pipe Clamp Cut away View

On the other hand, prefabricated and manufactured clamps are available in the market to suit various sizes and services. (see Figures 32.4A and 32.4B). Some of them are hinged to make the job easy. A look at the clamps will also suggest how to fabricate one. Similar products are available for leaks through various other pipe fittings such as union, coupling, elbow, and tee. See Figures 32.5A and 32.5B for an emergency repair product for leaky unions. They are simple and easy to install, and have generally nitrile rubber gaskets or liners inside. If required, the clamp can be welded in place after the leak is stopped.

Figure 32.5A

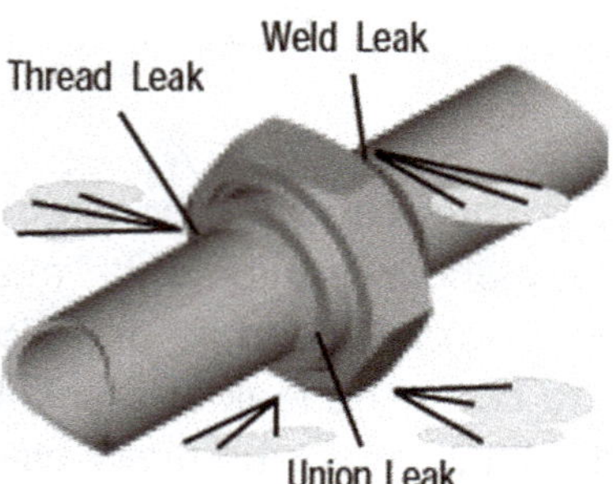

Figure 32.5B

Clamps for Leaky Unions

In some cases of high pressure lines, etc., clamps are provided with vent valves. The vent valve is kept open while the clamp is fixed. Soon after tightening the bolts, it is closed slowly.

LEAK SEALING

Another popular leak sealing job can be taken up on line at very high pressures and temperatures of almost 350 bar and 600°C. The procedure can be deployed on almost all services whether it is a leaky flange, valve gland, heat exchanger joint, pipe fittings, or a pipe. Present day online leak sealing practices are reliable and efficient. Experienced and professional agencies are offering these services which can be availed at a reasonable cost compared to the down time and wastage suffered due to leaks. Sometimes these joints can be left for considerable time before they could be taken off for a permanent repair. They save the plant from the loss of costly products and energy, prevent pollution, and ensure plant safety.

Basically, the technique is to install a metallic enclosure around the leak and inject a sealing compound into it at a higher pressure through a gun. The compound hardens and shuts the leak. For flanges and exchanger joints, the enclosure is a round clamp precisely machined to suit the periphery and the gap in between the flanges. For leaks on valve gland packings, a G-clamp is made. For leaks on pipes elbows, tees, etc., a clamp suitable to the fitting is fabricated. A wide range of sealants is manufactured to suit specific requirements.

GLAND LEAKS

A G-clamp is fitted around the stuffing box area and a hole is drilled into the stuffing box. Sealant that is similar to the original packing material is then injected with a high-pressure gun (see Figure 32.6). Safety both during the drilling and during the compound injection is ensured. This is the most widely used online leak sealing application and special G-clamps are available for inflammable stocks.

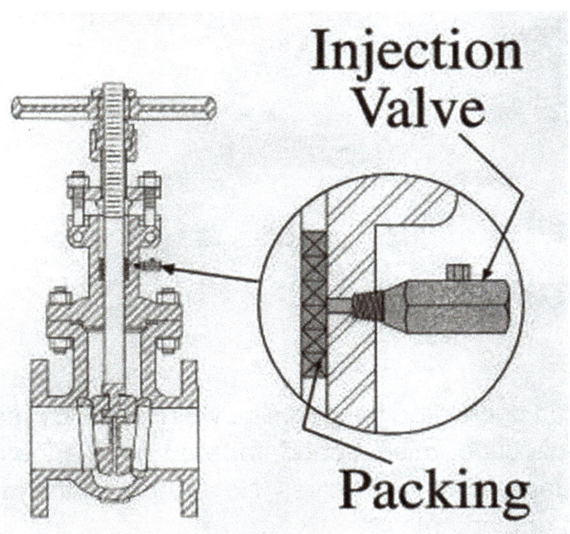

Figure 32.6
On-line Sealing of Leaky Glands

FLANGE LEAKS

Fortunately, flanges are standardized; hence, clamps can be prefabricated. A two-piece clamp is fixed around the leaky flange and it is tightened (see Figures 32.7A and 32.7B). Then sealant is injected through a number of points into the flange, which hardens soon enough (see Figure 32.7C). But if the flange gap is too little, a wire is inserted into its gap and then sealant is injected into the voids. However, this wire drawing procedure is only good enough for low-pressure services and is a rather short-term application.

Figure 32.7A

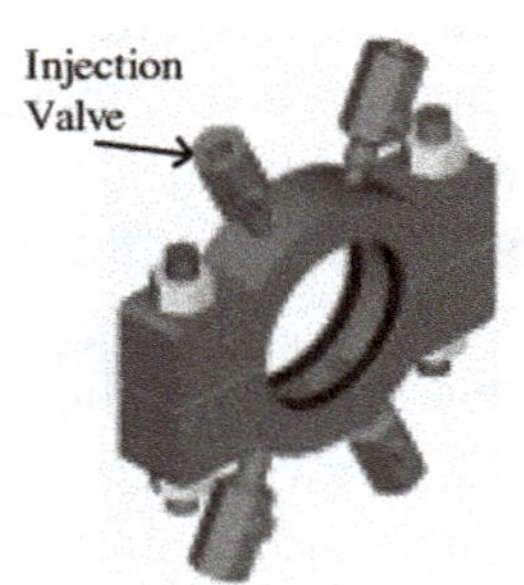

Figure 32.7B

On-line Sealing of Leaky Flanges

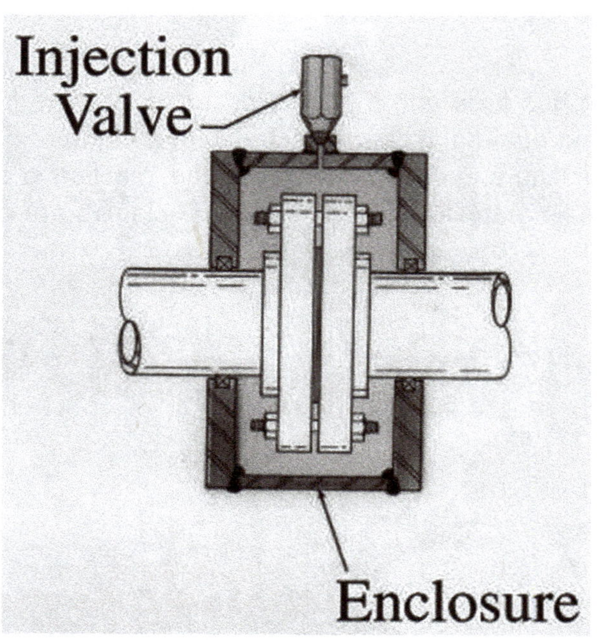

Figure 32.7C
Method of On-line Sealing of Leaky Flanges

This method can be adapted to leaking, pitted or cracked flanges, and weld leaks. By machining a suitable clamp, it can be adopted to out-of-round, mismatched, tapered gap flanges. This method can also be used on oval, round, or square bonnets of valves. Nevertheless, the method can be adapted to casing joint leaks on rotating equipment.

LINE LEAKS

Pipe fittings such as elbow, tee, or even straight pipe develop pinhole leaks. They may start off as unassuming small little ones, but soon develop into major problems. A pipe clamp suitable to the pipe size or fitting is fitted on to the pipe and tightened. Sealant is injected into the hole through a number of points which hardens and arrests the leak (see Figure 32.8). Even a broken pipe can be repaired, as shown in Figure 32.9.

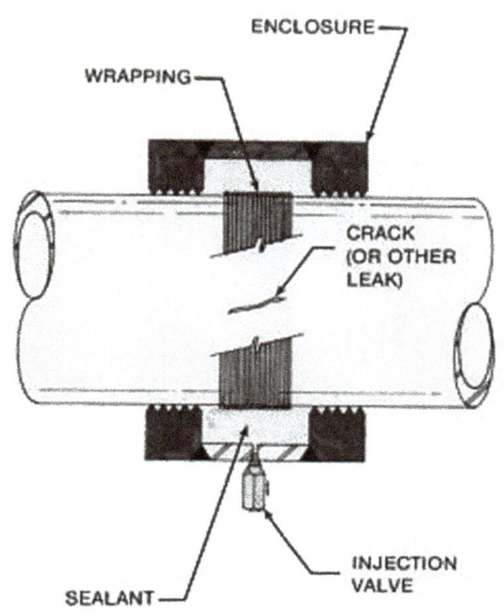

Figure 32.8
Method of On-line Sealing of Leaky Pipes

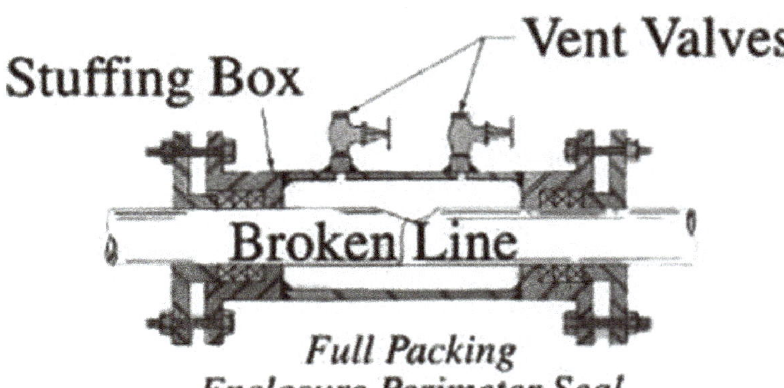

Figure 32.9
Repair of Broken Pipes

VALVE KILLING

Valves often fail. In engineering parlance, they "pass," which means that they are not able to give proper shut off. Sealant can be injected onto a badly passing valve seat, which hardens and shuts the valve (see Figure 32.10). However if the valve is opened, the sealant may flow into the system, but often drains out through the drain line.

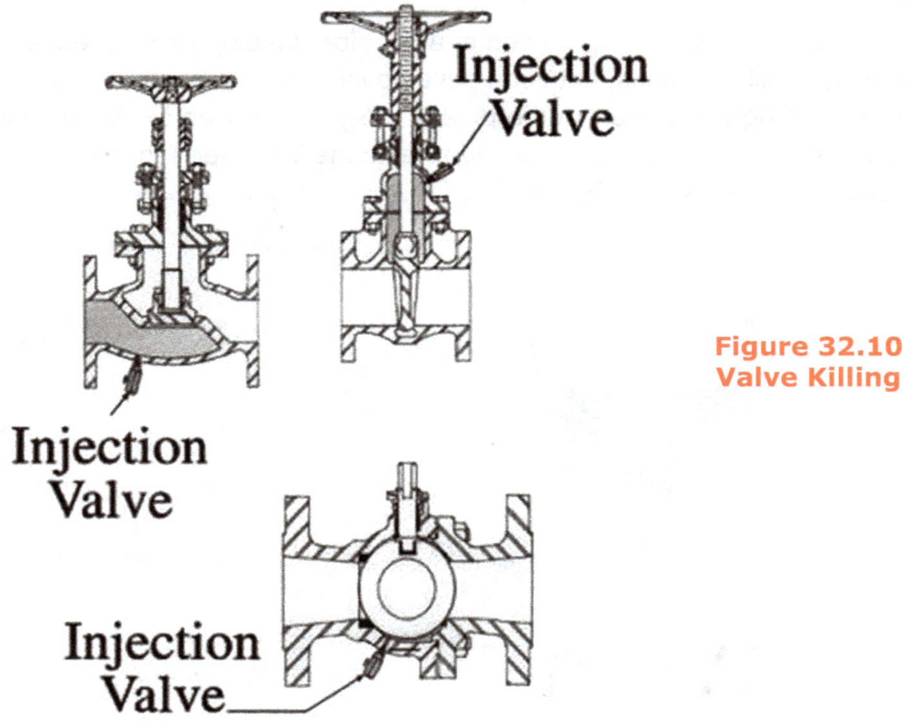

Figure 32.10
Valve Killing

On the other hand, in certain cases, particularly in the case of gate valves, they can be made operable by injecting sealant to the bottom center (see Figure 32.11).

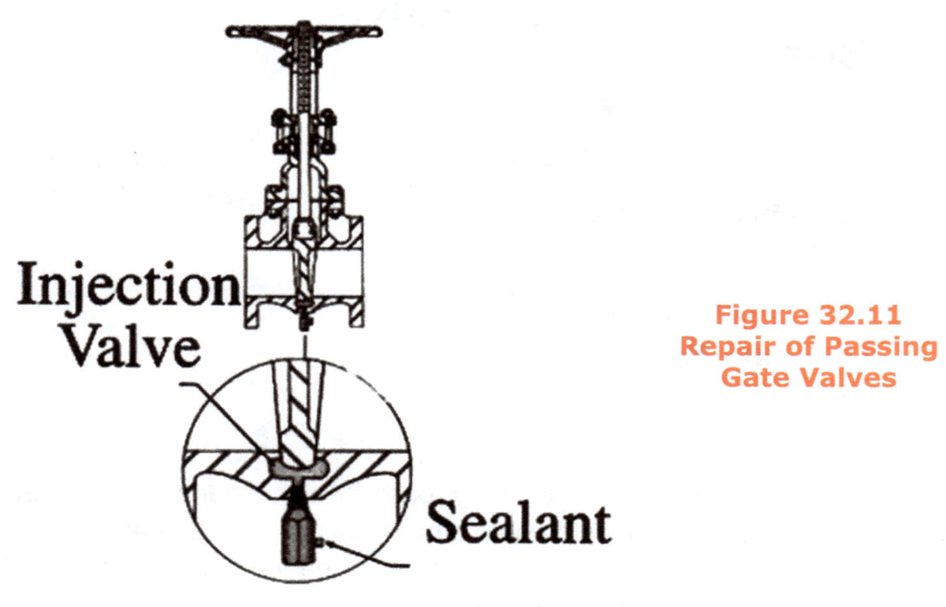

Figure 32.11
Repair of Passing
Gate Valves

The emergency pipe repairs discussed above are commonly used in piping systems. However, many leaks require an engineer's thought and imagination in order to devise a clamp or plug to suit a particular problem.

But remember, emergency repairs are often fraught under trying circumstances; the plant is running full and cannot take a shutdown, sometimes cannot even properly depressurize and drain. So it is essential to take every safety precaution in attending the job. A lot of good number of personal protective equipment has been developed to wear in specific circumstances. Safety should *never* take a back seat!

HOT TAPPING

Hot tapping is a method of making a branch connection, initiating a bypass, and allowing performance of a line stop without service interruption or product leakage and safely. Hot taps can be performed on almost any type of pipe from carbon steel to PVC containing almost any fluid. The system is normally available in 1/2 inch (15mm) through 60 inch (1500 mm) diameter and up to 1480 psig (100 bar) maximum operating pressure and temperatures as high as 700°F/370°C.

Typical hot tap installation consists of a tapping saddle, gate valve, and hot tapping machine. A tapping machine includes a drilling machine with a pilot drill and a hole saw or cutter (see Figure 32.12). All the parts of the system are sealed to outside atmosphere. After the installation is tested, the valve is opened. Drilling is advanced with the pilot drill and cutter through the valve, to the pipe. When the cut is completed, the cutter and pilot drill are retracted beyond the gate of the valve. The

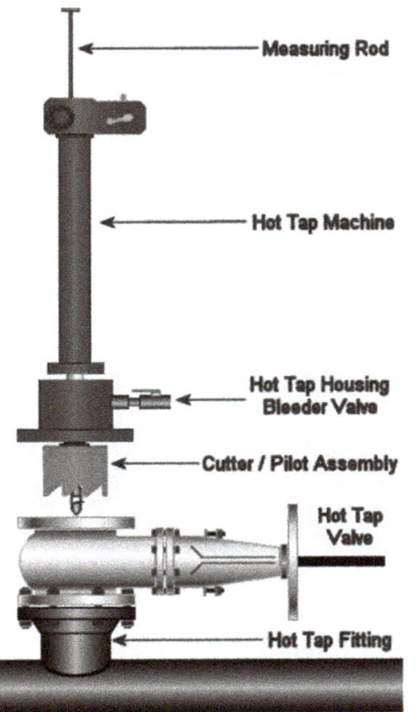

Figure 32.12
Hot Tapping

wired pilot drill catches the coupon or the cut pipe piece without letting it fall off. The valve is closed and the machine is removed.

Elbow taps, angle taps, and taps perpendicular to the pipe are all feasible, permitting new tie-ins to existing systems, the insertion of devices into the flow stream, and permanent or temporary bypasses.

LINE KILLING

Sometimes the process plants require a certain length of pipeline to be plugged or blanked off without stopping the plant operations. The reason may be that the line requires repair or renewal or modification. The job should be carried out online like hot tapping. The line stopping or line killing technique facilitates this blanking of the line by inserting a plug to shut off the flow while the pipeline is in service. In certain cases, a bypass line also can be introduced to continue the plant operations. Professional services are available to carry out these specialized jobs.

INSPECTION

The environmental and safety record of pipelines is excellent compared to other technology products. Pipeline systems are both the safest transportation mode and the most economical. We have millions and millions of kilometers of pipeline carrying everything from water to crude oil. The pipe is susceptible to attack by internal and external corrosion, cracking, third party damage, and manufacturing defects. Unfortunately, pipe lines do not give any audible early warnings like rotating equipment does. Hence, it is imperative that periodic inspections are carried out to detect defects and prevent damage.

INSPECTION METHODS

As the pipeline is fabricated and fixed, it is inspected by visual, radiography, magnetic particle, ultrasonic, and other inspection methods to evaluate the integrity of the piping and joints, and whether they are conforming to the laid down standards. Generally the pipe line is forgotten soon after. But the modern trend has been to bring various pipe lines within the schedules of inspection and preventive maintenance. Sophisticated and state of the art technologies are developed to inspect and predict the pipe line condition. Some of them are long-range ultrasonic testing or guided wave ultrasonic testing, magnetic flux leakage testing, and intelligent pigs.

The hydrostatic test is still one of methods often resorted to, though it can not always predict impending failure. A small pitting or corrosion will not stop the test to pass. You can read more on this subject in Chapter 31, *Testing of Equipment*.

The following discussion considers some non destructive testing methods of pipe lines. The most common methods of Non-Destructive Examination (NDE) include inspection at various points along the pipeline. This requires insulation to be removed at designated points over the length of the piping. NDE also requires the protective paint to be peeled off, the surface cleaned, and the pipe prepared suitable to the NDE method employed.

VISUAL INSPECTION

Visual inspection must always precede any other NDE method. Eliminate external defects, clean the suspected area, and if necessary grind it without sacrificing the detail. On occasion, magnifying glasses, mirrors, etc., may be used.

LEAK TESTING

Several techniques can be used to detect leaks in piping and pressure vessels. The traditional soap-bubble test is still used all over, there are several electronic listening devices, and ultrasonic leak detectors have been developed. Pressure gage measurements and liquid and gas penetrant techniques are also used.

DYE PENETRANT INSPECTION

Dye penetrant inspection (DPI), also called liquid penetrant inspection (LPI), is widely used to locate surface cracks in all non-porous materials (metals, plastics, or ceramics). This method is almost invariably employed for inspection of welds. This low-cost inspection technique is carried out frequently by the plant maintenance personnel themselves, rarely by third party inspection unless required by statutory regulations.

Liquid penetrant is normally supplied in three parts: cleaner, dye penetrant, and developer. Penetrant is applied to the test pipe by dipping, spraying, or brushing. Time is allowed for the dye to penetrate and excess penetrant on the surface is cleaned. Then a developer is applied. The developer draws penetrant out of the flaw to give a visible indication. Dyes can be fluorescent or non-fluorescent (visible). Specific chemicals are developed to suit a variety of industry demands. They are also available in aerosol cans.

PRE-CLEANING The test surface should be scrupulously cleaned. Dirt, paint, oil, grease, or any loose scale can obstruct the penetrant from reaching the defect or cause false indications. After mechanical cleaning is carried out, cleaner is applied, which evaporates any more grease or grime left on the surface. It makes the piece ready for further test.

APPLICATION OF PENETRANT The penetrant is then applied to the surface of the item being tested. Leave it for 10-to-30 minutes for the penetrant to soak into any flaws, depending on the material being tested and the size of the flaws. Thinner flaws require a longer penetration time. Any excess penetrant is then removed from the surface and care is taken not to disturb the penetrant trapped in real defects.

APPLICATION OF DEVELOPER White developer is applied to the sample. The developer should form a thin, even coating on the surface. The developer draws penetrant from defects out onto the surface for a visible indication. Colored stains on the white background indicate the positions and types of defects. The indication has a high visual contrast (e.g. red dye against a white developer background, or a bright fluorescent indication against a dark background). The developer draws the penetrant out of the flaw over a wider area than the real flaw, so it looks wider.

POST CLEANING If further inspection is required by this or any other method, it is necessary to post-clean the surface.

Unless the flaw is extensive, the cracked portion is ground and re-welded.

BORE SCOPE

A bore scope is a rigid or flexible tube with an eyepiece on one end and an objective lens on the other; they are linked together by a relay optical system in between. Bore scopes can be used to inspect the inaccessible interiors. They are often used for inspecting heat exchanger tubes.

THICKNESS GAGING OR ULTRASONIC FLAW DETECTION

Thickness gauging or ultrasonic flaw detection is normally carried out after the pipe line is shut down and cooled. An ultrasonic thickness gage can measure virtually all engineering materials with a very high resolution of 0.1 mm or 0.004 inches.

Piezoelectric transducers transmit bursts of sound waves at frequencies between 500 KHz and 100 MHz and receive the reflections. Thickness is determined by accurately measuring the time required for a short ultrasonic pulse to travel through the thickness of the material, and reflect from the back or inside surface. Though the time interval is only a few microseconds or less, modern electronics convert it to the actual thickness in inches or millimeters. A large variety of transducers have been developed to suit different industrial applications.

Ultrasonic flaw detection allows the detection of extremely small flaws deep in the piping. It is a nonhazardous and portable method. However, the method is manual and time consuming.

EDDY CURRENT INSPECTION

Eddy current inspection makes use of the principles of electromagnetism Eddy currents are created through a process called electromagnetic induction. They get their name from eddies that are formed when a liquid or gas flows in a circular path around obstacles when conditions are right. Eddy current inspection can be used for crack detection, thickness measurements, and coating thickness measurements. Small defects near the surface can be detected, but defects in plastics can not be detected.

RADIOGRAPHIC TESTING

Radiographic Testing (RT) is almost always specified for nondestructive examination (NDE) of inspecting welds on the pipe lines used for high pressure, high temperature, and critical services. It gives a virtual visual indication of the quality of weld and hidden flaws. Most often a radioactive source (Ir-192, Co-60, or in rare cases Cs-137) and occasionally an X-ray machine is used.

The weld joint or the piece of pipe to be inspected is placed between the source of radiation and the detecting device, usually the photographic film in a light tight holder. The film is exposed by the radiation source for sufficient time and then the film is developed. The radiograph is examined as a negative, because it is unnecessary to print it and some detail may be lost. Digital imaging is currently becoming popular where the radiographed picture can be viewed online.

However, only small areas can be radiographed at a time. During the course of this activity, the area should be cordoned off, holding up all other jobs there.

MAGNETIC PARTICLE INSPECTION

Magnetic particle inspection (MPI) is a relatively easy, nondestructive testing. Part surface preparation is not as critical as for other NDT techniques. However, MPI can be used only on ferromagnetic material such as iron, nickel, cobalt, or their magnetic alloys.

First, the test component is magnetized. If there are any defects on the surface or near, they create a leakage field. Iron particles — either in a dry or wet suspended form — are then applied on the surface of the magnetized specimen. The iron particles will be seen clustering around the flux leakage fields or at the cracks, which gives a clear visible indication of the defect.

Other technologies are emerging, such as acoustic emission testing, remote field testing, infrared/thermal testing, and long-range ultrasonic inspection.

PIPELINE PIGGING

Pipeline pigging is carried out for a variety of purposes apart from inspection of pipelines for internal defects such as dents, corrosion, and line plugging, etc. These purposes include cleaning up pipelines before commissioning; periodically removing foreign material, product leftovers, and dirt and water from the pipeline; brushing off liquids from gas pipelines; and separating products to reduce cross contamination. Pigging results in a proper flow of fluids in the line. It protects equipment from failure due to the ingress of unwanted material, increases the product yield, and reduces waste. The product can be properly separated and cross contamination can be avoided. Faster product changeovers can be accomplished with better quality. Present day intelligent pigs or smart pigs can continuously transmit data of the pipeline internal condition.

Pigs and spheres are pushed through the pipeline by the pressure of the flowing fluid. A pig usually consists of a steel body with rubber or plastic cups attached which seal against the inside of the pipeline. It is now inserted into a pig launcher. Pressure of the fluid in the pipeline pushes it along down the pipe. In the end, a pig catcher receives it. Different types of brushes and scrapers are available to suit specific requirements.

The pig may get stuck somewhere in the line if it is too small or too big, the wrong type, the wrong launch, or has bad pipe line and fittings. Lines with butterfly valves cannot be pigged.

Improper purging and lack of proper interlocking can create unfavorable conditions for pigging that can result in an explosion.

Pigging systems are invariably used before commissioning lines. These systems are regularly used in industries like pharmaceutical, food, paint, chemical, and petrochemical.

Nowadays pigs have become smarter and technically advanced. They use various techniques like magnetic flux leakage and ultrasonics, and continuously transmit data about the pipe line condition.

Originally metal discs were used in the oil industry to clean out wax build up in the pipe lines. Metal-to-metal contact made a squealing noise like the pigs. Hence, the name pigging! Pipeline inspection gage and pipeline inverse guide are other terms.

Pipeline Rigging

Lifting operations are inherently dangerous, even when proper training is conducted, equipment properly is maintained, employees work in a safe manner, and safety programs are conscientiously attended. *Accidents can still occur.* Crane operations are primarily mechanical. They can and do fail and the results can be catastrophic. You are the only controlling influence and can minimize hazard. Your life literally is in your own hands!

Pipelines fall into this more serious category. They are slender beings often lifted into position where there is no access. They are round and so they roll off. You need to lift a number of them at a time, which makes them slip. More often than not, pipeline supports are built after the pipe line is positioned. Therefore, you do not have anything to really hang on. When any part of the line needs to be removed, pipe fitters find, to their dismay, the support is not adequate for the rest of the line — it may hang, drop, or break loose. Furthermore, when you remove one section of pipe, another section may fall due to corrosion.

In such a scenario, rigging becomes an important activity of piping practice. There are a number of tools, lifting tackles, and resources available which are used for lifting, hoisting, and lowering pipelines and components, cranes, pulley blocks, come-alongs, etc.

HOISTS

There are two common types of hoists: the chain hoist and the wire rope type. As the names suggest, the chain hoist makes use of chains whereas the wire rope hoists make uses of braided wires. Chain hoists are often in the lower load versions with smaller lifts. (But not necessarily! Older versions still use heavier chains for heavier loads.) Longer reach and higher loads are invariably with wire ropes.

There are two types of chain hoists frequently used in piping practice; Come-a-long or ratchet lever hoists (Figure 34.1) and chain pulley block or chain fall (Figure 34.2).

A come-along is a hand-operated ratchet lever winch that is convenient and portable. It can be carried by one hand, fixed by one person, and operated by one person. It has a knob that can be set to lowering or lifting. It has a hand lever to operate it — to lower or lift, to tighten or loosen. The lever is ratcheted, which means that it slips in the other direction. A come-along is used for pulling joints together, stretching, lifting, lowering, and binding objects. Ratchet lever hoists have

Figure 34.1
Come-a-long

Figure 34.2
Chain Pulley Block

the advantage that they can usually be operated in any orientation. In piping, they are very handy as they can be fixed in inaccessible and confined places.

Come alongs are used most in aligning pipelines or realigning severely misaligned piping flanges. They are useful in positioning pipes or fittings in pipe racks.

While dismantling pipelines or fittings, you may find yourself in a precarious situation where the removed piece of pipe is the one which supports the other section. If this pipe is removed, the other sections of pipe may lose support and fall. Come alongs come into play now as they could now be placed as a temporary support, holding the other pipes while the present piece is repaired and fixed back.

The lifting capacity generally ranges from half to three tons with a travel length of six to twelve feet. As with all other lifting devices, it is not safe to exceed the weight limits. This is particularly relevant when the come-a-longs are used to pull the badly misaligned pipe lines, where the loads cannot be easily estimated.

> *The ratchet lever hoist or Come-A-Long was developed by Abraham Maasdam of Deep Creek, Colorado about 1919. It was later commercialized by his son, Felber Maasdam, about 1946.*

A chain pulley block uses a chain (called a load chain) with a hook in the end to lift heavy weights. The block itself has another hook to hang or support it. There is another smaller endless chain (called a hand chain) which is used to actually operate the hoist. Pulling this chain one way lifts the load, pulling the other length lowers the load. This is also known as chain fall or chain hoist.

Load chains are made of special alloy steel. They find wide application in lifting and dragging pipelines, fittings and other loads. They are particularly useful in confined places, in the open air, and overhead places for pulling and stretching work at any angle. They are available in 0.5 to 20MT capacities and are tested with 150% SWL. This does not mean that they can be loaded to that extent. Using any angle other than vertical may pose problems as either the load chain or hand chain may tend to get jammed.

CRANES

Then there are cranes, jib cranes, mobile cranes, and travelling cranes that use levers and pulleys to lift considerable weights. Indians, Romans, and Egyptians must have used some kind of these machines to build the massive structures a few thousand years ago.

A jib crane is the simple crane. Generally it is used on the shop floor, rarely in field piping work. Jib cranes are either wall mounted or pillar mounted, fully floating on bearings with a swivel facility. Normally, wall-mounted jib cranes have a swivel less than 180°, but cranes with 270° and 360° are also available with load capacities of 0.25 to 7.5MT. The swivel arms have reaches of 3 ft to 15 ft (1 mtr. to 5 mtrs. appx). Swivel arms are made suitable to accommodate trolleys with a chain pulley block for manual travel and lifting operations or an electric wire rope hoist for electric operations.

A relatively simple crane is the mobile crane. A telescopic boom (arm) or steel truss is mounted on a platform. The platform can rotate on its axis. Pulleys, levers, or hydraulic cylinders raise the boom. Generally a hook is suspended from the boom. Wire rope around the pulleys and drums does the actual lifting operations. Most of the cranes have telescopic tubes (known as boom) which are hydraulically operated to extend the reach. Mobile cranes are truck mounted; the trucks have traditional wheels for regular usage, railroad wheels for railroad application or specific locations, or caterpillar tracks used on rough terrain.

Mobile cranes are engaged extensively in fabrication, erection, and demolition of piping and piping components. They have telescoping tubes called booms which are extended or retracted hydraulically to increase or decrease the reach. The angle of the boom can be changed by another hydraulic means. All this arrangement can be rotated to turn sideways. When the boom is extended, the crane becomes unstable and may tilt. Hence, they are equipped with outriggers, which must be extended and loaded if the boom is extended. Outriggers (normally four on four corners) are extended horizontally from the chassis, then down vertically to level. They must be retracted vertically and horizontally while travelling. Travel speeds are slow. Still, the loads should not be swung sideways from the direction of travel as the crane may lose stability. These cranes invariably are equipped with counterweights to balance the center of gravity while hoisting. They are available with less than 10-tons capacity to a few hundred tons. Cranes are equipped with load charts which specify the loads that can be hoisted corresponding to boom length, angle, speed, on outriggers, on tires or tracks, etc. The specified loads must not be exceeded to avoid disastrous incidents.

A track-mounted crane or crawler has tracks just like the tanks used in war. It can be used on rough terrains; stability is greater, and cranes are available for higher operating loads. Furthermore, these cranes are stable on their tracks without the need for outriggers; they can travel with load. However, they are heavy and cannot travel farther than within the site. When required, they are dismantled and transported by trucks, rail cars, or ships and reassembled at the next site.

The overhead travelling cranes are used on the shop floor or a fixed field site. They travel point to point and have a trolley to make the lateral movement. Although the older type chain-driven and hand-operated cranes are still seen, most of the present day overhead travelling cranes are electric.

There are other cranes like stacker cranes, gantry cranes, and floating cranes which are used for specific jobs, but rarely in piping work.

EYE BOLTS, SHACKLES, AND HOOKS

Eye bolts (see Figure 34.3), shackles (see Figure 34.4) and hooks (see Figure 34.5) are useful for lifting loads without lashing them directly to a wire rope or chain. They can be attached to wire rope, fiber line, blocks, or chains.

Eye bolts are used in piping practice to lift some of the fittings if they have a threaded hole provided for the purpose or with the help of through holes. They are meant for vertical use; angular loading of less than 45° is not recommended. A standard eye bolt used with a 45° horizontal angle retains a mere 30% of its vertical capacity. Eyebolts must be tightened so that the shoulder is flush with the item being lifted. Please note that eye bolts are marked with their thread size, not with their rated capacities. Do not ever choke the sling through an eyebolt.

Shackles should be used for loads too heavy for hooks to handle. Shackles are available with screw pins or round pins, which should not be changed or substituted. Never replace the shackle pin with a bolt. Discard the shackle if the original pin is lost or damaged. Do not side load them.

Figure 34.3
Eye Bolt

Figure 34.4
D Shackle

Hooks with a provision for closing and locking should be used because the load can fall off the hook when it hits some obstruction, in the case of jerks, while dragging, or lifting at wrong angles. In extreme cases where the load needs to be lifted without such a locking arrangement, the hook should be moused. Mousing is a technique often used to close the open section of a hook to keep slings, etc., from slipping off the hook. Rope (several wraps of them), wire, or shackles are often used for this purpose. When hooks are overloaded, they straighten out slowly and the load may drop off. Such a hook should not be straightened and put back into service. Instead, cut it with a cutting torch and discard it.

Figure 34.5
Hook

FIBER ROPES

Wire slings, rope slings, and chains are used to actually connect the load to the crane or the pulley block. Ropes are no doubt still used to lift pipe lines. If the load is carefully selected, fiber ropes are still a better choice for lifting a bunch of pipes or heat exchanger tube bundle.

Fiber ropes are still used all over because of their easy availability and adaptability. Being soft in nature, they do not damage the load. However, their load carrying capacity is limited. The most common types of fiber line are manila, sisal, nylon, and Kevlar.

Manila rope is frequently used for lifting pipes and components because of its quality, relative strength, and resistance to wear and deterioration. It is a strong fiber from the abaca plant, with the fibers 1.2 to 4.5 meters long in the natural state.

Sisal rope is the next, made from two tropical plants — sisalana and henequen — that yield a strong, valuable fiber which is about 0.6 to 1.2 meters long in its natural state. Sisal rope is about 80 percent as strong as high-quality manila rope and it withstands exposure to seawater very well.

Nylon rope has a tensile strength that is nearly three times that of manila rope; it is waterproof, lighter, flexible, and elastic. It is better protected against shock and resists biological decay.

WIRE ROPES

Braided wire rope is made from large numbers of individual unbroken wires and braided. Eyes are made on both sides, spliced, and hand tucked or attached to a fitting such as sockets, clips, and thimbles. The braided sling is flexible and resistant to kinking and twisting. Each sling is rated for its capacity depending on its size, strength, and end fitting. Figure 34.6 shows various types of fittings attached to a sling.

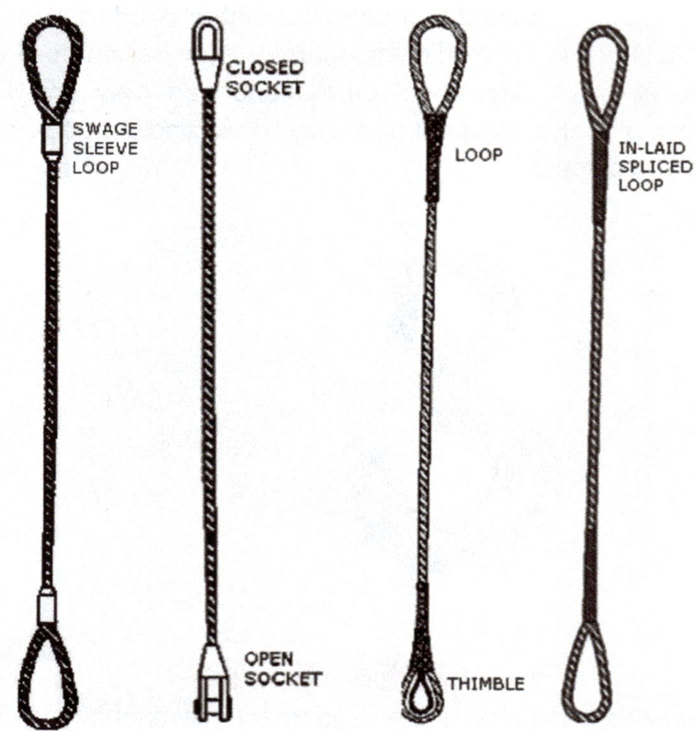

Figure 34.6
Wire Rope and Fittings

HITCHES AND MISSES

No matter what kind of sling you choose, or what kind of machine is used for lifting, there are certain common points that must be taken care of for a safe job. Safety only takes minutes, an injury can last a lifetime.

WEIGHT OF THE LOAD

Your first step should be to know the weight of the load. Check the physical condition of the load, feasibility to lift the load, and provisions for attachments. Select slings and other attachments like shackles, and eye bolts with sufficient capacity rating. Inspect them all before using and destroy all defective components. Then the difficult part is that pipe lines are round and long. If you are removing the pipe line or a fitting from the line, make sure that it is not clamped or bolted. If it is a valve or fitting, make sure that all the bolts are removed or the welding is completely removed. It is common in pipe line practice to remove all but a few bolts in place and rig it for removal. When the load is taken up, these bolts are removed. Metal is cut off from the leftover weld. Be careful while doing this step because the load may swing excessively in any direction. If unfortunately any leftover is forgotten, a really heavy load would be applied, which may break loose the slings, shackles, etc.

TYPE OF HITCH AND NUMBER OF SLINGS OR LEGS

The type of sling, hitch, and balancing will considerably affect the load-carrying capacity. The kind of hitch to be used depends on the kind of material to be lifted; the type, length, and safe load limit of the sling; whether lugs for slinging are available on the load; headroom; and other factors. Loads should be lifted from a point directly over the center of gravity. The Center of Gravity (CG) of the load to be lifted is a very important consideration, particularly in the case of pipelines because they are round and long. Sling the pipe and lift it a little. Check the balance and make corrections after lowering it. Check again and lift. If the balance is poor or, in other words, if the line is tilting to a side after lifting, even a little, the pipe may slip and fall. This is all the more important if a greater number of pipes are lifted in a single bundle. In all cases, make sure that the sling is tight and has a good grip on the load. The rated capacity limit of a sling depends on the hitch or method of applying a sling to the load, in addition to the strength of the rope and the efficiency of the attachment.

Every lift uses one of the three basic hitches: vertical, choker, and basket. Figures 34.7A, B, and C illustrate these hitches.

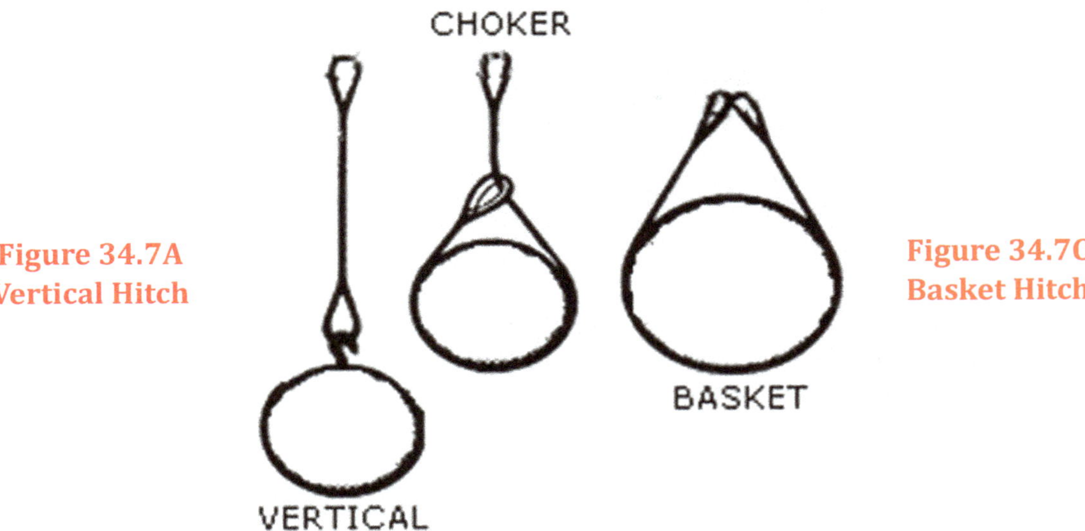

Figure 34.7A
Vertical Hitch

Figure 34.7C
Basket Hitch

Figure 34.7B
Choker Hitch

VERTICAL HITCHES

A vertical attachment is simply using a sling to connect a lifting hook or other device to a load. The full-rated load of the sling may be used, but never exceeded. This is rarely done on pipelines because they do not have lifting hooks and nobody would weld one unless it was extremely essential. However, many of the pipe fittings like valves and other devices, particularly the heavier ones, have a provision for eye bolts or lifting hooks (see Figure34.7A).

CHOKER HITCH

The choker hitch is the method most often used in pipeline lifting. Please read carefully the limitations imposed by this method. One end of a sling is passed around the load, through the loop or thimble of the other end, and then onto the hook of the crane. It creates an angular loading of the sling, and makes a small diameter bend in the sling at the choke point. Unfortunately the choker hitch reduces approximately 25 percent in the rated capacity of a sling. However, it provides better load control. A choker hitch is often used for lifting longer bundled loads such as a number of pipes (see Figure 34.8). However danger lurks as some pipes may slide off. Use of appropriate fiber rope would be a good idea. A better procedure would be to use a double-wrap choker hitch and with two slings. The double wrap compresses the load on all four sides and provides far better load control, whereas a standard choker hitch compresses the load from three sides only.

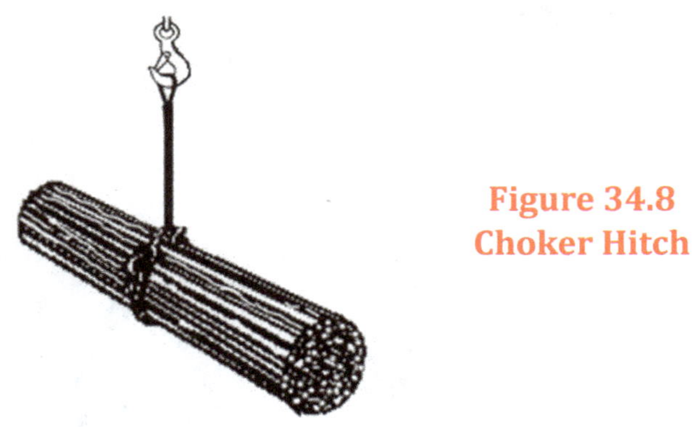

Figure 34.8
Choker Hitch

BASKET HITCHES

Basket hitches distribute the load equally between the two legs of a sling. But this method can roll the pipes off or imbalance them (see Figure 34.7C).

Figure 34.8 shows method of lifting a number of pipes.

SLING ANGLES AND CAPACITIES

The angle at which slings are used, and the number of legs used for lifting the load, can significantly affect their capacity. Only when a sling is used in a vertical hitch can the full lifting capacity of the sling material be utilized. All other angles reduce the capacity. The rated capacity of the sling decreases as the angle formed by the sling leg and the horizontal line decreases. The angle between the sling legs and the horizontal should be maximized. Suppose a sling has a working load limit of 500 lbs. It will have its full capacity in a vertical hitch. When the sling is choked (as in a choker hitch) — which is often done to lift a pipe with a single sling — the capacity goes down to 375 lbs. Slings rigged with this choker hitch achieve only about 75% of their potential capacity due to the stress created at the choke point. It would be risky to lift loads at an angle less than 30°. Choke it tight before lifting, not afterwards. Unchoked pipes roll, slip, and fall. Use the longest reach possible for completing the lift; this will provide the largest angle possible for minimum stress on the sling. **Chart 34.1** summarizes the sling capacities at various angles.

Chart 34.1
Sling Capacities

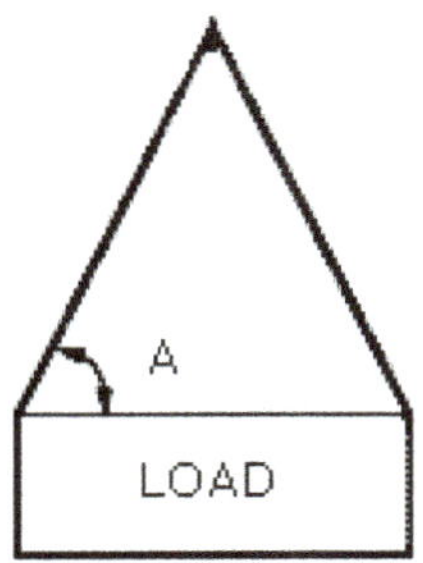

SLING ANGLE DEGREES	SLING ANGLE FACTOR	SLING ANGLE DEGREES	SLING ANGLE FACTOR	SLING ANGLE DEGREES	SLING ANGLE FACTOR
90	1.000	70	0.940	50	0.766
85	0.995	65	0.906	45	0.707
80	0.985	60	0.866	40	0.643
75	0.966	55	0.819	35	0.574

If the sling is used in a basket hitch with the legs at a 90° angle to the load, it would have a working load limit of 1,000 lbs. However, if it is used at a 60° angle to the load, the working load limit falls to 866 lbs. Do not ever exceed working load limits. Please understand that definition of Working Load Limit — the maximum load that shall be applied in direct tension to a new and undamaged sling or chain.

MORE SAFETY POINTS

When a wire rope is bent around any object, this bending action reduces its strength. The ratio of the diameter around which the sling is bent divided by the body diameter of the sling, known as the D/d Ratio is important. As the D/d ratio becomes smaller, this loss of strength becomes greater and the rope becomes less efficient.

Pulling or dragging the pipe or fitting will enormously reduce the load threshold. A sling at total vertical lift can take the full load within its capacity. But when a piece of load is dragged, although the sling takes about 90° between its ends, the load falls by 26%. At 30° the load falls to 51%. Never try to drag the load from a corner and pull it to the center. If you have to, which is often case as the materials are stashed in unreachable corners, give proper allowance of the load factor. A snapped sling has enormous power and it hits with a vengeance.

Keep hands and fingers from between the sling and the load.

Know where you're going to set the pipe in place! Check whether the new location can support the weight. Check out the available head room and obstructions on the way to the place, where the pipe would eventually be loaded. Directional movement should be made smoothly and deliberately. Avoid rapid movements in any direction. Remove slack from the sling, chain, or cable before lifting a load. Taking up slack with a sudden jerk, or raising or lowering a load with a jerk can exceed the working load limit of the sling.

Pipelines are long; they turn and twist while getting hoisted. Use a line, often a rope, to stabilize the load during the lift. Balance them before lifting. Be cautious while lifting a number of pipes at a time. Balance is particularity important when using a basket hitch where slippage of the load may occur.

Know your travel path in advance of the lift! Never carry loads over or past workers. Never allow a coworker between a fixed object and a load! Never allow riders on loads or hooks. Always ensure the safe travel radius of the crane. Move loads only after being signaled by the designated, qualified signaler. Signalers must keep line-of-sight with the operator and must watch the load.

Protect the load and sling from damage. Select the best sling material for the load and environment. Use softeners such as shims, burlap padding, wooden blocks, or sling protectors to prevent damage at corners or projections, particularly when lifting thin pipes, tubes, or heat exchanger bundles (see Figure 34.9).

Figure 34.9
Method of Lifting Thin Pipes

When lifting pipe with hooks, there is a practice of fixing the hooks at the ends of the pipe. Although it may seem an easy solution, it is inherently dangerous, as the pipe will slip and fall off at the slightest jerk and hooks will be damaged.

Lifting tackles such as slings, chains, shackles, and hooks should be given a visual inspection before each usage to determine their safety. An inspection should include such things as: broken wires, kinks or distortions of the sling body, condition of eyes and splices, reduction in diameter of the rope, cracks and corrosion, etc. If any lifting tackle is found to be defective, it must be cut and discarded.

Similarly make a visual check and test all hoist controls and brakes at the beginning of each shift. Always test brakes by making a short lift to ensure control. Check whether limits and latches are working well or not. Look around and make a visual check for any dangers lurking in the corners. Never attempt to operate a crane or hoist that is suspected to be unsafe. Never allow unauthorized persons to operate cranes.

Do not ever cross the rated load of the crane and do not operate at excessive speeds.

Always position the hook directly over the load before lifting the load off of the floor and lower loads directly below the hoist. Keep hoisting ropes vertical. Do not pull or push the load. Check for twisted, broken, or kinked cables or chains. Observe correct drum spooling as the hook is raised. Maintain two full wraps of cable on the hoisting drum. Inspect for deformed, cracked, or stretched hooks. Each hook must have a safety latch that automatically closes the throat of the hook.

While using travelling cranes, ensure that all loads are lifted high enough to clear obstructions before moving the bridge or trolley. Whenever possible, maintain a minimum clearance of one foot above loads and to the sides. Raise the load only to the height necessary to clear lower objects. Never pull a hoist by the pendant cable. Never leave the controls unattended while a load is suspended. If it becomes necessary to leave the controls, lower the load to the floor. If loss of electrical power occurs, place controls in the "OFF" position to prevent unexpected startup upon restoration of power.

Flexible Hoses

Flexible hoses, both metallic and non-metallic are extensively used all over the industry. Most of them are used primarily for vibration isolation or where the connecting upstream and/or downstream lines have considerable movement or change distance over a period. Flexible hoses are two kinds basically, non-metallic (Teflon, rubber, etc) and metallic. Lined metallic hoses are not uncommon. Rubber hoses consist of basic core tubing made out of rubber or plastic, (PTFE being popular) over which reinforcement with layers of braid is employed. Similarly, metallic hoses have a convoluted metal core with one or more metallic braids. The inner tube withstands the fluid while the outer cover strengthens it.

A corrugated metal hose tends to elongate when substantial internal pressure is applied; similarly, it would tend to contract when subjected to internal vacuum loads. Generally the braiding over the hose acts as the restraint. The braid as such does not affect the flexibility of the hose. However, the line or the components on either side of the hose must be properly supported. Additional braids can add to the working pressure but would reduce the minimum bend radius.

The bend radius is the minimum radius the hose can be curved without permanent deformation. Care must be taken when installing hoses so as not to exceed the specified bend. As a general rule of thumb, pressure drop in a corrugated metal hose is about three times over a comparable size standard steel pipe.

When selecting a hose, the following parameters need to be considered:

- Flow requirement
- Sizes of the hose and connected piping and required end fittings
- Required length of the hose
- Maximum and minimum temperatures
- Maximum and minimum pressures
- Corrosion conditions of the fluid and the required metallurgy of the hose
- Vibration levels and the movement of the connected piping

Flexible hoses are made with quite a good number of end connections such as fixed flanges, swivel flanges, male screw or female screw, union or certain couplings like quick or cam-lock, etc.

Where frequent connection/disconnection of hoses from the pipe lines is required some of the fast release couplings such as cam-lock or quick release couplings are used (see Figure 35.1).

Figure 35.1
Camlock Coupling

The following rules must be followed while connecting the hoses. Hoses are weaker than the weakest link in a system but would serve faithfully if installed and used properly.

- Do not allow the hose to be twisted while fixing any hose in any angle. Release all twisting and connect.
- Hoses should not be used to compensate for flange misalignment.
- Use two wrenches while connecting or disconnecting the hose.
- Do not compress, extend or stretch the hose. It is a good practice to allow a free length or slack of 5–8% in its total length.
- Hoses can take flexing in only one plane. Do not allow them to flex in all directions.
- Do not allow sharp radii in the hoses.
- Hoses should bend smoothly in the center.
- Strictly follow the minimum bend radius of the given hose. Fixing the hose with less than design will surely reduce its life.
- Do not use a hose as a sling or hang weights or loads on it.
- Please check that there is no damage on the inner or outer layers of the hose.
- Already bent hoses can not be straightened. Trying to do so may break the hose.

JUBILEE CLIP

Hose clamps are used to connect a hose to a pipe and one of the most popular ones is a jubilee clamp (see Figure 35.2). However, they are limited to use in moderate pressures. A hose clamp is basically a worm gear clamp where it has a band over which the pattern of worm gear is cut or pressed. The other end of the clamp has a fixed screw to match the worm gear pattern. The fixed screw has a screw driver slot or spanner head or both. The loose end of the band with worm gear teeth is inserted into the narrow gap of the fixed screw. When this screw is turned, the loose end is pulled into it and gets tightened and vice versa. Now this clamp is inserted over a hose with a pipe nipple inside and the screw is turned. The clamp tightens.

This clamp was invented in 1921 by L Robinson & Company (UK). They called it the "Jubilee Clip."

Figure 35.2
Jubilee Clip

CHICAGO COUPLINGS

The famous "Chicago Couplings" are almost universally used to connect hose lines of pneumatic tools to a compressed air source. Two couplers are used each with two lugs and a rubber gasket (erroneously but often called a washer). The couplers are firmly pressed into each other and twisted. Clips or tags on the coupler are designed to avoid accidental disconnection, but it is also a safe practice to wire or rope both the clamps together. Sizes over 1 inch may have four lugs instead of two. End connections are typically made to match the inside of the hose where it can be inserted and clamped (hose coupling — see Figure 35.3A) or the outside threads (male coupling — see Figure 35.3B), or made with inside (female coupling — see Figure 35.3C) to match the pipe fittings.

Chicago Couplings

Hose Coupling
Figure 35.3A

Male Coupling
Figure 35.3B

Female Coupling
Figure 35.3C

More Information

PRESSURE DROP IN STEAM PIPES

The first two charts in this chapter look at the pressure drop in steam pipes at different psi's and units. **Chart 36.1** considers pressure of 100 psi in imperial units, whereas **Chart 36.2** considers 7 bar steam pressure in metric units.

Chart 36.1
Pressure Drop in Steam Pipes at a Pressure of 100 Psi in Imperial Units

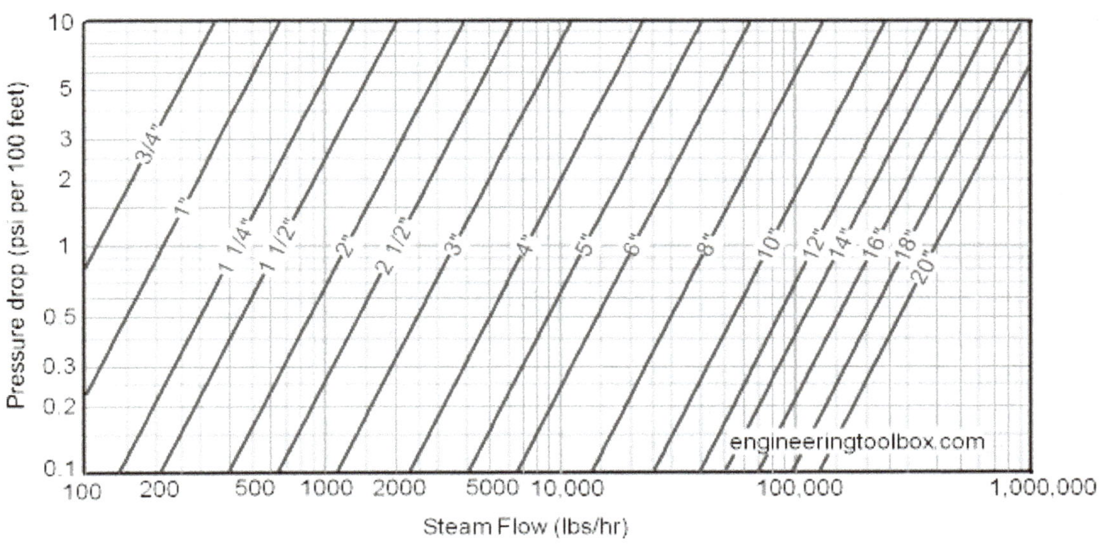

Correction Factors for Other Pressures

Pressure *(psi)*	0	5	10	30	60	90	100	110	150	200	250	300
Factor	6.9	5.2	4.3	2.4	1.5	1.1	1.0	0.92	0.70	0.55	0.45	0.38

- *21 lbs/hr = 1.26x10^{-4} kg/s*
- *1 psi (lb/in^2) = 6,894.8 Pa (N/m^2) = 6.895x10^{-3} N/mm^2 = 6.895x10^{-2} bar = 27.71 in H_2O at 62°F (16.7°C) = 703.1 mm H_2O at 62°F (16.7°C) = 2.0416 in mercury at 62°F (16.7°C) = 51.8 mm mercury at 62°F (16.7°C) = 703.6 kg/m^2 = 0.06895 atm = 2.307 Ft. H_2O*

Chart 36.2
Pressure Drop in Steam Pipes at a Pressure of 7 Bar Steam Pressure in Metric Units

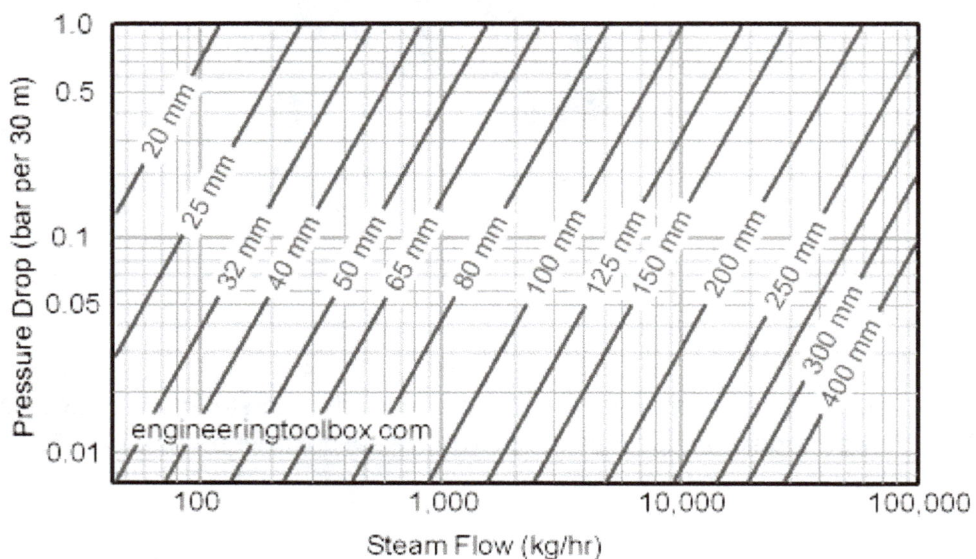

Correction Factors for other Pressures

Pressure *(bar)*	0	1	2	4	5	6	7	8.5	10	12	14	20
Factor	6.9	3.67	2.51	1.55	1.31	1.13	1.0	0.85	0.74	0.63	0.55	0.39

- *1 kg/h = 2.778×10^{-4} kg/s = 3.67×10^{-2} lb/min*
- *1 bar = 10^5 Pa (N/m^2) = 0.1 N/mm^2 = 10,197 kp/m^2 = 10.20 m H_2O = 0.9869 atm = 14.50 psi (lbf/in^2) = 10^6 dyn/cm^2 = 750 mmHg*

PRESSURE DROP AND FLOW VELOCITY FOR WATER FLOW

The pressure drop calculations are made with the D'Arcy-Weisbach_equation.

- Fluid : *Water*
- Pipe : *Steel Pipe – Schedule 40*
- Temperature : 68°F/20°C
- Density : 998.3 kg/m^3 (62lb/ft^3)
- Kinematic Viscosity : 1.004 10^{-6} m^2/s (0.01 stokes) (1.08E-5 ft^2/s)
- Pipe Roughness Coefficient : 4.5 10^{-5}

Chart 36.3 shows these calculations in Imperial units.

- *1 gal (US)/min = 6.30888×10^{-5} m^3/s = 0.0227 m^3/h = 0.06309 dm^3(litre)/s = 2.228×10^{-3} ft^3/s = 0.1337 ft^3/min*
- *1 ft/s = 0.3048 m/s*
- *1 psi (lb/in^2) = 6,894.8 Pa (N/m^2) = 6.895×10^{-3} N/mm^2 = 6.895×10^{-2} bar = 27.71 in H_2O at 62°F/16.7°C = 703.1 mm H_2O at 62°F/16.7°C = 2.0416 in mercury at 62°F/16.7°C = 51.8 mm mercury at 62°F/16.7°C = 703.6 kg/m^2 = 2.307 Ft. H_2O*

Chart 36.3
The D'arcy-Weisbach Equation (SI Units)

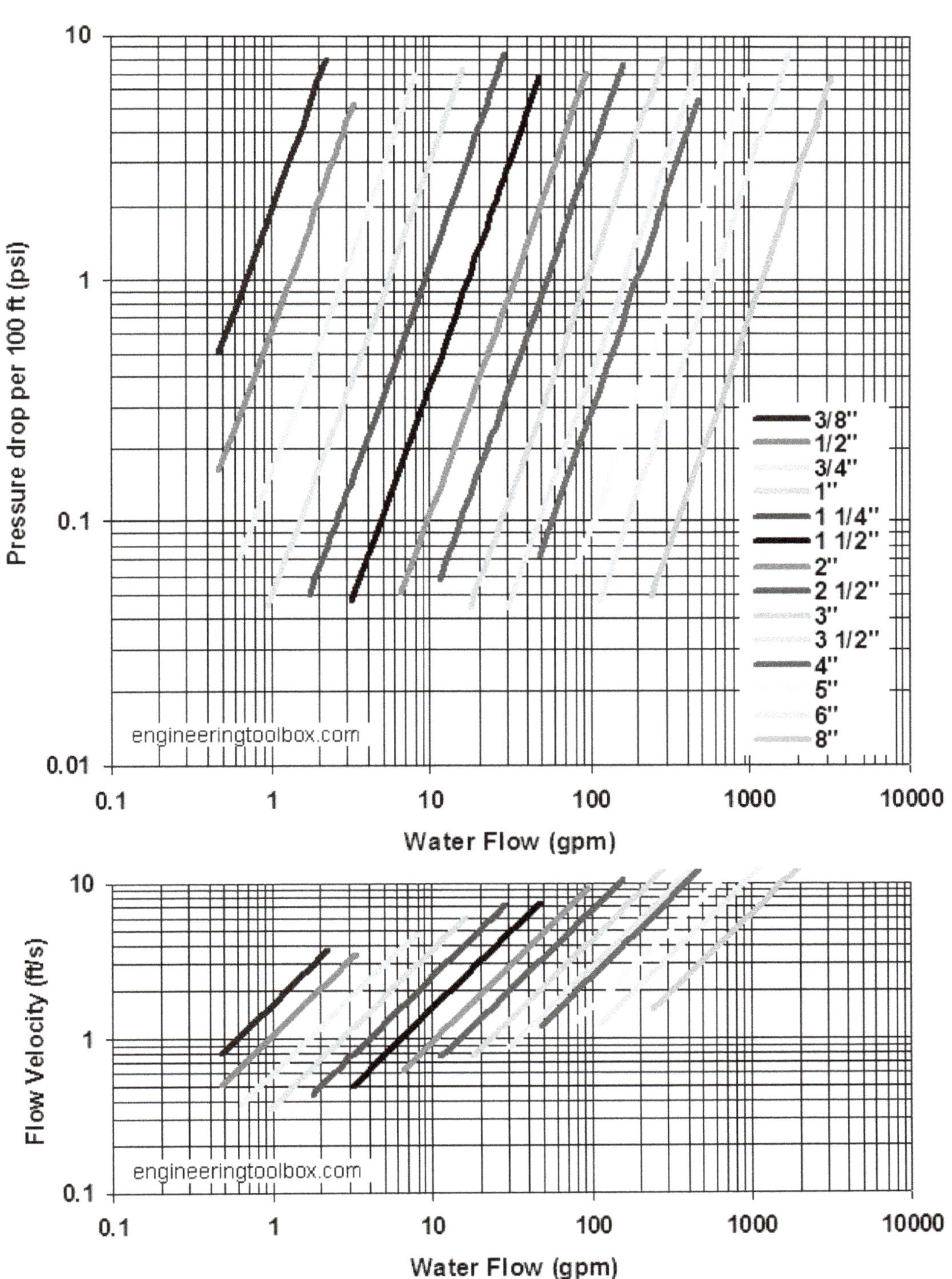

Chart 36.4
The D'Arcy-Weisbach Equation (Metric Units)

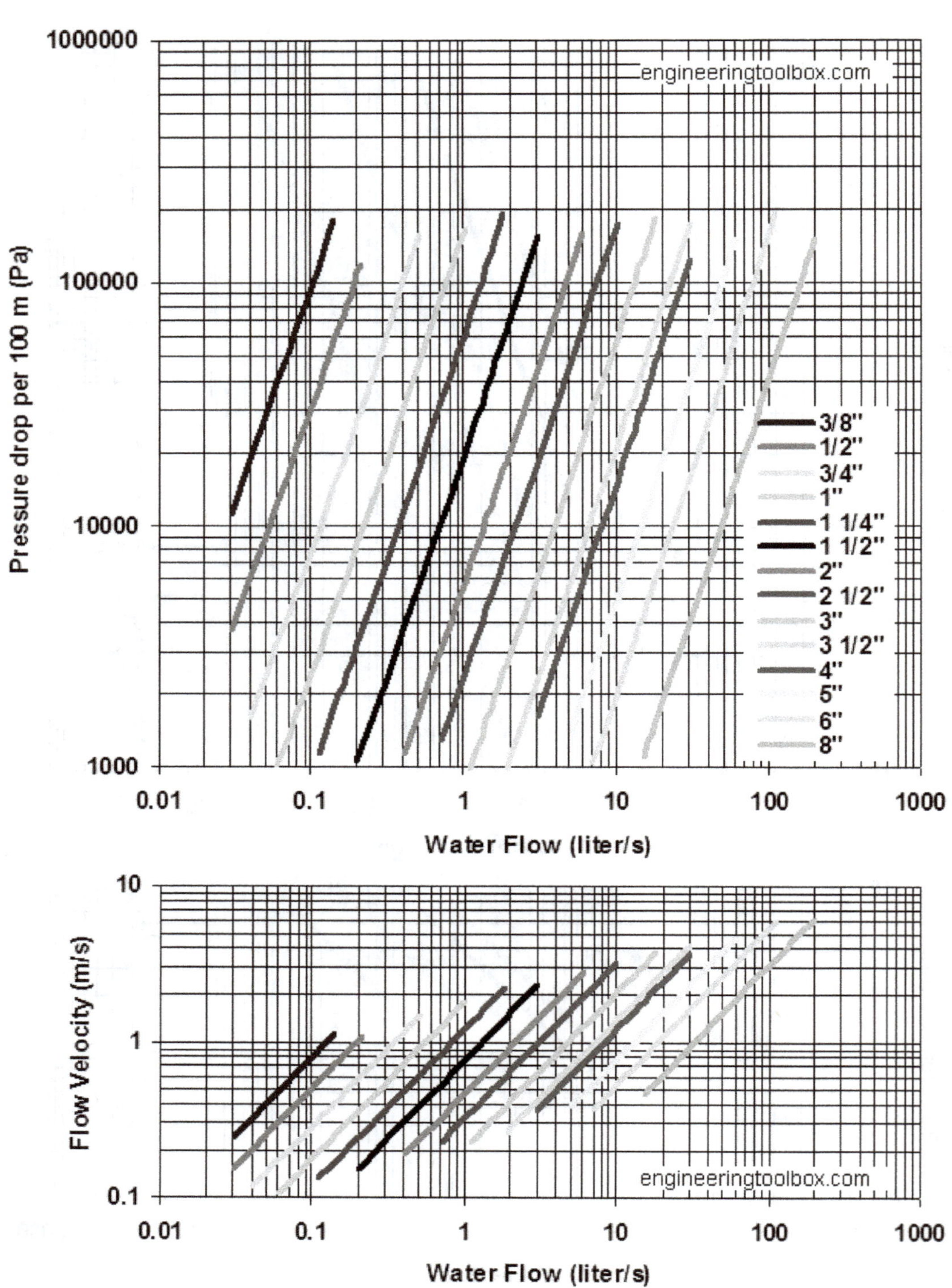

In turn, **Chart 36.4** shows these calculations in metric units.

- *1 Pa = 10^{-6} N/mm² = 10^{-5} bar = 0.1020 kp/m² = 1.02x10^{-4} m H_2O = 9.869x10^{-6} atm = 1.45x10^{-4} psi (lbf/in²)*
- *1 liter/s = 10^{-3} m³/s = 3.6 m³/h = 0.03532 ft³/s = 2.1189 ft³/min (cfm) = 13.200 Imp.gal (UK)/min = 15.852 gal (US)/min = 792 Imp. gal (UK)/h*
- *1 m/s = 3.6 km/h = 196.85 ft/min = 2.237 mph*

SPECIFIC GRAVITY AND WEIGHT

Chart 36.5 summarizes the specific gravity of common gases, and **Chart 36.6** *summarizes the weight of other materials.*

Chart 36.5
Specific Gravity of Some Common Gases

Gas	Specific Gravity[1)] - *SG* -
Acetylene (ethyne) - C_2H_2	0.90
Air[1)]	**1.000**
Ammonia - NH_3	0.59
Argon - Ar	1.38
Arsine	2.69
Benzene - C_6H_6	2.6961
Blast Furnace gas	1.02
Butadiene - C_4H_6	1.87
Butane - C_4H_{10}	2.0061
1-Butene (Butylene)- C_4H_8	1.94
Isobutene - C_4H_8	1.94
Carbon dioxide - CO_2	1.5189
Carbon monoxide - CO	0.9667
Carbureted Water Gas	0.63
Chlorine - Cl_2	2.486
Coke Oven Gas	0.44
Cyclobutane	1.938
Cyclopentane	2.422
Cyclopropane	1.451
Decane	4.915
Deutrium - D_2	0.070
Digestive Gas (Sewage or Biogas)	0.8
Ethane - C_2H_6	1.0378
Ethyl Chloride - C_2H_5Cl	2.23

Chart 36.5 Continued

Ethylene (Ethene) - C_2H_4	0.9683
Fluorine	1.31
Helium - He	0.138
Heptanes	3.459
Hexane	2.973
Hydrogen	0.0696
Hydrogen chloride - HCl	1.268
Hydrogen sulfide - H_2S	1.1763
Isobutane	2.01
Isopentane	2.48
Krypton	2.89
Methane - CH_4	0.5537
Methyl Chloride	1.74
Natural Gas (typical)	0.60 – 0.70
Neon	0.697
Nitric oxide - NO	1.037
Nitrogen - N_2 (pure)	0.9669
Nitrogen - N_2 (atmospheric)	0.9723
Nitrous oxide - N_2O	1.530
Nonane	4.428
Octane	3.944
Oxygen - O_2	1.1044
Ozone	1.660
Pentane	2.487
Phosgene	1.39
Propane - C_3H_8	1.5219
Propene (Propylene) - C_3H_6	1.4523
Sasol	0.42
Silane	1.11
Sulfur Dioxide - SO_2	2.264
Toluene-Methylbenzene	3.1082
Water Gas (bituminous)	0.71
Water Vapor	0.6218
Xenon	4.53

- [1)] NTP - Normal Temperature and Pressure - is defined as air at *20°C (293.15 K, 68°F)* and 1 *atm (101.325 kN/m2, 101.325 kPa, 14.7 psia, 0 psig, 30 in Hg, 760 torr)*
- Because specific gravity is the ratio between the density (mass per unit volume) of the actual gas and the density of air, specific gravity has no dimension.

Chart 36.6
Specific Weight of Some Other Materials

Product	Specific Weight - γ	
	Imperial Units (lb/ft³)	SI Units (kN/m³)
Ethyl Alcohol	49.3	7.74
Gasoline	42.5	6.67
Glycerin	78.6	12.4
Mercury	847	133.7
SAE 20 Oil	57	8.95
Seawater	64	10.1
Water	62.4	9.81

CORROSION

The last two charts address corrosion. **Chart 36.7** indicates acceptable combinations of more or less aggressive fluids and commonly used materials. Corrosion is complicated, depending on the combinations of materials and the fluids, the fluid temperatures, the surrounding environment and the galvanic currents in the constructions. This table must be used with care. Always check with the producer of the material. **Chart 36.8** summarizes the relation between corrosion and metal behavior.

Chart 36.7
Corrosion Resistance of Commonly Used Materials

CS	Carbon Steel	CI	Cast iron
302, 304, 316,416	Stainless Steel Grades		
B	Hasteloy B	C	Hasteloy C
D	Durimet	Ti	Titanium,
M	Monel	COB	Cobalt Based Alloy 6
Br	Bronze		

Corrosion Resistance:1- Good, 2- Be Careful, 3-Not Useable, na - data not available												
Fluid	Metal											
	CS	CI	302 304	316	Br	D	MI	B	C	Ti	COB	416
Acetaldehyde	1	1	1	1	1	1	1	na	1	na	na	1
Acetic acid, air free	3	3	2	2	2	1	2	1	1	1	1	3
Acetic acid, aerated	3	3	1	1	1	1	1	1	1	1	1	3
Acetic acid, vapors	3	3	1	1	2	2	2	na	1	1	1	3

Chart 36.7 Continued

Acetone	1	1	1	1	1	1	1	1	1	1	1	1
Acetylene	1	1	1	1		1	1	1	1	na	1	1
Alcohols	1	1		1	1	1	1	1	1	1	1	1
Aluminum Sulfate	3	3	1	1	2	1	2	1	1	1	na	3
Ammonia	1	1	1	1	3	1	3	1	1	1	1	1
Ammonium chloride	3	3	2	2	2	1	2	1	1	1	2	3
Ammonium Nitrate	1	3	1	1	3	1	3	1	1	1	1	3
Ammonium Phosphate	4	3	1	1	2	2	2	1	1	1	1	2
Ammonium Sulfate	3	3	2	1	2	1	1	1	1	1	1	3
Ammonium Sulfite	3	3	1	1	3	1	3	na	1	1	1	2
Aniline	3	3	1	1	3	1	2	1	1	1	1	3
Asphalt	1	1	1	1	1	1	1	1	1	na	1	1
Beer	2	2	1	1	2	1	1	1	1	1	1	2
Benzene (benzol)	1	1	1	1	1	1	1	1	1	1	1	1
Benzoic acid	3	3	1	1	1	1	1		1	1		1
Boric acid	3	3	1	1	1	1	1	1	1	1	1	2
Butane	1	1	1	1	1	1	1	1	1		1	1
Calcium Chloride (alkaline)	2	2	3	2	3	1	1	1	1	1	na	3
Calcium hypochlorite	3	3	2	2	2	1	2	3	1	1	na	3
Carbolic acid	2	2	1	1	1	1	1	1	1	1	1	
Carbon dioxide, dry	1	1	1	1	1	1	1	1	1	1	1	1
Carbon dioxide, wet	3	3	1	1	2	1	1	1	1	1	1	1
Carbon disulfide	1	1	1	1	3	1	2	1	1	1	1	2
Carbon tetrachloride	2	2	2	2	1	1	1	2	1	1	na	3
Carbonic acid	3	3	2	2	2	1	1	1	1			1
Chlorine gas	1	1	2	2	2	1	1	1	1	3	2	3
Chlorine gas, wet	3	3	3	3	3	3	3	3	2	1	2	3
Chlorine, liquid	3	3	3	3	2	2	3	3	1	3	2	3
Chromic acid	3	3	3	2	3	3	1	3	1	1	2	3
Citric acid		3	2	1	1	1	2	1	1	1		2
Coke oven gas	1	1	1	1	2	1	2	1	1	1	1	1
Copper sulfate	3	3	2	2	2	1	3	na	1	1	na	1
Cottonseed oil	1	1	1	1	1	1	1	1	1	1	1	1
Creosote	1	1	1	1	3	1	1	1	1		1	1

Chart 36.7 Continued

Ethane	1	1	1	1	1	1	1	1	1	1	1	1
Ether	2	2	1	1	1	1	1	1	1	1	1	1
Ethyl chloride	3	3	1	1	1	1	1	1	1	1	1	2
Ethylene	1	1	1	1	1	1	1	1	1	1	1	1
Ethylene glycol	1	1	1	1	1	1	1	na	na	na	1	1
Ferric chloride	3	3	3	3	3	3	3	3	2	1	2	3
Formaldehyde	2	2	1	1	1	1	1	1	1	1	1	1
Formic acid		3	2	2	1	1	1	1	1	3	2	3
Freon wet	2	2	2	1	1	1	1	1	1	1	1	na
Freon dry	2	2	1	1	1	1	1	1	1	1	1	na
Furfural	1	1	1	1	1	1	1	1	1	1	1	2
Gasoline	1	1	1	1	1	1	1	1	1	1	1	1
Glucose	1	1	1	1	1	1	1	1	1	1	1	1
Hydrochloric acid, aerated	3	3	3	3	3	3	3	1	2	1/2	2	3
Hydrochloric acid, air free	3	3	3	3	3	3	3	1	2	1/2	2	3
Hydrofluoric acid, aerated	2	3	3	2	3	2	3	1	1	3	2	3
Hydrofluoric acid, air free	1	3	3	2	3	2	1	1	1	3	na	3
Hydrogen	1	1	1	1	1	1	1	1	1	1	1	1
Hydrogen peroxide		1	1	1	3	1	3	2	2	1	na	2
Hydrogen sulfide, liquid	3	3	1	1	3	2	3	1	1	1	1	3
Magnesium Hydroxide	1	1	1	1	2	1	1	1	1	1	1	1
Mercury	1	1	1	1	3	1	2	1	1	1	1	1
Methanol	1	1	1	1	1	1	1	1	1	1	1	1
Methyl ethyl ketone	1	1	1	1	1	1	1	1	1		1	1
Milk	3	3	1	1	1	1	1	1	1	1	1	3
Natural gas	1	1	1	1	1	1	1	1	1	1	1	1
Nitric acid	3	3	1	2	3	1	3	3	2	1	3	3
Oleic acid	3	3	1	1	2	1	1	1	1	1	1	1
Oxalic acid	3	3	2	2	2	1	2	1	1	2	2	2
Oxygen	1	1	1	1	1	1	1	1	1	1	1	1
Petroleum oils	1	1	1	1	1	1	1	1	1	1	1	1
Phosphoric acid, aerated	3	3	1	1	3	1	3	1	1	2	1	3
Phosphoric acid, air free	3	3	1	1	3	1	2	1	1	2	1	3
Phosphoric acid vapors	3	3	2	2	3	1	3	1		2	3	3

Chart 36.7 Continued

Picric acid	3	3	1	1	3	1	3	1	1	na	na	2
Potassium chloride	2	2	1	1	2	1	2	1	1	1	na	3
Potassium hydroxide	2	2	1	1	2	1	1	1	1	1	na	2
Propane	1	1	1	1	1	1	1	1	1	1	1	1
Rosin	2	2	1	1	1	1	1	1	1		1	1
Silver Nitrate	3	3	1	1	3	1	3	1	1	1	2	2
Sodium acetate	1	1	2	1	1	1	1	1	1	1	1	1
Sodium carbonate	1	1	1	1	1	1	1	1	1	1	1	2
Sodium chloride	3	3	2	2	1	1	1	1	1	1	1	2
Sodium chromate	1	1	1	1	1	1	1	1	1	1	1	1
Sodium hydroxide	1	1	1	1	3	1	1	1	1	1	1	2
Sodium hypochloride	3	3	3	3	3	2	3	3	1	1	na	3
Sodium thiosulfate	3	3	1	1	3	1	3	1	1	1	na	2
Stannous chloride	2	2	3	1	3	1	2	1	1	1	na	3
Stearic acid	1	3	1	1	2	1	2	1	1	1	2	2
Sulfate liquor	1	1	1	1	3	1	1	1	1	1	1	
Sulfur	1	1	1	1	3	1	1	1	1	1	1	1
Sulfur dioxide, dry	1	1	1	1	1	1	1	2	1	1	1	2
Sulfur trioxide, dry	1	1	1	1	1	1	1	2	1	1	1	2
Sulfuric acid, aerated	3	3	3	3	3	1	3	1	1	2	2	3
Sulfuric acid, air free	3	3	3	3	2	1	2	1	1	2	2	3
Sulfurous acid	3	3	2	2	2	1	3	1	1	1	2	3
Tar	1	1	1	1	1	1	1	1	1	1	1	1
Trichloroethylene	2	2	2	1	1	1	1	1	1	1	1	2
Turpentine	2	2	1	1	1	1	2	1	1	1	1	1
Vinegar	3	3	1	1	2	1	1	1	1	na	1	3
Water, steam boiler feed	2	3	1	1	3	1	1	1	1	1	1	2
Water, distilled	1	1	1	1	1	1	1	1	1	1	1	2
Water, sea	2	2	2	2	1	1	1	1	1	1	1	3
Whiskey	3	3	1	1	1	1	2	1	1	1	1	3
Wine	3	3	1	1	1	1	2	1	1	1	1	3
Zinc chloride	3	3	3	3	3	1	3	1	1	1	2	3
Zinc sulfate	3	3	1	1	2	1	1	1	1	1	1	2

Chart 36.8
Metal Behavior-Corrosion

Non Chemical Environments

Ratings 0 = Unsuitable, 1 = poor, 2 = Fair, 3 = Fair to Good, 4 = good, 5 = good to excellent, 6 = excellent.

	Fresh Water	Sea Water	Steam	Steam Condensate	Air
Metal	Static/Turb	Static/Turb	Dry	Wet	City/Industrial
Grey Cast Iron-Plain or Low Alloy	4/3	4/3	4	4	3
Ductile Iron (High Strength)	4/4	4/2	4	4	3
Cast Iron Ni_Resist. (14% Ni, 7% Cu, 2% Cr bal Fe)	5/5	5/5	5	5	4
Ductile Iron Ni_Resist. (24 % Ni, bal Fe)	5/5	5/5	5	5	4
Mild Steel- Low Alloy steels	4/3	4/2	4	4	3
Stainless Steel Ferritic (17% Cr)	4/6	1/4	6	5	3
Stainless Steel Austenitic (18% Cr, 8% Ni)	6/6	2/5	6	5	4
Stainless Steel Austenitic (18% Cr 12% Ni, 2.5% Mo)	6/6	3/5	6	6	6
Stainless Steel Austenitic (20% Cro 29% Ni, 2.5% Mo, 3.5% Cu)	6/6	4/6	6	6	6
Hastelloy Alloy (55% Ni, 17% Mo, 16% Cr, 6% Fe, 4% W)	6/6	6/6	6	6	6
Inconel (78% ni, 15% Cr, 7% Fe)	6/6	4/6	6	6	6
Copper Nickel alloys (Up to 30% Ni)	6/6	6/6	5	6	5
Monel 400 (66% Ni, 30% Cu, 4% Si)	6/6	6/6	6	6	5
Nickel Commercial (99% Ni)	3/5	6/6	6	6	4
Copper and Silicon Bronze	6/5	4/1	5	6	5
Aluminium Brass	6/6	4/5	5	6	5
Bronze (88% Cu, 5% Sn, 5% Ni, 2% Zn)	6/6	5/5	5	6	5
Aluminium Alloys	4/5	0–5/4	2	5	5
Lead (Chemical/antimonial)	6/5	5/3	0	2	5
Silver	6/6	5/5	5	6	6
Titanium	6/6	6/6	5	6	6
Zirconium	6/6	6/6	5	6	6

Index

www.ingramcontent.com/pod-product-compliance
Lightning Source LLC
Chambersburg PA
CBHW080657220326
41598CB00033B/5232

* 9 7 8 0 8 3 1 1 3 4 1 4 3 *